D0987045

THE NEW POLITICAL SOCIOLOGY OF SCIENCE

SCIENCE AND TECHNOLOGY IN SOCIETY

Series Editors

Daniel Lee Kleinman
Jo Handelsman

The New Political Sociology of Science

Institutions, Networks, and Power

Edited by
SCOTT FRICKEL
and
KELLY MOORE

THE UNIVERSITY OF WISCONSIN PRESS

The University of Wisconsin Press
1930 Monroe Street
Madison, Wisconsin 53711

www.wisc.edu/wisconsinpress/

3 Henrietta Street
London WC2E 8LU, England

1 3 5 4 2

Printed in the United States of America

Library of Congress Cataloging-in-Publication Data
The new political sociology of science: institutions, networks, and power /
edited by Scott Frickel and Kelly Moore.
p. cm.—(Science and technology in society)
Includes bibliographical references and index.
ISBN 0-299-21330-7 (hardcover: alk. paper)
1. Science—Social aspects.
I. Frickel, Scott. II. Moore, Kelly. III. Title. IV. Series.
Q175.5.N456 2005
306.4′5—dc22 2005008260

CONTENTS

PREFACE

Many edited volumes begin life as a set of conference papers that get revised, edited, bundled together, and presented to readers as an organic product, the outcome of scholarly debate and synthesis. This collection is not one of those. It arose from our conversations with Daniel Kleinman about the structural inequalities flowing from globalization and neoliberal reforms, which appear to have complex and deepening influences across the sciences as well as among producers and consumers of knowledge. From the beginning we conceived the volume as an explicit attempt to infuse sociological and science studies scholarship with analyses of science policies and practices, the political and economic decisions behind them, and the ecological and social impacts that science continues to create downstream. In doing so, we intended to draw attention to questions of power, including why some knowledge doesn't get made, for example, or why some groups lack ready access to useful knowledge. To remain broadly relevant, we firmly believe that our scholarship must attend to the task of explanation. To that end, we invited contributors to tender individual or comparative case study analyses that explain why events and processes in science happen the way they do. All of them complied, most of them happily.

In our introduction, we spell out the basic contours of a new political sociology of science. In developing and organizing this framework, we hope to put into sharper focus the political and institutional dynamics that shape the funding, administration, and practice of science, doing so in a way that is engaged with broader social change processes as well as central elements of cultural science studies—particularly in its emphasis on meaning and networks. Contributors to this volume are working at these very intersections, and collectively their work speaks to the dynamic tensions that accrue from conceptual interaction. A similar dynamism shaped the evolution of this volume itself. Coediting was an intensely collaborative process, from early discussions of our vision for the volume, through several drafts of the introduction, to the final editing and indexing. There is little in this volume that does not bear the imprint of our combined hand, and so our names are listed in alphabetical order.

In the process of editing this volume, we were pleased to find that demand for critical institutional analyses of science is up, if the two well-attended and lively sessions we organized at the Society for Social Studies of Science conference in November 2003 and the American Sociological Association conference in August 2004 are any indication. We think they are.

The conference sessions provided opportunities for us and for contributors to the volume to flesh out the common—and distinct—themes and forms of analysis with each other and the audiences that attended these sessions.

ACKNOWLEDGMENTS

Many people deserve our sincere thanks in pulling this volume together, not least our contributors. We thank them all for their contributions and cooperation in abiding by the fairly strict set of deadlines we imposed on them between the project's first and final drafts. At the University of Wisconsin Press, series editors Daniel Kleinman and Jo Handelsman and press director Robert Mandel showed enormous faith in this project from beginning to end. We hope it meets their expectations. Once in production, David Herzberg, Erin Holman, and Matthew Levin shepherded the manuscript onward with skill and care. Jane Curran provided excellent copyediting, and we thank Blythe Woolston for preparing the index. Reviewers Elisabeth S. Clemens and Edward J. Hackett gave us rich and immensely useful commentary. Not least, our life partners, Rhys Williams and Beth Fussell, listened patiently and read drafts when they had more pressing things to do. We thank them for their love and support.

THE NEW POLITICAL SOCIOLOGY OF SCIENCE

I

Prospects and Challenges for a New Political Sociology of Science

SCOTT FRICKEL and KELLY MOORE

This chapter and the volume it introduces may be read as a scholarly response to contemporary politics in historical context. It maps a research program for sociologists and others concerned about the political impacts of recent transformations in market and regulatory arrangements on the development and use of scientific knowledge, and on scientific careers, fields, and policy regimes. It also seeks to better understand whether and how, in the wake of these shifts, knowledge production systems can respond adequately to the demands of those people—including some scientists—who join in struggle to make scientific knowledge more responsive to the needs of citizens. At root, it seeks new answers to an old question: what's political about science?

One answer was provided by British sociologist Stuart Blume. In his book *Toward a Political Sociology of Science,* Blume (1974:1) offered a critical analysis of his subject "founded upon the assumption that the social institution of modern science is essentially political and that, moreover, the scientific role is an integral part of the political system of the modern state." Empirical chapters addressed topics ranging from the structure of authority within science to the social and ideological conditions shaping scientists' tendencies toward unionization (in the United Kingdom) and politicization (in the United States), and from the role of elite scientists in the service of government to the limits to citizen participation in science policy decision making. If Blume's emphasis on scientists' professional and extra-professional roles within the social system of science was reminiscent of the North American brand of structural functionalism most closely associated with Talcott Parsons and his student Robert Merton, the interpretive style in which he made sense of his data was clearly sympathetic to the concerns of European critical theorists represented by Jurgen Habermas, Herbert Marcuse, and Steve and Hilary Rose. Blume concluded his study by noting, contra the Mertonian thesis of institutional autonomy, that "the social structure of modern science is highly dependent upon the social, economic, and political organization of society, and extremely sensitive to changes in this environment" (279).

Thirty years have passed since Blume's analysis of the politics of science was published. And while his conclusion about the interdependence of science, politics, and economy has become virtually axiomatic in contemporary science studies scholarship, the interconnections among the institutions he examined in deriving that claim have since undergone extensive transformation. Consider scientists' changing relationship to politics. In the 1950s, laypeople had virtually no role in decision making about sociotechnical issues, and scientists' involvement in politics occurred largely through advice to the government. By the early 1970s, much had already changed. The previous decade's wave of political activism included challenges directed at scientists' involvement in weapons research and the synthetic chemical industries, and new activists were beginning to challenge scientific and medical claims about women and minorities (Hoffman 1989; Bell 1994; Morgen 2002). At the same time, scientists themselves were becoming politically mobilized, using their expertise to challenge the government's and industry's uses of science and to call for the redirection of research toward "socially responsible" ends (Moore forthcoming). From the 1990s to the present day we have seen yet another shift, with the emergence of nonprofessional "lay experts" working with scientists to pursue and validate alternative theories and methodologies that are expressly value laden, from anti-evolutionists' promoting intelligent design theory in public school biology curricula to "popular epidemiologists" seeking answers to the illnesses that plague polluted communities (Fischer 2000; Hassanein 2000; Gross 2003; Schindler 2004; Brown 1987; Moon and Sproull 2000). By learning and modifying the language and tools of science, users and lay experts have contributed to scientific debate and in some cases have successfully challenged expert claims on a wide range of issues.

Increasingly complex engagements in formal and informal politics mark only one important dimension of a far broader set of changes confronting contemporary science. Like many other social activities in modern life, the production and use of scientific knowledge has become an intensely regulated and commercialized activity (Etzkowitz, Webster, and Healy 1998; Powell and Snellman 2004; U.S. Congress 2004; Kloppenburg 1988; Abraham and Reid 2002). To be sure, these changes have not been straightforward, nor are their implications certain. Rather, we now find a vastly richer array of interactions among scientists, citizens, government, and the private sector than existed in earlier decades. The stakes in understanding the interlocking structures of politics, law, and commerce are high. New or newly flexible institutional arrangements challenge older mechanisms for ensuring the credibility and authority of scientists' own claims as well as commonly held assumptions concerning the nature of intellectual work and property, the maintenance of legal and professional boundaries, and the trajectories and practices that characterize research. More broadly, the politicization, commercialization, and regulation of science carry profound implications for human health, democratic civil society, and environmental well-being.

Our goal in this chapter is to give readers a better sense of how research in the social study of science might become better equipped to understand these broad historical changes and their sociopolitical implications. To do so, we sketch an outline for a "new political sociology of science" (NPSS)—one inspired by Blume's earlier study but armed theoretically and methodologically to meet new challenges posed by the changing political and economic realities that structure the sciences of today and that will indelibly influence the organization and conduct of the sciences of tomorrow. Drawing on the sociologies of law, politics, social movements, economics, and organizations, NPSS demonstrates the ways in which institutions and networks shape the power to produce knowledge and the dynamics of resistance and accommodation that follow.

Repositioning the Political Sociology of Science

Major historical changes in the systems that produce, certify, and disseminate scientific knowledge have been accompanied, consciously or not, by a set of similarly broad intellectual changes as anthropologists, philosophers, sociologists, and others have crafted new methodological and conceptual tools—in some cases advocating entirely new vocabularies (Akrich and Latour 1992)—for the social study of "technoscience" (Latour 1987:174). The broader intellectual shift was first signaled by Thomas Kuhn's *The Structure of Scientific Revolutions* (1970 [1962]). Although derived from French conventionalist ideas, Kuhn's arguments were oriented toward a social-psychological theory of change in science, with his emphasis on cognitive dissonance generated by the accumulation of anomalies through "normal science." Taking their cue from Kuhn, sociologists in the 1970s began advocating a sociology of scientific knowledge that directly challenged the deeply held notion that science is unique among social institutions in gaining access to natural and universal truths (Barnes 1974; Mulkay 1979). Instead, researchers sought to demonstrate that scientific knowledge is social and particularistic and that the essential character of institutions and the cultural practices that support knowledge production in science are neither autonomous nor distinct. In relatively short order, Whiggish histories of science that related the stories of great men with great ideas, and the Mertonian emphasis on science as a self-regulating social system, fell from prominence as a decidedly more contingent, contextualized, and critical sociology and social history of science took their place.

Since then, research in this area has come to emphasize a strikingly anthropological and philosophical focus on how the material and epistemic stuff of science—facts and artifacts—are created. Drawn mainly from archival research, laboratory ethnographies, and discourse analysis, this eclectic, agency-oriented body of work describes the constructed nature of scientific knowledge through detailed case-study analyses that draw on the epistemology and methodologies of anthropology and cultural studies. Topically wide-ranging studies chart the epistemic journey from hypothesis to fact made possible by complex sociotechnical networks, language, and rituals. These studies show how machines, organisms, texts, data, people, and practices are brought together in "heterogeneous networks," the result of which is authoritative knowledge about and representations of nature (Callon 1995). From this perspective have come the important and related claims that the social and the natural are mutually constructed or co-produced categories (Latour and Woolgar 1986; Clarke and Oleson 1999); that the meanings scientists give to their subjects are locally contingent (MacKenzie and Spinardi 1995; Knorr-Cetina 1999); that stabilization of those subjects into obdurate facts, techniques, theories, or technologies are negotiated results emerging from the cooperative exchange of materials and coordination of practices (Fujimura 1996; Collins 1983; Pickering 1995; Casper 1998), as well as competition for credibility across geographical, organizational, and disciplinary domains (Clarke 1998; Gieryn 1999); and that it is this cultural apparatus—not some independently revealed "nature"—that supports and extends particular definitions of "what is true" or "what works best." The collective and negotiated character of credibility contests was illustrated in self-exemplifying fashion in edited volumes by Bijker and Law (1992), Clarke and Fujimura (1992), Galison and Stump (1996), Pickering (1992), and Star (1996). Published in the early and mid-1990s, these collections have been important contributions in cementing the dominant cultural and social constructionist approaches in social studies of science.

A deeper appreciation for the institutional bases of power in knowledge production is reflected in a second body of scholarship that has grown rapidly in the past decade or so. Research in this volume and elsewhere that we cast under the rubric of NPSS considers the intersection of rules and routines, meanings, organizations, and resource distributions that shape knowledge production systems. Sensitive to the arguments and incorporating insights gained from cultural science studies, but with an explicit focus on the structural dimensions of power and inequality in knowledge politics, NPSS extends the cultural schools' thick descriptions of how science works toward research explaining why science works better or more often for some groups than for others, and the ways in which social attributes such as race, gender, class, and profession interact with and condition those particular outcomes. While a number of edited volumes bring together elements of this approach to science and technology, these have focused on specific topical themes such as democracy (Kleinman 2000), industry-university relations (Restivo and Croissant 2001), feminism (Schiebinger, Creager, and Lunbeck 2001), or environmental health (Kroll-Smith, Brown, and Gunter 2000; Brown 2002).[1] In this volume, by contrast, we showcase research that represents a broad cross-section of substantive themes, arguments, and methodologies as a means of better identifying important programs and cross-currents within this still loosely integrated field.

The centerpiece of the NPSS project as we see it is the analysis of institutions and networks as they condition the availability and distribution of power in the production and dissemination of knowledge. Our basic understandings of these three central concepts are straightforward. *Power,* in the Weberian sense emphasized here, is the ability to influence others directly or indirectly, subtly or overtly, legitimately or illegitimately. Power is a dynamic and social condition whose character can be described empirically by the forms it takes, its distribution across societies, the mechanisms through which it is expressed, and the scope and intensity of its effects. By *institutions* we mean relatively durable sets of practices and ideas that are organized around social activities and that in various ways shape the contour and experience of daily life. Institutions embody routinized "ways of going on" that, even when largely taken for granted by individual members of society, nevertheless continuously shape or channel social choices, constraining certain courses of action and enabling others. *Networks* are dynamic configurations of relationships among individual and organizational actors. These configurations can operate within settings described entirely within a particular institutional setting, but we are most interested in the role that networks play in bridging or linking institutional domains and thus their operation as key mechanisms in the redistribution of power and in transformations of institutional arrangements. It is in the process of bridging that conflict is likely to take place, resulting in either solidification of or change in institutional practices.

Attention to the structural bases of power often (but not always) leads us to the investigation of controversies waged beyond the laboratory door—conflicts generated by ongoing and complex changes in society's dominant legal, economic, and political structures. In that realm, NPSS seeks answers to questions concerning, for example, the direct and indirect costs and benefits of profit-driven research, the implications of formal and informal conventions governing participation in decisions about research, and the processes through which such decisions are made and implemented. What knowledge gets produced? Who gains access to that knowledge? What kinds of knowledge is left "undone" (Hilgartner 2001; Woodhouse, Hess, Breyman, and Martin 2002)?

These are the kinds of pressing political questions that, generally speaking, more agency-oriented approaches are not well equipped to address. Constructionist approaches in social studies of science have been primarily descriptive, often showing how knowledge practices unfold at the local level. NPSS acknowledges the contingent and constructed character of scientific knowledge but also insists that construction processes are neither random nor randomly distributed. It strives to explain why collective efforts to change the way knowledge is produced succeed or fail. To be sure, the causal elements of our project do not avoid the epistemological problem that even modest explanations remain vulnerable to deconstruction. We note, however, that there are sustained, large-scale relationships that make some kinds of claims, outcomes, and processes far more likely than others. Critiques of science that eschew analysis of causality out of a concern that any causal statement will be shown to be wrong owing to its subject's endlessly "emergent" or "socially constructed" character are not suited to explain such broad patterns. Critique and description are valuable academic tools, but they are less helpful in cases where we try to understand why there are losers and winners in sociotechnical debates (Latour 2004). More importantly, as several chapters included in this volume demonstrate, a search for causes can become the basis for understanding those outcomes in ways that suggest new kinds of solutions.

While the research brought together in this volume can be read as a collective call to bring social structure (and structurally attuned concepts such as interests, regimes, and organizations) "back in" to social studies of science, NPSS is not retrogressive. It invokes a broader and decisively less deterministic analysis of the social relations of science than earlier political sociology of science that drew mainly from Marxist theory and its grounding in economic analyses (Arditti, Brennan, and Cavrak 1980; Aronowitz 1988; Noble 1977). It also explicitly avoids the structural-functionalist assumptions characteristic of earlier institutional analyses of science (Ben-David 1991; Hagstrom 1965; Merton 1973). Against both of these, we identify NPSS as an empirical project guided by a neo-Weberian emphasis on the relationships embedding scientific knowledge systems within and across economic, legal, political, and civil society institutions. In doing so we do not eschew culture (rituals, symbols, language, and other meaning systems) but understand it to be embedded in structured relationships of power (Bourdieu 1984, Lounsbury and Ventresca 2003; Schurman 2004).

We can begin to flesh out what this means by offering four general prescriptions that we hope will be useful for orienting and integrating new and existing research.

Key Elements in the NPSS Program

ATTENTION TO UNEQUAL DISTRIBUTIONS OF POWER AND RESOURCES

Our focus on formal power, or power situated in institutionally embedded social relationships and interests (Clemens and Cook 1999; Emirbayer 1997; Lounsbury and Ventresca 2003), is motivated by our concern with issues of equal access, responsibility, and inequality.[2] As Steven Lukes (2002:491) writes, understanding power helps us to know "whom to try to influence, whom to appeal to, and whom to avoid . . . and it helps to impute blame and praise." At any given moment, the structure of social relationships within which individual and organizational actors reside provides access to available resources and opens or closes possibilities for action—including action that seeks to change the distribution of power (i.e., politics). These arrangements confer formal power on specific actors by providing routinized channels and procedures for decision making and other forms of authority, making one's location within one or more fields of action an important factor in explaining outcomes. The resources that actors may use are also structured by pre-existing arrangements. Informal power is based on a similar principle: to the extent that one's social location provides access to social networks and other resources that can be used for bargaining, one can exercise power and engage in some form of politics.

The question of "who has the right to take self-determined and self-interested actions" is expected to vary over time and place (Scott 1995: 140). Members of certain social groups identified by ethnicity, class, race, gender, professional status, and age, for example, routinely enjoy advantageous positions that allow them to commandeer rules and resources and define situations. In other cases, more mutable resources—access to a key piece of information, for example, or, following Granovetter's (1973) terminology, connections to a "weak bridging tie"—may afford particular actors advantages that others in their group or network may not share. We thus see power, in part, as a variable function of actors' relative social location within more or less stable institutional configurations relative to the flexible networks that span those institutions; we see politics as collective action seeking to explicitly reproduce those configurations or, alternatively, to substantively change them.[3]

ATTENTION TO RULES AND RULE MAKING

Rules, whether legal or bureaucratic, are of central importance to NPSS because they often serve as decisive expressions of power and define the means by which debates can be carried out. One of the central features of such rules is that they set the baselines for the terms of debate, who has a right to be included in the discussions, and performance standards (Epstein 1998; Fischer 2000; Weisman 1998). Among the more influential baseline rules in science are those that involve protection of human subjects. Sixty years ago it was possible to treat human subjects as material subjects, since there were few protocols for protection and certainly no real legal recourse for those who were harmed during scientific research (Halpern 2004). Since then, human (and animal) subjects are under ever more rigorous systems of protection, including the elaboration of informed consent regulations, the required inclusion of varied populations in research studies, and requirements for ethics training for NIH grant recipients. These new rules have in different ways reconfigured the relationships between scientists and humans involved in biomedical and public health research.

Rules that are built into technological systems such as computer programs or patents structure power relations as well. At the most obvious level, they substitute for human decision making by restricting the kinds of information that can be processed and the order in which it must be considered, lessening human error and enabling effective communication and coordination. Yet they also have the effect of externalizing control away from human agents and limiting access to inside information. Privately owned systems compound these problems.

We encourage research that systematically examines the structure and fate of rules and procedural actions in science as a means of better understanding the conditions under which social relationships are routinized and become durable, and when these conditions become tumultuous and lead to significant rearrangements of power relations. By studying which procedures carry more influence than others, we also gain a better sense of the hierarchical or distributed arrangements within which rules and rule making are embedded. In this way, researchers can gain additional insight into such issues as what kinds of action or inaction is rewarded or sanctioned, what counts as knowledge, and how knowledge is used.

It would be a mistake to see this emphasis on rules as determinist.

What interests us as much as the structure of rules is how they are acted upon or ignored, resisted, or adapted for use in other settings. We also differ with structural functionalists' call to study how stability (the normal state) is disrupted by change (the abnormal state) and is then followed by renewed stability. We need not presume that stability and conflict are inversely correlated. Instead we can treat the relationship of those dynamic conditions as empirical questions to be resolved through research.

ATTENTION TO THE DYNAMICS OF ORGANIZATIONS

Studying organization in science is empirically important to NPSS because of the number and kind of organizations and organizational hybrids that have come to dominate sciences' social landscape. As in earlier decades, organizations are the sites in which nature is interpreted, processed, codified, and then presented to outsiders, but today those sites extend far beyond laboratories to include trade associations, patent offices, regulatory agencies, development banks, and social movement organizations, to name but a few. In the U.S. case (on which most of our studies draw), science is performed in what seems to be an increasingly diverse array of organizational settings. We need to understand these different settings and the "logics of action" through which organizations and organizational networks pattern the production of knowledge (Alford and Friedland 1991:243; Vaughan 1999).

There are important theoretical payoffs to studying organizations as well. One is that the Mannheimian (1991) question of why knowledge becomes organized in particular ways gains renewed relevance in light of the sweeping changes in science's relationships to the state, the market, and the public square. Another of equal importance is that organizational change often (but clearly not always) represents a shift in power relations. Scholars of organizations have identified many sources of organizational change; here we note three. The first is the explicit challenge to organizational arrangements made by organized activists and by the imposition of law (Epstein 1996; Edelman 1992; Sutton, Dobbin, Meyer, and Scott 1994). In these cases, moral, legal, or economic coercion compel compliance. Organizational change may also take place through imitation of successful models or the borrowing of templates for action from other settings (DiMaggio and Powell 1983; Clemens 1997; Lounsbury 2001). Finally, when the availability of valued resources shifts, we are likely to see changes in the structure of organizations.[4]

METHODOLOGICAL CONSIDERATIONS

For reasons just described, *organizations* offer strategic sites for the types of analyses we are advocating. Indeed, the majority of chapters collected for this volume cluster at the level of organizations and organizational networks. Yet as readers will notice, a fair number of chapters also offer more macro-level treatments of scientific fields (Woodhouse), social movements (Morello-Frosch et al.), or political culture (Kleinman and Vallas) and one provides a micro-level ethnographic study of identity conflict in science (Henke). This raises a question about the appropriate level of analysis in studies that seek to draw conclusions about the intersection of human action and larger-scale structures. In our view, the level at which one pursues research is not in itself a critical issue, except insofar as that decision influences the kinds of evidence marshaled and the kinds of questions that can be answered. What is more important is recognizing that social life is meaningfully organized on different scales and that researchers may usefully work across those scalar differences. Thus, research undertaken at any level of analysis needs to account for how different institutions intersect with and shape the organization of human agency, from local laboratories (Kleinman 2003) to intellectual movements (Frickel and Gross 2004) to nation-states (Mukerji 1989).

Comparative approaches are also especially valuable for building causal explanations. They permit the researcher to evaluate the relative importance of theoretically important relationships, processes, actors, or mechanisms. Following Weber (1949:183), we seek neither purely idiographic nor entirely nomothetic forms of explanation, but to build probabilistic causal statements about why we observe distributions of power in the politics of science. Especially fruitful are comparisons between successes and failures. For example, inventions that succeed are usually considered "innovations," but we cannot be confident about what causes inventions to become innovations without attention to failed cases. Explicit use of other kinds of comparisons (over time, geographic area, groups of actors, scientific subjects, mechanisms of aggregation and distribution, etc.) provides a similar advantage, illuminating how institutional arrangements satisfy the needs and interests of some groups and leave others out.

One of the most distinctive methodological bases of NPSS is its *attention to aggregation and distribution processes*. To be sure, agent-oriented approaches to the social study of science are also concerned with these processes, but their focus on the continual reconfiguring of heterogeneous assemblages contrasts with our concern with which social groups in and outside the laboratory benefit or not from scientific knowledge. The mechanisms of aggregation and distribution that concern us most are those that operate among laboratories, markets, governments, and universities.

Finally, we believe research should pay closer attention to *scope conditions*. That is, we need to give explicit consideration to whether and how research claims relate to other cases. We should strive to make clear whether our claims are applicable to only one case, are widely applicable, or, more typically, concern a limited number of situations. This is not a call for eliminating case studies or for privileging a specific kind of evidence collection or analytic strategy, but rather a reminder that specifying scope conditions is an essential component of theory building and is crucial for identifying the empirical applicability of research.

Overview of Chapters

The chapters brought together in this volume represent contributions from both younger and more established scholars working in a variety of substantive and disciplinary traditions that share several if not all of the major features of the new political sociology of science we sketched above. Within the domain of the life sciences, this volume provides more or less evenly distributed attention to agricultural, biomedical, biotechnological, environmental, and molecular biological research and policy, but only one chapter—Martin's analysis of defense research—falls entirely outside that domain. While not our original intention, the substantive foci of the chapters mirror the substantive focus of the field as a whole over the past twenty years.

In part, this heavy concentration of social scientific research in the life sciences reflects the dramatic increase in the amount of biological research undertaken in the United States and around the world and the general perception, no doubt shared by many in science studies, that life sciences are important means by which understandings of personhood, citizenship, gender, and race are becoming reconstituted and redefined (Clarke, Shim, Mamo, Fosket, and Fishman 2003). We would be remiss after all our talk about bringing structure back into science studies not to acknowledge the relative ease of access—conceptually and institutionally—that life sciences seem to afford scholars who want to understand how science is social. But at the same time, we also hope that readers will see the value of political sociology of science in studying fields not centered in biology and in science as it is organized and practiced in countries other than the United States.[5]

The book is organized into three parts that roughly parallel science's complex entanglements with the economy, civil society, and the state. As readers will surely note, while our contributing authors have situated their analyses primarily within one of these major institutions, nearly all emphasize convergences, hybridities, conflicts, and contradictions that arise when institutional logics interpenetrate. Ample evidence in the following chapters that institutional logics are being constituted in new ways reinforces our belief that our current historical period is one of intense change in which knowledge producers are playing ever-more critical roles in shaping public and political life. However, this does not mean that traditional ways of conducting research or organizing knowledge are disappearing or becoming irrelevant.[6] Where there is change, there is also resistance to change. To understand the implications of this intensifying dynamic, our research needs to remain dually attuned to how these enabling and constraining forces interweave. The chapters in part 1 examine different dimensions of the intertwining of private- and public-sector science and identify some of the potential consequences of those processes for the state and civil society.

THE COMMERCIALIZATION OF SCIENCE

We lead this section with a chapter by Daniel Lee Kleinman and Steven P. Vallas, who build on their theory of "asymmetrical convergence" for conceptualizing what they see as a general trend toward isomorphism in academic and industry biotechnology. The authors argue that universities, once viewed as sites of collegiality and autonomy, are increasingly viewed as places where scientists are compelled to respond to the constraints of the commercial world, while scientists working in science-intensive firms find their workplace environments to be characterized by the sophisticated equipment, freedom to pursue ideas, and collegiality that are traditionally thought to be available only in academia. Grounding their framework in neo-institutionalist sociology of organizations and economic sociology, Kleinman and Vallas explore many of the "contradictions, anomalies, and ironies" that emerge from this blurring of deeply embedded institutional norms and practices. The most important of these is that the convergence process is highly uneven, with the economic logic of markets holding greater comparable sway in university laboratories, promotion and tenure committee meetings, and classrooms than is the penetration of academic logic into industry research settings. This study also has notable implications for science and technology studies. In contradistinction to the past two decades' emphasis on how virtually everything is socially (or otherwise) constructed (Hacking 1999), they insist on the continued value of studying how "the *already constructed* can shape social phenomena of all varieties, especially and including the technoscientific field." The next chapter addresses similar issues, while also demonstrating how the integration of constructionist concepts that are the indirect target of criticism for Kleinman and Vallas, when properly contextualized, can enhance understanding the transformations underway in contemporary universities.

Jason Owen-Smith is also interested in the uneven and sometimes contradictory effects of deepening commercial engagement on universities. In chapter 3 he describes academic research commercialization as a dual process, first of shifting hierarchical orders within academia, or what he calls the logic of "grafting," and second as extensions or "translations" (pace Latour 1987) that reposition the academy vis-à-vis industry in ways that facilitate the constitution of novel actors and arrangements. The former process highlights observed disparities in the power of actors, while the latter highlights features of a system in motion. Drawing on case studies of commercialization attempts at Columbia University, Boston University, and the University of Wisconsin, Owen-Smith argues that attempts to graft the formal rules and hierarchical orders of one institutional logic onto another temporarily offers opportunities for actors to force novel connections in response to ambiguity. Over time, however, these connections are rendered durable through the redistribution of resources and, more importantly, through the reconstitution of horizontal webs of connection. While the contemporary period may represent a window of opportunity in which new kinds of relationships linking universities to the broader society are rapidly emerging, that window is not likely to remain open indefinitely as new institutional configurations become cemented in place and a new knowledge economy settles in.

The influence of the market on science is complex. In chapter 4, Steven Wolf looks to agriculture to demonstrate that the unevenness of those influences extends well beyond biotechnology. Wolf's study focuses on how the changing structure and composition of reference networks are potentially eroding farmers' contribution to the production of new technical knowledge in agrofood systems. Reference networks are composed of organizations that pool empirical records of performance of production technologies applied in varied settings in order to inform managerial decision making (e.g., how much of what kind of fertilizer to use on a particular type of farm field). Such databases serve to transform farmers' experiential knowledge into codified knowledge that can be archived, aggregated, circulated, and privatized. Wolf argues that since the 1990s, reference networks have undergone reorganization in ways that emphasize commercial relations with agribusiness firms at the expense of farmers' professional (collectively organized) capabilities. The emergence and coordination of highly specialized agronomy consultants employed by agrochemical input dealers and proprietary digital animal husbandry databases have weakened farmers' traditional interdependence and marginalized farmers within the decentralized process of knowledge generation. Thus, like Kleinman and Vallas and Owen-Smith, Wolf finds that although farmers continue to participate in knowledge production, their inclusion in heterogeneous networks is also simultaneously a hierarchical relationship that constrains their ability to control valued resources, including proprietary rights to data from their own fields.

In chapter 5, David J. Hess exploits a different entry point than the previous chapters, arguing for "a more deeply historicized sociology of scientific knowledge." He advocates using comparative case studies that span long periods of time (decades, rather than months or years) and insists that we remain mindful of the historical character of the "external" institutions, actors, cultures, and materials that shape research programs, as well as how those programs simultaneously undergo change. To illustrate this framework, Hess compares the fate of two cancer therapy research programs. One program has focused on antiangiogenesis drug-based therapies. The research network constituting this field established early connections with industry, and as regulatory changes encouraged the proliferation of university-industry relationships, the drug-based approach was advantageously positioned to capture interest, money, marketing, and results. The second case involves cancer therapy research that focuses on shark cartilage and other natural or food products. Unlike the deep corporate coffers that fed drug-based antiangiogenesis research, the food-based therapies have been supported by a handful of "old-fashioned" researchers and some social movement actors, have remained mired in controversy, and have not experienced even a fraction of the growth and profitability of the drug-based field. Hess rejects the simple argument that differential access to funding explains the divergent histories of these two fields, however. Instead, he finds a more satisfying explanation involves understanding broader historical changes in the institutional scale of cancer therapy research laboratories, the increasing differentiation of actors and roles as scientists and clinicians, the increasing cultural investments in biomedicine that emphasize research on molecular-level mechanisms, biochemical pathways, and gene expression, and the increasing denaturalization of the products that biomedicine produces, away from nature-based regimes and toward synthetic (and often toxic) ones.

Taking a cue from the previous chapter, in chapter 6 Edward J. Woodhouse asks us to further complicate the role of industry in driving what kinds of knowledge are produced. He compares the development and enthusiasm surrounding two techniques for rearranging atoms and molecules to better serve human purposes: nanotechnology, which has been met with optimism and lavish funding, but whose dangers (and benefits) are poorly understood, and "green chemistry," an environmentally friendly alternative to the standard, highly toxic, molecule production, which has received dramatically less attention and funding. Conventional analyses of these divergent fates would treat industrial managers, market demand, and government regulators as the central actors in the story. Woodhouse acknowledges their importance but reminds us that is it ultimately chemists themselves who develop scientific processes, not their bosses and regulators. He argues that scientists' continued development of toxic "brown chemistry" and inattention to the potential dangers of nanotechnology are due to more complex interactions between markets, government, and, most importantly, the lack of availability of alternative knowledges in textbooks, at conferences, and in publications. In a fitting segue into our focus in part 2 on the politics of science in civil society, Woodhouse concludes that scientists need to take a more active role in decision making about whether to produce harmful or potentially harmful products, but he recognizes that scientists, as much any other people, are not always able to see alternatives because of the webs of routines and resources in which they are embedded. It is nonscientists, he argues, and especially the public, who must take some of the responsibility by demanding alternatives from scientists.

SCIENCE AND SOCIAL MOVEMENTS

Part 2 includes five chapters that identify some of the synergistic accomplishments that can develop when citizens cross into the domain of science and when scientists carry their concerns as public citizens with them to work. Chapter 7, by Scott Frickel, examines the organizational politics of the Environmental Mutagen Society (EMS), a small professional organization that was instrumental in the early-1970s in institutionalizing the field of genetic toxicology. Frickel's interest lies in exploring the ways in which organizations condition the relationship between scientific research and scientist activism. He argues that a priori definitions of "activism" in science occlude the ways in which conventional scientific practices can become contentious within the context of rapid institutional change. In this case, drawing clear distinctions between the work of genetic toxicology researchers and environmental groups—a conventional boundary maneuver—allowed the EMS to nurture less obvious but arguably more effective variants of environmental activism within the organization's expanding domain of influence. Frickel concludes that conventional behavior in pursuit of contentious goals can sometimes produce significant sociopolitical change, but to fully appreciate this, researchers need to systematically attend to questions about the political significance of seemingly mundane action. This will require a renewed sensitivity to how activism in science is constituted, where strategic and contentious politics are most likely to be deployed, and what are the consequences for the production and organization of knowledge.

An analysis of the relationship between scientific work and environmentalism is developed further in Christopher R. Henke's contribution in chapter 8. Henke's intention is to identify some of the factors that "enable and constrain scientists as agents of environmental change." To do so, he goes (literally) into the field, where he finds agricultural scientists seeking to convince farmers in California's Salinas Valley to adopt a simple and inexpensive soil test to help them better monitor and thus reduce crop field nitrate levels. It turns out that this seemingly straightforward solution to the serious problem of agricultural water pollution is fraught with conflict and subtle complexities as scientists' relationships with growers describe an edge space between being advocates for farmers and California agriculture, on the one hand, and advocates of environmental improvement, on the other. Henke shows how scientists struggle to negotiate their own ethical, professional, and environmental interests in interplay with the exigencies of the farm political economy and threats posed by the regulatory state. Henke argues that despite diverse structural constraints and sociocultural pressures, agricultural scientists find ways to preserve some degree of autonomy as relatively independent agents who defy cynical treatment by academic and environmentalist critics as "pawns of industry." In a provocative conclusion, he extends Woodhouse's call for increased activism by scientists to ethnographers of science—agents of change he believes are advantageously positioned to make important contributions to decisions about environmental regulation.

Following Frickel's and Henke's focus on scientists as political actors, the next three chapters address ways in which nonprofessionals are organizing to induce change in knowledge production systems. Rachel Morello-Frosch and her coauthors are interested in the origins, tactics and successes of what they call embodied health movements. EHMs, these authors argue in chapter 9, have emerged in response to the scientization of policy debates that have turned political and moral issues into scientific questions that threaten to prevent nonscientists from having a voice in policymaking. They show that asthma, Gulf War Syndrome, and environmental breast cancer activists have leveraged their own bodily experiences to effect significant change in levels of research funding, law, media coverage, and approaches to research. The growth and effectiveness of EMHs can be traced not only to their ability to exploit new interest in the bioethics of research but to their ability to leverage scientific uncertainty, and to build alliances with scientists at the same time that they challenge research questions, methods, and treatment options.

In the past twenty years, an impressive body of science studies scholarship on the relationship between power and knowledge has grown. In chapter 10 Brian Martin argues that however important this scholarship might be for academics and policymakers, it is usually of little help to citizens who wish to develop ways to create alternative visions of science, or a science in which citizens have more active roles in funding, creating, and using technoscience. Eschewing critique in favor of providing strategies for action, Martin explores visions of alternative science based on the concepts of "science for the people" and "science by the people." Each of the four visions he offers describes a distinctive role for citizens and scientists, giving more or less power to either group. Enacting such visions can be accomplished using several strategies, including pressure-group politics, state-led transformation, "living the alternative" by engaging in "science by the people," and through grass-roots empowerment. Martin uses the example of defense technology to show how these strategies have shaped and can shape power relations among states, citizens, and scientists.

Our final chapter in part 2, by Kelly Moore, is also concerned with the origins and political significance of nonprofessionals' involvement in scientific knowledge production. Her goal is to identify some main institutional origins and challenges to science of three kinds of "participatory science": professional-initiated, amateur, and grass-roots science. She traces the growth of nonprofessionals' involvement in the production of knowledge to legal changes in the 1960s and 1970s that required citizen participation in environmental policy and the development of protocols for human subjects, and to social movements that treated the knowledge of ordinary people as legitimate and essential for the accurate and just creation of scientific knowledge. Drawing on examples from public health research, ornithology, ecological restoration, and air pollution activism, Moore argues that each form poses specific challenges to our understanding of what constitutes science. Citizen-initiated reforms have also presaged new roles for scientists in legal and regulatory arenas, a topic we turn to in the book's third section.

SCIENCE AND THE REGULATORY STATE

In chapter 12, Steven Epstein examines a wave of recent reforms at NIH and FDA that signal a paradigm shift in biomedical research in which the assumption of a "standard human" is rapidly being replaced by a new biomedical model that recognizes important differences in human variation, and incorporates those differences in research design and the development of clinical therapies. Now, in order to receive federal funds or licensing, academic researchers and pharmaceutical companies are required to incorporate women, racial and ethnic minorities, children, and the elderly into their studies as research subjects. Epstein uses this case of policy change to explore the "politics of categorization" whereby actors moving between science, the state, and social

demonstrates legal compliance but stops far short of the serious institutional change that higher-credit courses taught by full professors in regular classroom settings would indicate.

The final chapter in this section and the volume, by Maren Klawiter, examines ongoing struggles within biomedicine to redefine risk and reconfigure clinical drug users. She describes "pharmaceuticalization" as a process that has involved industry and regulatory efforts to legitimate the extension of pharmaceutical drugs from treating diseases to treating disease risk. Situating her analysis within the context of cancer prevention, Klawiter relates a series of struggles that took place during the 1990s over tamoxifen (brand name: Nolvadex), a drug that industry promoted as a technology of disease prevention but whose efficacy would require the reconfiguration of healthy women as end users. Klawiter argues that important changes in the regulatory regime and therapeutic environment have reorganized the institutional field in which the production of biomedical knowledge about pharmaceutical technologies is made credible to doctors, patients, regulatory agents, and biomedical researchers. Ironically, the reconfiguration of institutional actors has strengthened the voices of patients, consumers, and end users while at the same time enhancing the power as well as the vulnerability of the pharmaceutical industry.

Conclusion

We began our introduction by noting the connection between the contemporary geopolitical condition and the fast-growing body of NPSS research that seeks to understand the changing place of science within the larger political economy. Another useful lens for gauging the arguments we offer are analyses that emphasize the ritualistic, discursive, and symbolic dimensions of science and science policy. Our contention is that without explicitly linking social action to structural conditions and forces, social studies of science have experienced and will continue to experience difficulty addressing science's role in shaping and responding to some of the most fundamental social issues of our time.

The irony here is that over a period in which the interconnections of politics and science have become more pervasive, the field has developed largely through attention to the interactions and discursive worlds of scientists themselves. While valuable, this theoretical and methodological turn toward scientific life has directed attention away from the ways in which scientific research is embedded in and entwined with systems of economic, political, and legal power. We believe that an unintended consequence of the relative inattention scholars have given to the structural features of power in science has been a failure to develop the field's capacity to shape debate in science and science policy (Martin 1993). Thus, we hope that our arguments will be read as a double engagement—with real-world politics and with the narrower politics constituted as academic debate. We think they are not unrelated and that finding new ways to deepen those interconnections are worthwhile intellectual and political goals.

It is in this spirit that we have sought to sharpen the justification for an analytical focus on science as it interpenetrates markets, states, and civil society. As the chapters that follow attest, disquisitions within these broad domains reveal much about the complex and changing distributions of power associated with knowledge production and about the varied ways that rule-making procedures and organizational networks combine to alter institutional arrangements or to resist such change. As important as these three domains are, perhaps more instructive for the larger NPSS project is recognizing the themes and questions that cut across them and help unify the fifteen chapters in this volume. Although readers will likely find several to choose from, four issues repeatedly rose to our attention. We pose them here as prescriptive questions to guide future research: How do intersecting or overlapping logics shape the content and conditions of knowledge production? What are the primary formal and informal mechanisms of institutional change in science? Which ones matter most in shaping the trajectory of scientific research and its broader dissemination and use? What impacts do those mechanisms have on access to knowledge and technology? The authors in this volume answer these questions in different ways. Taken together, these chapters provide an intellectual scaffolding for solidly grounded and analytically focused analyses of what is political about science—and why.

Acknowledgments

We kindly thank Elisabeth S. Clemens, Steven Epstein, Neil Gross, David Hess, Rachel Schurman, and Rhys H. Williams for insightful comments on an earlier version of this chapter.

Endnotes

1. Some culturally based postcolonial and feminist analyses of science and technology (e.g., Downey and Dumit 1997; Harding 1998) share our concern with understanding why science benefits some groups more than others. We differ with these scholars in our understanding of formal power relationships, organizations, and networks as the bases of inequality.

2. Contra rational choice theorists (e.g., Coleman 1990), who understand interests as being fixed, clearly spelled out, and acted upon with a relatively full complement of distinct choices that promise to enhance self-satisfaction, we retain the notion that people have interests that may be based on altruism as well as or instead of personal benefit, and their interests are shaped by the institutions in which they find themselves.

3. Besides being relational, we recognize too that power comes in many different forms—for example, as individual acts of resistance or as discursive or symbolic representations (Foucault 1978; Scott 1990)—but we see these as generally less consequential for larger-scale social and political change.

4. To be sure, not all organizational change affects power relationships in significant ways. Some changes may be symbolic rather than substantive (Meyer and Rowan 1977), for example, or may be short lived.

5. In its U.S. focus, this volume decidedly does *not* represent the international breadth of relevant NPSS research.

6. After all, the emergence of supra-state alliances has not meant that nation-states have withered away (Weiss 1998), nor has the advent of e-mail activism diminished the influence of political parties or interest groups. We suspect that the same is true for science. For all the talk of interdisciplinarity, disciplines carved into the organizational structure of North American and European universities in the late nineteenth and early twentieth centuries remain the primary axes for producing and certifying public knowledge; the addition of new citizen voices on NIH review panels has not drowned out the choirs of experts and public officials making science policy.

References

Abraham, John, and Tim Reid. 2002. "Progress, Innovation, and Regulatory Science in Drug Development: The Politics of International Standard-Setting." *Social Studies of Science* 32:337–370.

Akrich, Madeleine, and Bruno Latour. 1992. "A Summary of a Convenient Vocabulary for the Semiotics of Human and Non-human Assemblies." In *Shaping Technology/Building Society: Studies in Sociotechnical Change*, ed. W. E. Bijker and J. Law, pp. 259–264. Cambridge, MA: MIT Press.

Alford, Robert, and Roger Friedland. 1991. "Bringing Society Back In: Symbols, Practices and Institutional Contradictions." In *The New Institutionalism in Organizational Analysis,* ed. W. W. Powell and P. J. DiMaggio, pp. 232–263. Chicago: University of Chicago Press.
Arditti, Rita, Pat Brennan, and Steve Cavrak. 1980. *Science and Liberation.* Boston: South End Press.
Aronowitz, Stanley. 1988. *Science as Power: Discourse and Ideology in Modern Society.* Minneapolis: University of Minnesota Press.
Barnes, Barry. 1974. *Scientific Knowledge and Sociological Theory.* London: Routledge & Kegan Paul.
Bell, Susan. 1994. "Translating Science to the People: Updating the New *Our Bodies Ourselves.*" *Women's Studies International Forum* 17:9–18.
Ben-David, Joseph. 1991. *Scientific Growth: Essays on the Social Organization and Ethos of Science.* Berkeley: University of California Press.
Bijker, Wiebe E., and John Law. 1992. *Shaping Technology/Building Society: Studies in Sociotechnical Change.* Cambridge, MA: MIT Press.
Blume, Stuart S. 1974. *Toward a Political Sociology of Science.* New York: Free Press.
Bourdieu, Pierre. 1984. *Distinction: A Social Critique of the Judgment of Taste.* Cambridge, MA: Cambridge University Press.
Brown, Phil. 1987. "Popular Epidemiology: Community Response to Toxic Waste-Induced Disease in Woburn, Massachusetts." *Science, Technology, & Human Values* 12:78–85.
———, ed. 2002. *Health and the Environment.* Thousand Oaks, CA: Sage.
Callon, Michel. 1995. "Four Models for the Dynamics of Science." In *Handbook of Science and Technology Studies,* ed. S. Jasanoff, G. E. Markle, J. C. Petersen, and T. Pinch, pp. 29–63. Thousand Oaks, CA: Sage.
Casper, Monica. 1998. *The Making of the Unborn Patient: A Social Anatomy of Fetal Surgery.* New Brunswick, NJ: Rutgers University Press.
Clarke, Adele E. 1998. *Disciplining Reproduction: Modernity, American Life Sciences, and the Problems of Sex.* Berkeley: University of California Press.
Clarke, Adele E. and Joan H. Fujimura. 1992. *The Right Tools for the Job: At Work in Twentieth-Century Life Sciences.* Princeton, NJ: Princeton University Press.
Clarke, Adele, and Virginia L. Oleson, eds. 1999. *Revisioning Women, Health, and Healing: Feminist, Cultural, and Technoscientific Perspectives.* New York: Routledge.
Clarke, Adele E., Janet K. Shim, Laura Mamo, Jennifer Ruth Fosket, and Jennifer R. Fishman. 2003. "Biomedicalization: Technoscientific Transformations of Health, Illness, and U.S. Biomedicine." *American Sociological Review* 68:161–194.
Clemens, Elizabeth S. 1997. *The People's Lobby: Organizational Innovation and the Rise of Interest Group Politics in the U.S., 1890–1925.* Chicago: University of Chicago Press.

Clemens, Elisabeth S., and James M. Cook. 1999. "Politics and Institutionalism: Explaining Durability and Change." *Annual Review of Sociology* 25:441–466.

Coleman, James S. 1990. *Foundations of Social Theory*. Cambridge, MA: Harvard University Press.

Collins, Harry M. 1983. "The Sociology of Scientific Knowledge: Studies of Contemporary Science." *Annual Review of Sociology* 9:265–285.

DiMaggio, Paul J., and Walter W. Powell. 1983. "The Iron Cage Revisited: Institutional Isomorphism and Collective Rationality in Organizational Fields." *American Sociological Review* 48:47–160.

Downey, Gary Lee, and Joseph Dumit, eds. 1997. *Cyborgs and Citadels: Anthropological Interventions in Emerging Sciences and Technologies*. Santa Fe, NM: School of American Research Press.

Edelman, Lauren. B. 1992. *Legal Ambiguity and Symbolic Structures: Organizational Mediation of Civil Rights Law*. Chicago: University of Chicago Press.

Emirbayer, Mustafa. 1997. "Manifesto for a Relational Sociology." *American Journal of Sociology* 103:281–317.

Epstein, Steven. 1996. *Impure Science: AIDS, Activism, and the Politics of Knowledge*. Berkeley: University of California Press.

———. 1998. "History and Diagnosis of 'Scientific' Medicine." *Social Studies of Science* 28:489–495.

Etzkowitz, Henry, Andrew Webster, and Peter Healy, eds. 1998. *Capitalizing Knowledge*. Albany: State University of New York Press.

Fischer, Frank. 2000. *Citizens, Experts, and the Environment: The Politics of Local Knowledge*. Durham, NC: Duke University Press.

Foucault, Michel. 1978. *The History of Sexuality*. Vol. 1: *An Introduction*. Trans. R. Hurley. New York: Vintage.

Frickel, Scott. 2004. *Chemical Consequences: Environmental Mutagens, Scientist Activism, and the Rise of Genetic Toxicology*. New Brunswick, NJ: Rutgers University Press.

Frickel, Scott, and Neil Gross. 2005. "A General Theory of Scientific/Intellectual Movements." *American Sociological Review* 70:204–232.

Fujimura, Joan H. 1996. *Crafting Science: A Sociohistory of the Quest for the Genetics of Cancer*. Cambridge, MA: Harvard University Press.

Galison, Peter, and David J. Stump. 1996. *The Disunity of Science: Boundaries, Contexts, and Power*. Stanford, CA: Stanford University Press.

Gieryn, Thomas F. 1999. *Cultural Boundaries of Science: Credibility on the Line*. Chicago: University of Chicago Press.

Granovetter, Mark S. 1973. "The Strength of Weak Ties." *American Journal of Sociology* 78:1360–1380.

Gross, Matthias. 2003. *Inventing Nature: Ecological Restoration by Public Experiments*. Lanham, MD: Lexington Books.

Hacking, Ian. 1999. *The Social Construction of What?* Cambridge, MA: Harvard University Press.

Hagstrom, Warren O. 1965. *The Scientific Community*. New York: Basic Books.
Halpern, Sydney A. 2004. *Lesser Harms: The Morality of Risk in Medical Research*. Chicago: University of Chicago Press.
Harding, Sandra. 1998. *Is Science Multicultural? Postcolonialisms, Feminisms, and Epistemologies*. Bloomington: Indiana University Press.
Hassanein, Neva. 2000. "Democratizing Agricultural Knowledge through Sustainable Farming Networks." In *Science, Technology, and Democracy*, ed. Daniel Lee Kleinman, pp. 49–66. Albany: State University of New York Press.
Hilgartner, Stephen. 2001. "Election 2000 and the Production of the Unknowable." *Social Studies of Science* 31:439–441.
Hoffman, Lily M. 1989. *The Politics of Knowledge: Activist Movements in Medicine and Planning*. Albany: State University of New York Press.
Kleinman, Daniel L., ed. 2000. *Science, Technology, and Democracy*. Albany: State University of New York Press.
———. 2003. *Impure Cultures: University Biology and the World of Commerce*. Madison: University of Wisconsin Press.
Kloppenburg, Jack R. 1988. *First the Seed: The Political Economy of Plant Biotechnology, 1492–2000*. Cambridge: Cambridge University Press.
Knorr-Cetina, Karin. 1999. *Epistemic Cultures: How the Sciences Make Knowledge*. Cambridge, MA: Harvard University Press.
Kroll-Smith, Steve, Phil Brown, and Valerie J. Gunter. 2000. *Environment and Illness: A Reader in Contested Medicine*. New York: New York University Press.
Kuhn, Thomas S. 1970 [1962]. *The Structure of Scientific Revolutions*. Chicago: University of Chicago Press.
Latour, Bruno. 1987. *Science in Action: How to Follow Scientists and Engineers through Society*. Milton Keynes, UK: Open University Press.
———. 2004. "Why Has Critique Run Out of Steam? From Matters of Fact to Matters of Concern." *Critical Inquiry* 30:225–248.
Latour, Bruno, and Steve Woolgar. 1986. *Laboratory Life: The Construction of Scientific Facts*. Second Edition. Princeton, NJ: Princeton University Press.
Lounsbury, Michael. 2001. "Institutional Sources of Practice Variation: Staffing College and University Recycling Programs." *Administrative Science Quarterly* 46:29–56.
Lounsbury, Michael, and Marc Ventresca. 2003. "The New Structuralism in Organizational Theory." *Organization* 10:457–480.
Lukes, Steven. "Power and Agency." 2002. *British Journal of Sociology* 53:491–496.
MacKenzie, Donald, and Graham Spinardi. 1995. "Tacit Knowledge, Weapons Design, and the Uninvention of Nuclear Weapons." *American Journal of Sociology* 101:44–99.
Mannheim, Karl. 1991 [1936]. *Ideology and Utopia*. London: Routledge.
Martin, Brian. 1993. "The Critique of Science Becomes Academic." *Science, Technology, & Human Values* 18:247–259.

1

The Commercialization of Science

2

Contradiction in Convergence

Universities and Industry in the Biotechnology Field

DANIEL LEE KLEINMAN and STEVEN P. VALLAS

Understanding the emerging knowledge economy should be a central concern of science and technology studies. In this chapter, we hope to contribute to that understanding and to simultaneously illustrate the virtues of a broadly organizational and institutional approach to the study of what Pierre Bourdieu called the "scientific field." We focus on what we term "asymmetrical convergence" (Kleinman and Vallas 2001)—the process by which the codes and practices of industry and the academy are increasingly traded across the boundary between the two. We aim to strengthen existing conceptualizations of that process through analysis of data collected through interviews with academic and industry employees doing work in what can broadly be termed *biotechnology*.

In a way that seems very healthy to us, the field of science and technology studies has moved from being overshadowed by a limited number of approaches and analysts and has become richly fragmented in recent years, with a wide array of orientations being used to explore an equally diverse range of subject matter. At the same time, the legacy of the broad array of work that might generally be termed *social constructivist* continues to deeply shape the contours of the field and the work done in it. Concepts like actor network (e.g., Callon 1986), boundary work (e.g., Gieryn 1999), and social worlds (e.g., Fujimura 1988) have been widely drawn upon, and names like Trevor Pinch (e.g., Pinch and Bijker 1989), and especially Bruno Latour (e.g., 1987), continue to ring loudly in science and technology studies. And well they should. What we have learned from the waves of scholarship that dominated science and technology studies since the cracking of the old "institutional" paradigm should not be understated,[1] and that technoscience, the social/technical boundary, and nature itself are constructed cannot be denied. At the same time, in the rush to look at all phenomena as socially constructed, something has been lost (see Kleinman 2003). We have tended to overlook the ways in which the *already constructed* can shape social phenomena of all varieties, especially and including the technoscientific field.

By arguing for attention to the already constructed, we are not implying that the social world is static. Indeed, the focus of this chapter is on transformation: the simultaneous change of the academic and high-tech industry fields. However, this transformation and ultimate blurring is shaped by deeply entrenched norms and practices—broadly, cultures—that constrain actors in their efforts to remake their respective fields. In undertaking an analysis of what we have termed *asymmetrical convergence,* we draw inspiration from a set of literatures that until very recently have had only the most marginal influence on science and technology studies: the new institutionalism in organizational analysis (e.g., DiMaggio and Powell 1983; Schneiberg and Clemens forthcoming), economic sociology (e.g., Saxenian 1994), and new class theory (e.g., Gouldner 1979).

Our theory of asymmetrical convergence contends that an uneven and contradictory process of convergence is occurring, increasingly bringing together previously distinct institutional fields. Social codes and practices from one domain (already constructed and deeply institutionalized) are imported into the other, yielding new structures of knowledge production that defy existing characterizations. On the one hand, science-intensive firms find it useful to invoke academic conventions, such as the publishing of journal articles, sponsoring of intellectual exchanges, and supporting curiosity-driven research (though in complex and often contradictory ways that articulate with corporate goals). On the other hand, academic institutions increasingly resort to entrepreneurial discourses and practices[2] (though these are often reinterpreted and recast in terms more palatable to the professoriat; see Owen-Smith, this volume). The result, we contend, generates contradictions, anomalies, and ironies that violate long-held normative understandings yet increasingly pervade both university and industrial laboratories.

The process of convergence we are witnessing is not equivalent to simple isomorphism where two previously distinct domains directly mirror one another. Instead, each domain adopts elements of the other in ways that are not always straightforward, and integrates these with aspects of its own domain, creating two new hybrid regimes. Each adopts the codes and practices of its counterpart domain in ways that aim to facilitate the realization of its distinctive objectives, creating something like an *inverted mirror* of the other, where each is a contradictory amalgam that conjoins distinct sets of normative orientations in a novel matter.

Importantly, this process is marked by the ongoing coexistence of conflicting logics within institutional fields that are dominated by one or the other of these logics. Thus, the movement of entrepreneurial norms into the university can contribute to behaviors (e.g., data hoarding) that are fundamentally at odds with widely held views of the purpose of the university. At the same time, industry scientists may press their employers to institute practices that provide no obvious benefit to firms (e.g., early scholarly publication or seminars at firms including university scientists). The result often introduces tension and conflict into the organization, as different actors debate ways of resolving such disparities (for an example within manufacturing settings, see Vallas 2003). In this sense, the convergence process is contradictory, and these contradictions reflect the fundamentally different histories and objectives of industry and academia. We call the process *asymmetrical convergence* because although the emerging hybrid regimes are constructed of codes and practices from both sides of the divide between industry and academia, "in the last instance" it is the logic of profit that is shaping this process.[3] Industry adopts attributes of academic culture in the interest of increasing profitability, and academia draws on private-sector codes and practices for either directly commercial purposes or indirectly because of the legitimacy universities gain by adopting elements of commercial culture.

The data we present below provide an image of institutional change that is saturated with contradiction and irony. Universities, once viewed as sites of collegiality in which researchers had a measure of autonomous control of their work process, are increasingly viewed as places where scientists feel compelled to respond to the constraints of the commercial world and where administrators place a high priority on "entrepreneurialism" among their faculty. At the same time, work in many science-intensive firms is characterized by a cooperative spirit and freedom from managerial pressures that researchers in today's university seldom enjoy. Two consequences seem to result. First, and ironically, scientists often find that the conditions and resources needed to support traditional academic norms—the most sophisticated equipment, the most generous budgets, the greatest distance from entrepreneurial pressures—are most readily available in corporate laboratories. Second, as respondents in each domain become more aware of the other's modes of operation, distinct and divergent logics increasingly coincide within *both* university *and* corporate laboratories, yielding an often-contradictory and uneven process of convergence in which the academic and industrial science are much less distinct than before. Important changes are indeed affecting the structure of knowledge production in the United States, but understanding them adequately will require broader and more inclusive perspectives than researchers have thus far employed.

We begin this chapter by outlining the essential tenets of our theory of asymmetrical convergence. Briefly alluding to previous approaches, we tease out key propositions that we contend characterize the restructuring of knowledge production that is currently underway. Then, after describing the methods used to generate our interview data, we present an empirical analysis that focuses on the changing structure and culture of scientific research. We begin with a discussion of trends apparent in academic science and then consider parallel developments found among industrial laboratories. Our evidence suggests that organizational logics commonly associated with one institutional domain are today found on both sides of the university/industry divide. We conclude by offering some informed speculation on the major sources of variation in this process of asymmetrical convergence.

Engaging the Theory: The Process of Asymmetrical Convergence

In the early 1980s, amid rising concern for U.S. economic competitiveness, postwar understanding of the public role of scientific and academic research began to undergo significant change. Once viewed as a *relatively* autonomous realm, university research was increasingly construed as a source of economic and technological innovation. Universities, endowed with newfound powers of patenting and intellectual property protection as the result of both developments in biology and changes in federal policy (especially the Bayh-Dole Act of 1980), were encouraged to play an increasingly central role in the process of capital accumulation. Amid rising fiscal constraints on public spending (and with social entitlements placing limits on public support for higher education), university administrators increasingly looked to market-based sources for much-needed resources. The result, many suggested, involved a historically significant shift in the very logic that traditionally informed university research (Slaughter and Leslie 1997; see also Owen-Smith and Powell 2001; Powell and Owen-Smith 1998; Etzkowitz and Webster 1998).

Early research on university-industry relationships (UIRs) focused on the apparent erosion of traditional academic freedom and autonomy that accompanied the increase in formal contractual relationships between academic biologists and biotechnology companies (see, for example, Blumenthal et al. 1986; Shenk 1999). Much of this literature voiced concern over the ways in which joint ventures of various types between universities and corporations, or academic efforts to foster licensing arrangements or patent protection, threatened both the free flow of knowledge and the autonomy of scientific research. At about the same time, a parallel literature on knowledge workers in the corporate sector arose, where industry was perceived as providing an increasingly favorable, autonomous realm in which scientists and engineers might enjoy the trappings of flexible and collaborative work situations (Saxenian 1994; Albert and Bradley 1997; Powell 2001).

Each of these literatures is troublingly ahistorical in its analysis. As we have argued and shown elsewhere (Kleinman and Vallas 2001; Kleinman 2003), although the American university has changed since the late nineteenth century, it never mapped well onto the ivory tower image with which analysts often represent it. For one thing, the formal connections between academia and industry, which have prompted such extensive discussion since the 1980s, are by no means entirely novel (Noble 1977:110; Geiger 1993:284; McMath et al. 1985:189; Lowen 1997:75).

Looking at the organization of industrial research, the historical pattern is again more complex than more recent analysts allow. According to historian David Hounshell, throughout most of the twentieth century, research directors of corporate labs worked to give industrial researchers "a semblance of an academic research environment" as a way to attract university trained scientists (1996:26, 27; see also Mees 1920). Such efforts were unevenly successful throughout the twentieth century.

While formal university-industry relations and industrial scientist autonomy are not entirely novel, what *is* new is the emergence of a broad structural trend in which previously distinct organizational fields have begun to converge or co-evolve. As we have argued elsewhere (Kleinman and Vallas 2001), commercial codes and practices increasingly pervade academia, even as academic codes and practices are imported into science-intensive firms. The result of such a commingling of norms and practices—a process we term asymmetrical convergence—is a hybrid knowledge regime in which each field is a fractured mirror of the other. Thus, university science is increasingly marked not only by deeply institutionalized norms concerning the value of basic knowledge but also by an essentially commercial culture in which competition and entrepreneurship are buzzwords, and intellectual property is the coin of the realm. In similar fashion, knowledge-intensive firms are driven not only by the profit imperative but also by conventions that support research autonomy and the value of scholarly publication.

The convergence we observe in academic and industrial science is explained by several factors. Among them is the mobility of scientists and other professionals across the divide between the two realms—a pattern not yet sufficiently studied, but one that seems likely to reorder inherited prestige hierarchies among professional scientists (Smith-Doerr 2004; Leicht and Fennell 1998). In addition, corporate concerns for legitimacy in the eyes of investors pressure firms to establish their scientific currency, whether by hiring "star" scientists or by publishing in the most prestigious journals (Stephan 1996). At the same time, universities seek to legitimate their activities, especially as public institutions come under growing pressure to demonstrate their "relevance"—a goal often interpreted in terms of the academic contribution to private-sector economic growth. Despite these pressures and the process of convergence they promote, each field retains characteristics linked to its broadly accepted position in the U.S. social formation. Thus, to varying degrees universities retain commitments to educating citizens and curiosity-driven research, and knowledge-intensive firms rarely lose sight of commitments to product production and profit. As we noted at the outset of this chapter, we term the convergence process "asymmetrical" in recognition of the greater power wielded by the logic of capital and the goal of economic competitiveness that is the guiding premise of public policy toward scientific research. More specifically, firms adopt academic norms and practices ultimately in the name of profit, while universities adopt profit-making practices in the face of neo-liberal fiscal policies that have made governmental resources insufficient for universities' needs.

To enhance our understanding of asymmetrical convergence, we undertook a set of semi-structured interviews with scientists, support staff, and administrators (N=80) doing biotechnology-related work in industry and academia. Our interviews were undertaken between the fall of 2001 and the summer of 2002, and approximately half were with respondents from six universities—three in California and three in Massachusetts. The other half of our interviews were with a parallel sample of workers from fourteen biotechnology companies. Again, roughly half were from Bay Area firms and half from Massachusetts-based firms. We selected the San Francisco Bay Area and Boston and vicinity because both areas have high concentrations of knowledge-intensive industries (Saxenian 1994), and both represent important centers of research in biotechnology. All of the university researchers we interviewed conduct broadly similar work in biotechnology-related biology; the universities that employ them include both public and private Research I institutions.[4] All firms in our study do diagnostic or medical-related biotechnology; none do agriculture-related biotechnology. The sample of firms was stratified to assure variation by firm size and age. Thus, we interviewed at both small start-up firms and large, well-established and highly prominent corporations. Our respondent samples were stratified to reflect the full array of positions in the university (assistant professor, associate professor, full professor, postdoctoral fellow, graduate student, technician, and administrator) and in firms (PhD scientists leading research groups, more junior PhD scientists, research associates/ technicians, and managers). Interviews were semi-structured and ranged from forty-five to ninety minutes. Questions fell into five topic areas: background, organization of work or management, information flow or intellectual property, gender in the work environment, and careers and occupational culture.

Findings

THE CHARACTER OF SCIENTIFIC WORK IN ACADEMIC BIOTECHNOLOGY

The existing literature on university-industry relations (UIRs) has often focused on the importance of direct and formal relations between university scientists and companies. And indeed, the university scientists we interviewed talked about joint ventures, partnerships, and agreements between their departments and private industry. Only a very few, however, indicated that they currently had a working relationship with a private sector firm. This finding is consistent with national-level studies that have consistently found that only a minority of researchers receive industry funding (see work cited in Kleinman and Vallas 2001: 455). We suggest that while these formal UIRs are not unimportant, their actual presence in a given academic setting is not the most important part of the story. UIRs may signal an institutional change and may even serve to prompt further shifts in academic culture. Yet, as our interview data suggest, even in the absence of direct, formal transactions between industry and academia, broad changes in the culture of university science are underway that are at least partially independent of UIRs as such (see also Kleinman and Vallas 2001; Slaughter and Leslie 1997).

This point begins to emerge in comments made by one senior scientist at an elite Boston university. Summarizing the shift he saw in the normative milieu, he said:

> [W]hen we first started this company . . . I had to take a lot of abuse from people. . . . The implication was that you were doing it for the money and there was, there's a certain culture in academia that [laughs] . . . you're not *supposed* to make money, and if you want to make money it's, it's bad. You know it's a negative motivation. But I think that's all changed now.

Continuing, this man suggests that while academic scientists commonly disparage industrial science, many of his colleagues have come to see virtues in the pursuit of commercially relevant research.

Academic administrators seem to be at the cutting edge of these changes, serving as a major impetus to changes in academic culture. Feeling the weight of budget cuts at state and federal levels and pressure from elected officials to make academic science economically important, deans, provosts, and presidents encourage formal university-industry relations. Beyond formal UIRs, these administrators appear to be a major force in changing academic culture. One dean described the changes occurring in academia as follows:

> Right now as a university we're going through a fairly [major] search and re-evaluation of who we are and what we do, and how well we're

> doing it in view of budget cuts, how we should react and what I've tried to convince my colleagues is that it would be reasonable to think of a university as a manufacturer of capital goods. We manufacture minds, ideas, patents in some cases, and these are the capital goods that industries are built around, and I think if you view yourself as a manufacturer of capital goods, then your success is actually having your capital goods placed in a production environment somewhere, so we build students, they go off, they get jobs, they do great stuff. That's really what we're for. . . . That's why we're here.

This dean is not only describing changes; he is also telling his faculty how things *must* change.

The response of a second administrator we interviewed makes clear the pressure he is under to change academic culture: "*We are not given the privilege any longer of doing research just because we're curious about an answer*. . . . Because nowadays I think it's absolutely critical that we justify the use of taxpayer money based upon the fact that it has some potential to have impact on people." This dean suggests that "number of patents, number of companies . . . and the impact on the economy" will soon become standard criteria in tenure evaluation, if they have not already. Academic administrators increasingly feel the need to legitimate hiring and promotion decisions in a way that is consistent with a logic of capitalism, not just the university.

Especially since the academic labs in our sample reported relatively few direct transactions with corporate or for-profit organizations, the comments reported above point to a change in the *culture* of the university that seems broader and deeper than the UIRs that previous studies have stressed. Again, we do not mean to deny that formal UIRs are a significant aspect of the new knowledge economy, nor that they have played a role in reshaping academic culture or the world of high-technology industry. However, in the massive attention to UIRs, often overlooked are more subtle, *in*direct processes at work that induce academic laboratories to become increasingly, if unevenly, isomorphic with their corporate counterparts. Understanding such processes requires that we ask how and to what extent the infusion of commercially derived norms and practices is altering the work situations of university scientists in ways that are at least partly independent of direct formal relationships between firms and the university.

The contradictory character of the convergence process in the university is nicely illustrated by the simultaneous belief in traditional academic norms and the widespread use among our respondents of an entrepreneurial rhetoric. Thus, scientists we interviewed stressed their autonomy in defining their research. As one scientist characteristically noted: "Well, I'm the person in charge of the lab and my intellectual interests are the overriding concern." In addition, all of our academic scientists described their research as "basic." At the same time, many of the academic biologists we interviewed draw forcefully on a neoliberal discourse that stresses competitive and highly instrumentalized, exchange-like relations to characterize scientist-scientist interactions.

When asked about collaboration and cooperation, respondents commonly described their labs and departments as highly cooperative, but as often as not, they suggested that collaboration beyond one's lab is undertaken for narrowly instrumental reasons: to get needed information or materials.[5] Noted one respondent, an assistant professor: "if you're interested in some question and somebody has a reagent or a mouse model or animal model which might be very useful for you, and you have that, and so then you collaborate with them to get [the information or research material you need]." We were told over and over again not of an environment of mutual exchange, benefit, and cooperation, but of narrowly constructed collaboration created to advance individual agendas.

Ironically, some of our university scientists saw industry as an environment that is more likely than academia to foster real scientific collaboration (on this point, see Smith-Doerr 2004). Describing a company in which he was involved, one of our respondents said: "So in a company there's much more of a team spirit and excitement about getting something done, and so you can put pieces together in ways to accomplish things that could never be done in an academic lab." As we describe below, this view is echoed by several of our industrial respondents who previously worked in academia.

In the matter of intellectual property protection, the spread of an entrepreneurial logic in academic biology is also evident. Robert Merton stressed the idea that a distinctive attribute of academic science was its aversion to private property. Common ownership, he said, is an "integral element of the scientific ethos" (1973:273). Analysts and critics of UIRs stress the supposed increase in emphasis on private property—patenting and intellectual property considerations—in university science. This focus, according to critics, threatens to inhibit the flow of information and materials among academic scientists. However, our data show it is not just *direct* relations between university scientists and commercial enterprises (and their correlative nondisclosure or intellectual property agreements) that are changing the character of the university, as many analysts suggest. Major constraints on the sharing of information and materials that we observed stem from larger, *indirect* influences impinging on the culture of academic science. It is not only formal UIRs that have led to restrictions in the "free flow" of information and research materials. In addition, we are witnessing a sharpening competition for professional distinction. Maintaining scientific priority—a lead in the race to make a significant discovery—is more important than ever. Combined with the entrepreneurial ethos enveloping academic science, this competition has erected potent barriers to the sharing of knowledge among scientists in the same or similar fields.

Reflecting the contradictory character of the asymmetrical convergence process, overall our university respondents articulated a deep commitment to the Mertonian norm of scientific communism (Merton 1973), even as many violate it in practice. One established scientist at a Boston-area university is emblematic of this commitment. He said: "I mean it, it's been a, sort of a strong position of mine from the very beginning that I consider anything that I publish, reagents, anything, part of the public domain and send it out without any questions about what people are going to do." At the same time, reflecting the view of others, one respondent noted that "people are really not sharing things the way they used to and . . . it's becoming more competitive." This respondent captures other respondents' descriptions of their practices:

> A lot of people you know would claim that this has to do with closer interactions . . . with companies and financial interests and everything and I don't believe that. I think, I think the driving force in this are egos and career advancement and that . . . people keep those things to promote themselves, not to get rich. . . . It's mostly self-protective and it doesn't have to do with financial interests. It has to do with credit, advancement, grants, prestige, all those things and that's why I think the simple answer is that the field has become highly competitive.

Importantly, our respondents do not point to increases in patenting as an explanation for the attenuated character of the norm of information sharing. As a junior university scientist put it, restrictions on the flow of information in academia are not about patenting, but about not being "scooped." This professor said: "you can't afford to sort of let it go [because] somebody else could keep it."

In brief, our data suggest that the infusion of a commercial or entrepreneurial ethos in academic science is indeed multiplying the number of constraints that impede the flow of information, but in ways that involve broad cultural influences more often than direct (formal) influences (for similar findings, see Marshall 1997:525 and Campbell et al. 1997). Our data begin to suggest that if the commercialization of academic science impedes the flow of information (and we believe it does), this effect is a mediated one that operates as much through subtle changes in academic culture as direct and explicitly legal arrangements or financial transactions. University scientists increasingly act according to the logic of the market and industry, independent of direct relations with private firms.

The data presented in this section point to several important conclusions concerning what we have termed asymmetrical convergence. First, insofar as our data are representative, a commercial culture appears to be infusing academia and increasing the acceptability of profit as a motive for scientific work and the prospects for evaluation of merit of academic scientists according to private-sector criteria. Second, while our scientists contend they autonomously shape their research (an opportunity viewed as a traditional privilege of the "academic role"), they do not view their work as a collective endeavor. Instead, competition for resources has promoted or reinforced a kind of entrepreneurialism in academic biology. While modern academic science may have always included entrepreneurial practices to a degree, we suspect that these practices have intensified in recent years as available support from the federal government and private foundations to fund increasingly expensive biological research has not kept pace with demand (Adams, Chiang, and Starkey 2001:74).[6] Finally, our interviews suggest that a commitment to "scientific communism" sits uneasily with a belief in the need to engage in the hoarding of information and materials. Importantly, insofar as we are witnessing increasing restrictions on the flow of information and materials in academic biology, these do not appear to reflect direct commercial links between academic biologists and private firms, but a competitive culture that in many ways mirrors the market.

THE CHARACTER OF SCIENTIFIC WORK IN INDUSTRIAL BIOTECHNOLOGY

In this section, we explore two issues that figure centrally in scholarly discussions of knowledge work in industry. First, we present data on the structure of work in biotechnology firms. Next, we explore problems of worker cooperation and information sharing. Our data suggest that the character of workplace cooperation and autonomy as well as the nature of information sharing in biotechnology firms is shaped, in part, by normative pressures that push biotechnology firms to look increasingly, if unevenly, like academic settings.

Previous researchers have found that with the increasing intermingling of corporate and university activities and personnel, high-tech firms often avoid the bureaucratic character of entrenched corporations, adopting a more collegial organizational culture (Smith-Doerr 2004). Our data conform to this pattern. Indeed, reflecting the descriptions of many of our private-sector interviewees, one of our respondents stressed the "minimum of bureaucracy" at the firm where he works. But it is not just the relative absence of bureaucracy that characterizes this scientist's work experience. It is the ironic ways in which his flexible work environment offers more of the freedom he seeks at work than did his previous academic position. Thus, he says:

> [There is more] time that one can concentrate on the science, [compared to] my experience as an assistant professor . . . [despite the] so called academic freedom. . . . From my experience one saw more and more a push [in academia] to do experiments that would create the next grant funding. That sort of antagonism between what will create grant funding and what might be in the interest of the scientists exists in academia just as similar antagonisms exist in companies. With the fight for funding in universities, one saw less and less of the opportunity for individual creativity and more and more of a move in biological sciences to what one could call the big projects where individuals don't have much say anymore in what they do. They were just playing a small role in part of a much bigger project.

Looking at his experience, this scientist drew conclusions much like ours:

> I think the work in the companies is becoming very much like the work of the academics. They're converging. I think a lot of academicians who are supposed to always pride themselves on being pure scientists are looking more and more for the commercialization of their ideas and that I believe is being [encouraged] by the universities for their licensing departments.

Graduate students we interviewed also pointed to the changing character of the university as a reason to seek private-sector employment. As one West Coast graduate student noted:

> I don't think I could lead a[n academic] lab. I would not want the responsibility of being in control of directing an entire research effort, and being in control of other people's lives in terms of funding. I mean . . . if you're a PI and you can't get funding then you can't pay your graduate students and you can't pay people relying on you. I think maybe in industry I could hopefully get into a position where I wouldn't have to be in such authority.

Here again is an ironic inversion of the traditional, if ultimately never realized, norm: this graduate student is considering employment in industry, *precisely to escape the managerial or entrepreneurial pressures he encounters in the university*. For these respondents and others, industry seems to offer the freedom and autonomy widely considered characteristic of academia.

These views notwithstanding, the stereotypical view of scientific research in corporate settings holds that scientific research is conceived, designed, and conducted along lines ultimately envisioned by corporate managers (cf. Dubinskas 1988). The unevenness of the spread of academic culture into industry and the sometimes uneasy way industrial and academic practices meld in the biotechnology industry mean that there is an element of validity in this view. Indeed, several of the firms in our sample had adopted measures designed to focus their research activities more tightly on the discovery of saleable products than they had previously done. Managers and scientists recognize the potential contradiction between the profit imperative of industry and the autonomy imperative of academia. They are engaged in a difficult balance in which management aims to make scientists' desire for autonomy and control serve the interest of profit and not undermine it.

At most of the firms we studied, scientists retained substantial measures of autonomy in the conduct of their work and were often at liberty to pursue research of their own design. In the more tightly controlled firms, scientists were expected to gain formal approval for their proposals (a hurdle that could at times be informally circumvented). Yet at several of the other firms—both small start-up firms and large, highly prominent corporations—the organizational culture and practices assumed forms that drew freely from academic norms and conventions in many important respects.

Indeed, at many of the biotech firms in our sample, one would be hard pressed to distinguish organizational routines from those found within academic laboratories. This was especially true at one large commercial enterprise in the Bay Area. Here is an excerpt from the interview transcript:

Q: Do you have the opportunity to suggest projects you'd like to work on, simply because they seem intellectually interesting to you?

A: Yes, we have kind of the mandate in our department, and it might be different in others. We're more basic research focused where we can either work on what is a drug candidate, a therapeutic problem that's totally obvious that people at [the firm] want to study, or you can work on just about anything else provided it will be published in *Science* or *Nature*. So you know that's hard to say because you don't know what that's going to be until you get there but if you're doing that kind of glamorous, visibly recognized science, it doesn't matter what it is because that's something we also care about is our reputation within the scientific community for doing cutting-edge research, and you're supported in doing that. So I feel the freedom to say you know this is really hot, and we really need to study this. No it's not a drug, but it's something important.

Here, the firm quite clearly provides lavish support for basic (even curiosity-driven) research, despite the commercial goals it must achieve. Importantly, the firm supports basic research of the most highly visible variety and not more routine edifice-building research.

But there is unevenness in the spread of academic norms in corporate settings,[7] and at firms in which academic norms were less thoroughly institutionalized, scientists enjoyed correspondingly lower levels of autonomy and control over the conduct of their research. Most common is an arrangement in which general research areas and goals are set by management, and scientists then design or lead projects that fit within those priorities. A typical comment in such settings was that "I think I have a lot of freedom to do whatever I'm interested in, as long as it's within . . . [the] broad focus of the company and I get to pick my own . . . research directions and . . . it's the freedom I really like."

The managers we interviewed seemed quite conscious of the need to retain elements of autonomy for their research scientists while at the same time maintaining an overall focus on the development of profitable therapeutic products. Several respondents spoke of an ongoing dialogue among bench scientists, the director of research, and lead managers. Said one scientist at a small Boston firm:

> [The director of research] comes up with ideas that he kind of throws out there. Some of, most of, them I have not accepted. I also come up with ideas that I throw at him. And my ideas are based on what I see is

> a need in the . . . subgroup. So that's pretty much two-way about how [we decide topics of research.]

At the largest firms, there was more variation, with some firms adopting formal programs of control and accountability over scientific work. Yet even in such relatively controlled settings, there tends to be little managerial oversight of day-to-day work activities. Less experienced scientists may receive guidance from their directors, in ways that resemble the mentorship of an academic lab, as this director described:

> I would say [the choice of research methods is] mostly up to them, but I do see a lot of these people I've hired have not really had experience doing this before so I read a lot of stuff and I say maybe we could try it like this and I show them the paper and they'll go read the paper and then they'll either decide yeah let's do it that way or we'll do it a better way.

We do not wish to overstate the extent of scientist autonomy and intellectual freedom in these biotechnology firms. None of our firms precisely reflects the academic myth of scientist autonomy, and there is clear variation in the control that firms allowed scientists in the selection of topics for research. Still, at many of our firms, managers clearly made substantial concessions to professional autonomy, in keeping with norms typically assumed to guide the practices of academic scientists. As we discuss below, where firms embrace academic norms, their policies appear to reflect legitimacy pressures coming from their own scientists, prospective employees, investors, and other firms.

In contrast to the circumstances we found in academic biology, where an ethos of competitive relations often obstructed the flow of information across laboratories and universities, within many industrial labs we found a set of relations that often adhered more closely to the academic ideal than was the case in actual university settings. Commonly, the companies we studied were able to engender a cooperative set of social relations among their research personnel that stressed the value of collaborative work and the sharing of information and techniques across different ranks and departments within the firm. Work processes were generally informed by an ethos of mutual support, and information within the firm flowed quite openly. Said one scientist, in an observation that was common in many of our industrial labs:

> I think there's a feeling that everyone's sort of in this together, that . . . the success of one project only helps the success of other projects and so at least from what I've seen so far . . . any time someone can offer some valuable input on a project, they will, and it's used.

Such a normative environment did not spontaneously emerge, of course. Rather, we found that science directors and human resource managers seemed to make a conscious effort to engender such a culture in their labs. Said one human resources director: "The open environment [that] we provide is one that [allows] for collaboration, the hallways and the bathrooms, wherever you are, to try to get people to engage in conversation, talk out ideas, and it's a very open environment that way." Likewise, a manager in another firm spoke of his company's conscious effort to create an open, collaborative culture in its labs:

> Well—we do things to try to enhance [information sharing]. . . . We have monthly research meetings which are not to present data but to present plans for research, to organize research going forward. . . . Even in the new facility we have areas that we call collaborative areas which are just off the lab. People can go and sit down and talk about projects or . . . read, or discuss. It's important because everybody needs to know what is going on in other projects so that there's not redundancy. . . . [So,] because people have worked on many different things, there is sort of an interdependence.

Of course, this openness is limited: it is within firms, not across them. Still, the implication is an ironic one: in these labs, managerial initiatives within profit-oriented firms have led to cooperation and collaboration, whereas the laissez-faire freedom of the university has often engendered atomization and autarchy instead. Managers maintain academic norms both to satisfy their scientists *and* to enhance the prospects for profitability.

For many of our industry scientists, the collaborative character of industrial work situations provided a marked contrast with what they had previously encountered in the university setting. One scientist, who was fairly new to working in industry and was very enthusiastic about the opportunities for collaboration his company offered, said:

> I saw more opportunity for, for the type of things that I was interested in, in industry. You know in academics . . . you tend to get a little isolated in what you're working on and it involves quite a bit of effort to initiate collaborations. . . . Whereas in a company, they need to have the

> whole research outfit within the company so that you have all these different [forms of] expertise that you would ever need there.

Several scientists at commercial firms reported that their decisions to work in industry instead of academia were linked to precisely this desire for teamwork and to their feelings of isolation in academia. As one research director noted, "one thing that was really missing in [academic] science is the opportunity to work on a team." The irony here is that so-called academic norms are more likely to be realized in corporate settings than in universities, and in some cases, this reality drives scientists from academia to industry.

A central feature of the scholarly world is publication. On this score, the long-held concern has been that commercial enterprises will be loathe to allow publication of their results, for doing so places privately owned knowledge in the public domain. To be sure, several of the firms in our study looked askance at such a prospect and made few, if any, provisions for journal publication. However, this orientation applied to a minority of the firms in our sample. Driven in part by the scholarly orientations of the scientists they recruit, and partly by the commercial benefits they derive from publication of their research (McMillan, Steven, Narin, and Deeds 2002), many of our firms actively embraced the academic convention of journal publication, making ample provision for scientists to publish the results of company research.

Our interviews made it clear that the norm of journal publication was strongly favored by the scientists employed at the biotech firms. Said one scientist at a large Bay Area corporation:

> We want to be successful scientists and we're, in many ways I would say, academics at heart and we know that . . . when you're excited about what you do, you want to tell people about it and you want to get their ideas and so you have to be open and if you're not talking to people on the outside, you get . . . kind of stuck in one way of thinking. It's just . . . part of the whole scientific process . . . If you're not publishing, if you're not going to conferences, then of course . . . *you're not a real scientist* . . . [Something we] care about is our reputation within the scientific community for doing cutting-edge research and [here], you're supported in doing that.

Another scientist at a small firm in Massachusetts emphasized a more practical view, but one that equally supported the practice: "Personally, you can keep current in your publications and your CV. We all know that with a Ph.D. you're *supposed* to publish."

Many of our respondents pointed to the benefits that firms seemed to derive from lending support to the publication of scientific results. The research director at a Bay Area firm, for example, said:

> When a company is new and trying to establish itself and establish a reputation, publications help the company establish credibility and it also establishes you . . . as part of a community of scientists and you attract people to come here to work, to give lectures here . . . we're close to some universities so we get a lot of people who give seminars, and publications are all a part of that, adding to the culture of the company.

On its face, pursuit of scholarly publication does not seem like it should be a pivotal activity for profit-seeking enterprises. However, even beyond academia, scholarly publication is recognized as a crucial cultural currency in science. In industrial biotechnology, although our interviews indicate that pursuit of publication and participation in scholarly exchange at scientific meetings is not unencumbered, it is widespread, and our respondents give two reasons for this. First, scientists coming from academic environments *expect* to engage in these practices. Managers recognize these expectations and believe that permitting publication and scholarly-like communication will aid in the recruitment, motivation, and retention of the best scientists. Publication and information exchange establishes a firm's legitimacy in the eyes of scientists within the firm. Second, managers at many firms contend that the pursuit of journal publication—with the proviso of legal protection in advance—provides an effective means with which they can establish or enhance their corporate legitimacy in the eyes of key players in the industry: venture capitalists, stockholders, other firms, universities with whom they wish to interact, and influential scientists.

The data we highlight in this section lead us to several conclusions. First, while the drive of "knowledge for knowledge's sake" is often asserted to be the impetus for flexibility, cooperation, and the free flow of information that early analysts viewed as the crux of the academic endeavor, our data suggest that under certain conditions the pursuit of profit may be more likely to foster these dimensions of scientific practice than do academic norms. Second, our interviewees imply that deeply institutionalized notions of science (ironically, academic science) appear to play a substantial role in pushing industrial science to look, in many ways, like the academic ideal. That is, managers, industrial scientists, and investors all have ideas—fundamentally taken for granted—about what science should look like, and managers and private-sector scientists push to enact this vision.

Discussion

This study suggests that a complex process is underway in which academic codes and practices are increasingly woven into industry, and vice versa. On the university side of the equation, our data challenge the idea that direct, formal ties with industry are threatening the traditional autonomy once the preserve of academic science. Recognizing that there have always been limits on such autonomy, at a general level our biologists retain a considerable degree of autonomy. At the same time, however, we see considerable growth of an entrepreneurial and competitive spirit among many university biologists. The practices that result from this spirit (e.g., restrictions on the flow of information) are often not the direct result of formal university-industry relations. Still, they do reflect the spread of private-sector values into the academic domain. Although our data cannot definitively address the specific conduits through which such influences flow, they do suggest that administrators in particular are acting upon the signals that many legislators and patrons have sent concerning the new mission of the university. The emphasis on pursuing commercial ventures, on seeking external revenues, and on establishing lines of research that address patentable topics—these and other priorities—are abundantly apparent in the university environments we studied.

It is important to reiterate the point that these pressures are not entirely new. Still, such entrepreneurial pressures appear to have accelerated in recent years, apace with the increased cost of "big biology," at a time when available resources for academic science cannot keep up with scientists' demands for these funds. Although our discussion has not addressed the relationship between scientific discipline and asymmetrical convergence, our data begin to suggest that such pressures are likely to be unevenly felt across different scientific specialties. Smaller fields with resources adequate to meet demand are less likely than larger fields with intense resource demands to adopt an entrepreneurial spirit and competition-related practices.

Our study further suggests that as commercial enterprises have enlisted scientific knowledge as a key factor of production, they have at the same time invoked institutional logics and practices traditionally associated with the university. At least one explanation for the cooperative nature of work organization in biotechnology firms is the professional culture in which firms are at least partly embedded and the values that scientists import into biotechnology firms. The result is again ironic: mindful of the greater material resources that firms enjoy, many of our industrial scientists report that their employers provide a greater opportunity to realize academic values than does the university. Especially in light of the respondents we interviewed who left the university precisely because of its entrepreneurial emphasis, it seems safe to conclude that in some settings at least, *companies have become more "academic" than academia itself.*

Predictably, many biotechnology firms have policies in place to protect intellectual property rights. While this certainly leads to restrictions on the flow of information between firms and other organizations—recall some of the incidents reported by our university respondents—many of our firms actively encourage scholarly publication by their scientists, and many of the scientists view publication as an important outlet for their work. Firms appear to have two primary motivations for encouraging publication. On the one hand, doing so enables them to conform to the values that scientists are likely to import into the firm, thus recruiting, retaining, and motivating the most sought-after scientists through appeal to academic norms (Albert and Bradley 1997). On the other hand, some of our firms, although certainly not all, saw in scholarly publication a powerful means of establishing or enhancing organizational legitimacy in the eyes of scientists and investors.

Although we cannot develop the point in the present context, we believe that a key element in the emerging, albeit uneven, convergence between university and industry is the convertibility of academic capital into economic capital and back. Indeed, successful alliances between university departments and commercial enterprises have as their object precisely this goal. Where one partner in a university-industry initiative may derive enhanced opportunity for publication, the other may derive augmented revenues. Ironically enough, *the firms* may be more interested in the publications, while *the universities* are primarily interested in the accumulation of revenues. Of course, for firms, accumulation of academic capital is ultimately a means to economic capital.

Although our data suggest that the general outlines of a novel and inherently contradictory knowledge regime are beginning to come into view, it is important to note the existence of critical sources of variation in the outcomes we observed. One such variation involves significant inter-firm differences in attitudes toward publication. Though much more research is needed on this point, our data begin to suggest the presence of a U-shaped pattern affecting the level of support for academic culture and practices among corporate labs. Simply put, provisions for academic practices such as publishing and relative autonomy for scientists were most apparent among two types of firms: smaller, privately owned start-up firms that were still establishing themselves within the industry and larger, relatively prominent firms with recognized positions and solid revenues. The *least* support for traditional academic practices, by contrast, seemed to occur among middle-sized, publicly traded firms that had not yet established an ongoing market position, whether through the reputations of their scientists or any revenue-generating products. In other words, it was within firms that held the most precarious market positions, yet were subject to growing pressure to enhance their commercial performance, that academic norms seemed to hold little currency. Under other conditions, industry seemed quite capable of assimilating academic practices to commercial ends, finding important elements of congruence between the accumulation of capital and of academic prestige.

A second source of variation in the outcomes we observed often occurred in conjunction with shifts in the power of particular actors within the organization. Again, much more research is needed on this point, but in a number of cases we found evidence of ongoing debate among parties holding different orientations toward the value of basic or applied research. Some of our biotech firms appear to have undergone a sudden shift from a culture that encouraged discovery-oriented research toward a more applied or pragmatic outlook. Additionally, our interviews unearthed evidence of dissension concerning a given organization's emphasis—a point that was especially apparent in academic laboratories and between administrators and faculty. The point here is that the knowledge regime that seems to be emerging is not a simple, static, or uniform structure but is instead a volatile, precarious, and shifting arrangement that our research suggests is the product of cross-cutting pressures from disparate elements in the environment.

If, as our research begins to reveal, biotechnology provides an instance in which institutional boundaries are rapidly being reconfigured and in some cases even dismantled, our findings may hold theoretical significance far beyond the field of biotechnology as such. Institutionalist theories of organizational structure have until recently tended to speak the language of stasis, durability and uniformity, and legitimation through conformity to overarching normative demands (Meyer and Rowan 1977; DiMaggio and Powell 1983). The case of biotechnology, an important element within the knowledge economy, reminds us that multiple and often-conflicting codes and practices can coexist over time, generating unstable and inherently contradictory organizational practices. Contrary to the assumptions that informed the early institutionalist literature, our research suggests that isomorphism need not be accompanied by consensus, stability, and consistency. Although commercial and scientific norms are increasingly conjoined, for example, their confluence need not generate a homogeneous field but instead may give rise to *hybrid* and inherently contradictory institutional regimes (Vallas 2003) that reflect the competing demands that impinge on firms or organizations in different ways, promoting isomorphism in some areas and diversity in others. Recognizing the diversity that is often latent in institutional fields and the conflicting logics that it produces may in the end help us better understand the shifting character of organizational life and the ways in which organizations respond to their wider environments.

Although our data cannot fully address the point, we believe that the structure of knowledge production in the life sciences is the outcome of crosscutting influences that often pit the inclinations of administrators, legislators, and corporate actors against the institutionalized cultures of professional scientists and faculty members. Mindful of the greater power of commercial interests—a point that, as we noted earlier, leads us to speak of an asymmetry in the convergence we detect—we also note that scientists are capable of mobilizing lines of action that can modify, mediate, or contest aspects of the emerging knowledge regime. Indeed, the coordinates of such conflict constitute an important area for future research. How does the mix of orientations toward the commercialization of science vary across the different tiers of higher education? Are the outlooks of administrators in public institutions driven more by pressures from state legislatures, by growing interactions with corporate personnel and managerial thinking, or by concerns to shore up the public legitimacy of their universities? And how precisely does the commercialization process affect the public credibility that university science enjoys? Arguably, the legitimacy won through economic development must be set alongside growing threats to the claim of objectivity and concern for the public good. These and other questions remain important avenues that future research will need to pursue.

In one sense, our work is much like the earlier generation of broadly social constructivist studies. We are interested in the construction of a new knowledge regime and in the way new boundaries are being constituted. However, our work does not show the de novo construction of a fact, a field, or a phenomenon. Nor would we argue that construction or (re)construction is possible at every moment. We are at a particular historical juncture. It is a period of crisis in the sense of the term used by institutional political economists. Here, the rise of science-intensive firms has made possible the reconstruction of industry, while fiscal crises have made possible the reorientation of academia. Beyond the factors that make the construction of this hybrid field possible—which really serves as the background to our study—our data shed light on the ways in which established norms, practices, and structures shape how a hybrid culture is being constructed. This is a construction process in which actions are shaped by existing norms and practices even as these recombine in novel ways. Agents are not free to institute practices as they see fit. Instead, the movement of personnel across the industry-academic divide, and legitimacy pressures tied to deeply seated notions of what a scientist is and what the commercially viable should look like, are driving the construction of this convergent yet contradiction-ridden field. Actors on both sides must work to balance established notions of what universities and firms are, even as they twist and transform them.

This chapter draws on interviews with workers in biotechnology-related fields in academia and industry as a means to explore questions raised by our theory of asymmetrical convergence. Our findings begin to illustrate the need for research that traverses the university/industry divide and so is able to explore the structural and normative mechanisms that increasingly conjoin these previously relatively distinct institutional domains. Needed are additional studies of this sort that focus on other scientific fields and center on particular facets of such institutional interaction.

Acknowledgments

This chapter was originally prepared for presentation at the meeting of the Society for the Social Studies of Science meetings in Atlanta, Georgia, in October 2003. We thank the participants in our study and those, including Sam Gellman, who helped us gain access to research sites. In addition, we gratefully acknowledge funding from the Fund for the Advancement of the Discipline (American Sociological Association/National Science Foundation), the Georgia Tech Foundation, and the Graduate School of the University of Wisconsin–Madison. Finally, we thank Abby Kinchy and Raul Necochea for their research assistance and Elisabeth Clemens, Scott Frickel, Kelly Moore, Woody Powell, Karen Schaepe, Marc Schneiberg, Laurel Smith-Doerr, and Robin Stryker for their comments on earlier versions of this chapter. The authors contributed equally to the conceptualization of this chapter.

Endnotes

1. Here we refer to the work of the "father" of the sociology of science Robert K. Merton. Merton viewed science as a central social institution and focused a large part of his attention on those factors that provided the basis for the smooth functioning of science and gave science its distinctive character. See Merton (1973).

2. When we refer to entrepreneurial discourses and practices, we mean to capture the sense that faculty increasingly operate in highly competitive (not cooperative), self-interested, and individualistically oriented ways. We are not using entrepreneur as Shumpeter did to refer to actors who combine the familiar in novel ways.

3. Owen-Smith (this volume) offers a different, but largely compatible, analysis of the emergence of the contemporary university as a hybrid. Wolf's chapter in this collection nicely illustrates how actors can be involved in a process of co-construction that is asymmetrical. Finally, Woodhouse (this volume) captures the disproportionate capacity of industry to shape social outcomes.

4. Importantly, the experiences of the scientists we interviewed are likely to be different than the experiences of faculty at non–Research I institutions. Such variations among universities constitute an important issue for future research. See Owen-Smith 2003.

5. Owen-Smith (this volume) suggests that the current transformation of U.S. universities is associated with increased collaboration, but he says nothing about the quality of this collaboration. Increased collaboration may mean more instrumental, narrowly focused collaboration, and if this type of collaboration is currently associated with economic innovation, this may be a benefit that will not persist over time.

6. A colleague who read our chapter challenges this claim, noting the massive budget increases at the National Institutes of Health during the 1980s and 1990s as well as industry support for university biology research. We recognize these increases, but our sense is that the available dollars do not meet the demand for support. Certainly this is the perception of our university respondents

and many of our colleagues in the biological sciences. They are constantly anxious about securing the funds needed to keep their research programs going.

7. That is, some academic norms are uniformly adopted across firms, others are partially adopted by most firms, and still other academic conventions are found in some firms but not others.

References

Adams, James D., Eric P. Chiang, and Katara Starkey. 2001. "Industry-University Cooperative Research Centers." *Journal of Technology Transfer* 26: 73–86.

Albert, Steven, and Keith Bradley. 1997. *Managing Knowledge: Experts, Agencies, and Organizations*. Cambridge: Cambridge University Press.

Blumenthal, David, Michael Gluck, Karen Seashore Louis, Michael A. Stoto, and David Wise. 1986. "University-Industry Research Relations in Biotechnology: Implications for the University." *Science* 232 (June 13): 1361–1366.

Callon, Michel. 1986. "Some Elements of a Sociology of Translation: Domestication of the Scallops and the Fishermen of St. Brieuc Bay." In *Power, Action, and Belief: A New Sociology of Knowledge,* ed. John Law, pp. 196–233. Sociological Review Monograph no. 32. London: Routledge and Kegan Paul.

Campbell, Eric G., Brian R. Clarridge, Manjusha Gokhale, Lauren Birenbaum, Stephen Hilgartner, Neil A. Holtzman, and David Blumenthal. 2002. "Data Withholding in Academic Genetics: Evidence from a National Survey." *Journal of the American Medical Association* 287:473–480.

DiMaggio, Paul, and Walter Powell. 1983. "The Iron Cage Revisited: Institutional Isomorphism and Collective Rationality in Organizational Fields." *American Sociological Review* 48:147–160.

Dubinskas, Frank. 1988. "Cultural Constructions: The Many Faces of Time." In *Making Time: Ethnographies of High-Technology Organizations,* ed. Frank A. Dubinskas, pp. 3–38. Philadelphia: Temple University Press.

Etzkowitz, Henry, and Andrew Webster. 1998. "Entrepreneurial Science: The Second Academic Revolution." In *Capitalizing Knowledge: New Intersections of Industry and Academia,* ed. A. Webster H. Etzkowitz and P. Healey, pp. 21–46. Albany: State University of New York Press.

Fujimura, Joan. 1988. "The Molecular Biological Bandwagon in Cancer Research: Where Social Worlds Meet." *Social Problems* 35(3): 261–283.

Geiger, Roger. 1993. *Research and Relevant Knowledge: American Research Universities Since World War II.* New York: Oxford University Press.

Gieryn, Thomas F. 1999. *Cultural Boundaries of Science: Credibility on the Line.* Chicago: University of Chicago Press.

Gouldner, Alvin. 1979. *The Future of Intellectuals and the Rise of the New Class.* New York: Oxford University Press.

Hounshell, David. 1996. The Evolution of Industrial Research. In *Engines of Innovation: U.S. Industrial Research at the End of an Era,* ed. Richard S. Rosenbloom and William J. Spencer, pp. 13–85. Boston: Harvard Business School Press.

Kleinman, Daniel Lee. 2003. *Impure Cultures: University Biology and the World of Commerce.* Madison: University of Wisconsin Press.

Kleinman, Daniel Lee, and Steven Vallas. 2001. "Science, Capitalism, and the Rise of the 'Knowledge Worker': The Changing Structure of Knowledge Production in the United States." *Theory and Society* 30(4): 451–492.

Latour, Bruno. 1987. *Science in Action.* Cambridge, MA: Harvard University Press.

Leicht, Kevin, and Fennell, Mary. 1998. "The Changing Context of Professional Work." In *Annual Review of Sociology,* ed. John Egan. Palo Alto, CA: Annual Reviews.

McMillan, G. Steven, Francis Narin, and David L. Deeds. 2000. "An Analysis of the Critical Role of Public Science in Innovation: The Case of Biotechnology." *Research Policy* 29:1–8.

Marshall, Eliot. 1997. "Secretiveness Found Widespread in Life Sciences," *Science* 276 (April 25): 525.

McMath, Robert C., Jr., Ronald H. Bayor, James E. Brittain, Lawrence Foster, August Giebelhaus, and Germaine M. Reed. 1985. *Engineering the New South: Georgia Tech, 1885–1985.* Athens: University of Georgia Press.

Mees, C. E. Kenneth. 1920. *The Organization of Industrial Scientific Research.* New York: McGraw-Hill Book Company, Inc.

Merton, Robert K. 1973[1942]. *The Sociology of Science.* Chicago: University of Chicago Press.

Meyer, John, and Brian Rowan. 1977. "Institutionalized Organizations: Formal Structure as Myth and Ceremony." *American Journal of Sociology* 83:340–363.

Noble, David. 1977. *America by Design: Science, Technology, and the Rise of Corporate Capitalism.* New York: Oxford University Press.

Owen-Smith, Jason. 2003. "From Separate Systems to a Hybrid Order: Accumulative Advantage Across Public and Private Science at Research One Universities." *Research Policy* 32:1081–1104.

Owen-Smith, Jason, and Walter Powell. 2001. "Careers and Contradictions: Faculty Responses to the Transformation of Knowledge and Its Uses in the Life Sciences." In *The Transformation of Work: Research in the Sociology of Work,* vol. 10, ed. Steven P. Vallas. Greenwich, CT: JAI Press.

Pinch, J. Trevor, and Wiebe E. Bijker. 1989. "The Social Construction of Facts and Artifacts: Or How the Sociology of Science and the Sociology of Technology Might Benefit Each Other." In *The Social Construction of Technological Systems: New Directions in the Sociology and History of Technology,* ed. Wiebe E. Bijker, Thomas P. Hughes, and Trevor J. Pinch, pp. 17–50. Cambridge, MA: MIT Press.

Powell, Walter. 2001. "The Capitalist Firm in the Twenty-First Century: Emerging Patterns in Western Enterprise." In *The Twenty-First-Century Firm: Changing Economic Organization in International Perspective,* ed. Paul DiMaggio, pp. 33–68. Princeton, NJ: Princeton University Press.

Powell, Walter, and Jason Owen-Smith. 1998. "Universities and the Market for Intellectual Property in the Life Sciences." *Journal of Policy Analysis and Management* 17(2): 253–277.

Saxenian, Annalee. 1994. *Regional Advantage: Culture and Competition in Silicon Valley and Route 128.* Cambridge, MA: Harvard University Press.

Schneiberg, Marc, and Elisabeth S. Clemens. Forthcoming. "The Typical Tools for the Job: Research Strategies in Institutional Analysis." In *Bending the Bars of the Iron Cage,* ed. Walter Powell and Dan Jones.

Shenk, David. 1999. "Money + Science = Ethics Problems on Campus." *Nation,* March 22, 11–18.

Slaughter, Sheila, and Larry Leslie. 1997. *Academic Capitalism: Politics, Policies and the Entrepreneurial University*. Baltimore: Johns Hopkins University Press.

Smith-Doerr, Laurel. 2004. *Women's Work: Gender Equality vs. Hierarchy in the Life Sciences*. Boulder, CO: Lynne Rienner.

Stephan, Paula E. 1996. "The Economics of Science." *Journal of Economic Literature* 34:1199–1235.

Vallas, Steven P. 2003. "Why Teamwork Fails: Obstacles to Workplace Transformation in Four Manufacturing Plants." *American Sociological Review* 68(2): 223–250.

3

Commercial Imbroglios

Proprietary Science and the Contemporary University

JASON OWEN-SMITH

The last thirty years have witnessed dramatic increases in the depth and variance of university involvement in commerce. Academic institutions play more complicated commercial roles than ever before. University endeavors range from the prosecution and marketing of intellectual property, to active venture capital investment, to intimate involvement in the far-flung contractual networks that are the knowledge economy's center of gravity. The increasing commodification of academic research and development (R&D) has spawned new professional groups, shifted faculty career trajectories, reworked academic stratification hierarchies, and sparked transformations in the organizational infrastructure on many campuses. This latter transformation is particularly important because technology-licensing offices maintain a permeable boundary between the academy and industry.

Alterations to the institutional and organizational arrangements that underpin the traditional academic research mission have prompted a wide range of responses. They have been characterized as "revolutionary" (Etzkowitz, Webster, and Healey 1998) and as key components of the social "shockwave" that is transforming the academy (Kerr 2002). Economically oriented pundits laud the legislation (the 1980 Bayh-Dole Act) that sparked dramatic acceleration in university research commercialization as a "golden goose" responsible for the nation's economic resurgence.

> [S]uffice it to say that the sole purpose of the Bayh-Dole legislation was to provide incentives for academic researchers to exploit their ideas. The culture of competitiveness created in the process explains why America is, once again, pre-eminent in technology ("Innovation's Golden Goose" 2002).[1]

The same outcomes (e.g., faculty's increased concerns with exploiting their profitable findings and a proprietary culture of competitiveness in the halls of the academy) generate deep concerns about the health of the university itself on the part of more reformist critics (Press and Washburn 2000; Bok 2003). As Sheldon Krimsky (2003) frames the issue, proprietary science, private gain, and a culture of secrecy and competitiveness in the academy stand to destroy the primary benefit open science offers to a democratic society—a cadre of autonomous and passionately dedicated experts whose secure positions enable them to "speak truth to power."[2] The loss of an academic platform for concerns involving the public good is, in this view, a devastating side effect of increasing commercialization.

Whether appreciative, admonitory, or agnostic, observations about academic research commercialization share common points of departure and thus manifest little disagreement over the causes of the alterations they observe. Estimations of the consequences of commercial endeavors, though, diverge. Debates about academic commercialization rest, at least implicitly, on a notion that academic and proprietary science represent distinct and contradictory institutional regimes, and that the university's role as a citadel of science requires a unique place in civil society.

Collapsing distinctions between public and proprietary science shift arrangements and practices on campus, reposition universities relative to other institutions, and reforge the diverse linkages that constitute the modern academy. Such shifts necessarily alter both the university and the system of which it is a part. Commercialization's effects derive from systematic transformations in the relationship between basic and commercial science on campus and shifts in the university's broader connections to society. Whether such changes are taken to be beneficial or detrimental depends less on the processes by which realms converge than on expectations of academe's characteristic features and their resilience in the face of the market. Commercial engagement may corrode the academy's core, but the features that are at risk are not natural; they are outcomes of distinctive and contingent historical processes. Concerns about the academy's health and prognostications about its future must be grounded in the systematic changes that occur at the intersection of academic and commercial spheres.

Strong distinctions between academic and proprietary knowledge rest on a Mertonian conception of science as an institution characterized by a system of strong norms and distinct rewards (Merton 1968, 1973; Dasgupta and David 1987, 1994). Here, science and its organizational home (the university) contribute to the social order precisely by remaining separate from it.

Commercial engagements flout this division. Changes in the university can reverberate to alter adjacent structures, and most of civil society is adjacent to academe. Understanding commercialization's effects requires that we appeal to both the stability and distinctiveness of academic arrangements *and* to the knowledge that these structures are fragile and can be damaged by contradiction. If shifting fault lines within the university are to matter for society as a whole, we must imagine that internal transformations will resonate beyond the citadel's walls. In short, debates about research commercialization span institutional and agentic conceptions of science while situating academic institutions in a complex, changeable network that establishes their character.

University science holds an important place in multiple orders of meaning and can be understood along multiple dimensions. In what follows, I explore the implications of research commercialization in terms of the intersection between a view that emphasizes the stability and autonomy of academic science and an approach that highlights the university's reliance on the diverse networks in which it is embedded. The former view understands commercialization as a process that merges two separate institutional orders to create a hybrid system where new opportunities for mobility reshuffle the status hierarchy that has governed academic science for half a century. The latter considers commercialization in terms of the changes it makes in the university's relationship to the market, the government, the citizenry, and other constituencies. Instead of emphasizing integration and changing opportunities, this lens offers insight into the effects that shifting social position has on expectations for the university and its science.

Academic research commercialization offers a viable site to examine institutions and hierarchies as outgrowths of the same phenomena that constitute and stabilize individual and collective actors. The American research university is simultaneously an institution of and an actor in societal transformation. Its stability and efficacy result from dense connections to other actors and institutions. Deepening commercial engagement generates a hybrid university in at least two senses: by grafting (e.g., collapsing two distinct sets of rules for creation and dissemination of new innovations in the same location) and by translation (e.g., by extending an entity's reach across separated fields of endeavor, in a supple and contingent network whose solidity constitutes the actor itself) (Latour 1993). I explore these dual notions of hybridity in the context of a transforming university embedded in an established political economy with an eye toward demonstrating a conceptual interconnection between the stable, institutionalized, hierarchical social orders that convey power, and the more flexible, effortful, extensive networks that underpin construction.

Ontology, Power, and Change

Political economies of power and being are fundamentally relational. Power results from relative position in an objectified hierarchical order where the distribution of resources and opportunities favors those at the top (Wright 1984). Ontology is a largely political accomplishment of enrolling and stabilizing diverse participants into a constellation of relationships (Latour 1987; Mol and Law 1994). While the first confronts participants with the necessities of a reified structure, the second constructs real actors through situated, pliable, and far-from-inevitable extensions.

Possibilities for change, too, vary across these two different conceptions. Hierarchical orderings—particularly those that match structurally-based resource inequity with ascriptive distinctions and expectations—are stable and capable of resisting all but the most fundamental change. The structures and institutions of power are durable sources of constraints, interests, and differential opportunities. Change occurs, but it is most likely to do so at the margins (Phillips and

Zuckerman 2001) or in interstices generated by disjuncture and overlap (Friedland and Alford 1991; Sewell 1992; Clemens and Cook 1999). In durable systems of power, novelty springs up from the crevices.

In contrast, flux is the natural state of extensive networks. Stability is effortful but necessary to the appearance of objectivity and the mobility of things. Beyond continuous translation, resilience can result from processes that obscure contingency and shroud underlying linkages. If social ontologies reside in the relational performances of actors, then the attributes of systems are only as stable as the weakest of their assembled components and their capacities for self-repair. Within an established social order it is easy to equate destabilization with danger.

These two dimensions of relationality bracket considerations of outcomes for the commercializing research university. Explanations for observed variations in commercial and academic success on campus emerge from examinations of contingent processes of transformation. Attending to the timing and trajectory of universities' entry into new competitive arenas offers insights into both the opportunity structures and perils of a changing field. Yet transformations that simply reshuffle the hierarchical ordering of a single class of incumbents miss the point that universities exist in a field choked with established players of many sorts. While a view of change in terms of grafting emphasizes timing and point of entry to explain mobility in suddenly destabilized systems, changes via extension remind us that tracking connections across classes of actors is also essential and that the effects of such connections need not be felt symmetrically (Kleinman and Vallas, this volume). Diverse networks and institutionalized stratification orders, then, offer alternative views of the same process, and those views are complementary rather than exclusive.

In what follows I turn to discussions of the ways in which increasing academic research commercialization can be understood in terms of shifting hierarchical orders within the existing system of academia, and at the same time as a set of extensions and translations that reposition the academy and constitute novel types of actors and arrangements. Before grounding these seemingly contradictory conceptions of power and construction in a consideration of the transfigured, "entrepreneurial" university, I sketch an argument that has some potential to resolve the tensions inherent in simultaneous focusing on durable, constraining systems of power and contingent, accomplished networks of being.

Linking Topologies of Power and Emergence

Knitting together disparate theoretical traditions exceeds this chapter's scope. Nevertheless, I take a few steps to ease discomfort with the attempt at bricolage to come. Linking social ontology and stratification requires conceptualizing history and dynamics. The former suggests a source of observed disparities in the power of actors, while the latter highlights features of a system in motion. History suggests the importance of trajectories, and institutional dynamics generate windows of opportunity to leave them. Contingency suggests not only localization in time and space but also the possibility that things might have been otherwise, that any particular arrangement emerges from specifiable processes.

Real things emerge from contingent and situated procedures undertaken in the context of existing interests, relations, and actors (Latour 1999). Contestation occurs during stabilization because emerging entities gain their observable attributes through the pattern of their relationships, but those relationships also alter existing configurations. But ontological performances and power exercises can flop. Some entities are stillborn (Latour 1996), and mandated changes can meet resistance or mere symbolic compliance from established players (Smith-Doerr, this volume). Things are no less real for having been constructed, but construction always occurs in a space littered with prior successes and failures.[3] It is precisely that reality, the very apparent, even intransigent thingness of these artificial entities that provides the crosscutting sets of interests, possibilities, and potential sources of resistance for future enrollment efforts. Things may emerge from diverse networks, but that emergence does not occur in a clear field.

Stabilized entities resist change. Opposition is all the more significant when the contingent sources of existing arrangements are obscured. Whether we think in terms of the black-boxing of facts and artifacts, the processes of rationalization that transform normal situations into normative conditions (Meyer and Rowan 1977; Zucker 1977), or the routinization of organizational actions (March and Simon 1958), processes that obscure contingency in an air of natural inevitability reduce multiple orders to a single constellation of taken-for-granted arrangements (Law 1994). It may well take more than a laboratory to raise a world (Latour 1983; Kleinman 2003), but it is all too easy to equate a world not of one's own making with a world that is not, in fact, made.[4] That equation is wrong. Actors, objects, interests, and institutions are conditionally constituted and, once stabilized, structure their own and others' futures.

Grafting and translation offer alternate views of similar phenomena. Grafting opens possibilities for change within a sector by creating new possibilities for mobility that can shuffle the hierarchies that support fields of power and interest. Translation envisions change as resonance, where forging new linkages fundamentally alters the characteristics of all components in a system. Here horizontal relationships reconfigure the meaning and characteristics of existing and emerging actors. The evolution of a distinctively "entrepreneurial" university partakes of both types of processes. Ever more complex commercial engagements alter both the academic stratification system and the cross-sectoral linkages that make an autonomous, "ivory tower" university sensible.

University Hybridities

Increasing commercialization sparks both vertical and horizontal change. Proprietary approaches to the development and dissemination of scientific and technical findings are increasingly important formal components of the university research mission, and commercial success can be leveraged into academic mobility (Owen-Smith 2003).

The university is increasingly conceptualized as an "engine of economic development," a "creator and retailer of intellectual property," and a source of competitive advantage rather than a citadel of science (Feller 1990; Chubin 1994; Slaughter and Rhoades 1996). That conceptual shift repositions the university relative to its constituencies and, by doing so, alters the characteristics of both.

Changes at work in the research university are simultaneously transformations to the institutional arrangements and stratification orders characteristic of "the academy" and to the broader cross-sectoral linkages whose stabilized patterns opened and maintained the possibility of an academic, ivory-tower university in the first place.[5] The sections that follow offer broad-brush empirical support for the simultaneous existence of these dual shifts before turning to a preliminary consideration of where these linked transformations leave the contemporary "entrepreneurial" university.

GRAFTING SEPARATE SYSTEMS INTO A HYBRID ORDER

If change in commercially engaged universities results from tensions at the interstices of contradictory institutional orders, then the interface between academic and proprietary approaches to research opens a window onto transformation.[6] The distinction between these systems is most easily grasped in terms of their characteristic outputs. Patents are the coin of the proprietary realm, while publications are the primary credit in the academician's ledger. Arie Rip's (1986) evocative analogy linking patents to fences and academic publications to funnels speaks to the distinctive characteristics of each.

Patents and publications represent alternate institutions that validate and disseminate the outcomes of contemporary research efforts. Both publications and patents are markers of accomplishment and means to disseminate information. Publications, however, are funnels precisely because their success depends on the breadth of future use. Articles represent an author's formal release of control over the uses to which a finding is put.[7] "Ownership" is largely a matter of peer recognition and priority. Returns to publication accrue in reputation, and rewards are intimately linked to others' uses of an author's findings. In short, articles have (and to be successful must have) no presumption of exclusivity. In contrast, patents are fences in the sense that they demarcate a "plot" of knowledge legally owned by its inventor. Property ownership conveys a bundle of rights. Excludability (the right to prevent others' use of your property) and appropriability (the right to capture economic returns from the use of your property) are central, and their various implications have generated much of the hue and cry that has accompanied academic commercialization.

Patent ownership also establishes priority. Returns to patenting are primarily pecuniary. Ownership is recognized through a bureaucratic process and thus divorced from recognition and the use of inventions by others. While still tethered to fecundity, rewards associated with patenting are a matter of controlling and monitoring others' uses of findings in order to extract a portion of the revenues that come from sale of a product based on the technology (royalty income) or payment for the right to practice a protected invention (licensing fees).

During the Cold War era, academic and proprietary approaches to science and engineering remained (at least rhetorically) separate. Publications were the territory of academics, and patents were concentrated in industry. The increasing commercial engagement of universities—and particularly the huge upswing in academic patenting, which now accounts for 5 percent of all U.S.-owned patents (National Science Board 2002)—heralds significant blurring of the institutional boundaries between the academic and commercial spheres.[8] More importantly for an understanding of shifting arrangements within the academy, increased patenting signals the importation of divergent mandates, constituencies, rewards, and practices to the academy.

The changes that accompany such transitions may be subtle but pervasive, adding novel criteria to the evaluation of universities and their scientists and altering established standards for judging success (Owen-Smith and Powell 2001). At a supra-organizational level, widespread university patenting has opened a new arena for academic competition as an established set of players from one highly stratified competitive field (academic research) enter another (proprietary research). The explicit integration of public and private science orientations in the academic research mission has fractured the established status order that structured possibilities for success in "traditional" academic competitions. By altering the rules of the game to open an alternate route to achievement, research commodification catalyzes the emergence of a hybrid institutional system characterized by positive feedbacks across commercial and academic uses of science (Owen-Smith 2003).

STRATIFICATION BY ACCUMULATIVE ADVANTAGE

The academic stratification order certainly represents a constraining structure that is not made by any single scientist or university. In the winner-take-all world of priority races, grant competitions, and reputational rewards, success breeds success. Merton (1968) characterized this stratification process in famously biblical terms: "for whosoever has, to him shall be given, and he shall have more abundance." Within a system governed by the Matthew Effect, where increasing returns to success are driven by peer evaluation based on reputation, chances for mobility are limited, and the detrimental effects of transformations may be felt disproportionately at the bottom of existing hierarchies.[9]

Nevertheless, opportunities and resources that enable mobility can arise in institutional channels distinct from the peer-review system that reinforces existing status differentials. Commercial engagement and particularly patenting—by virtue of its separation from academic standards of validation and its potential to return unrestricted income flows to universities and researchers—offers just such an alternative.[10] Ownership and marketing of intellectual property (IP) supports efforts to generate blockbuster royalty streams and increases trends toward equity ownership in faculty start-up companies (Association of University Technology Managers 1999). Returns to commercial engagement on campus can catalyze academic success to the extent that the financial payoffs are folded back into the academic mission.

Such royalty streams are vanishingly small on most campuses and do not approach the costs of R&D on even the most successful ventures (Mowery et al. 2001).[11] Nevertheless, income streams that are not directed to specific projects and uses can offer benefits greater than their magnitude suggests. As Michael Crow, the former executive vice provost of Columbia University, noted in a recent article regarding Columbia's failed attempt to extend the life of a lucrative and fundamental biomedical patent, "This is an income stream that is absolutely critical to us. It is the single most important source of free and clear funding. Everything else comes with a string attached" (Babcock 2000). In another interview later the same week, Crow expanded his claim, noting that this free and clear funding stream enables the university to "do some things none of our other resources allows us to do" (Pollack 2000).

Other universities evince similar beliefs. Stanford University and the Wisconsin Alumni Research Foundation (WARF) use royalties to fund university-wide fellowship competitions that support faculty and student research. Likewise, Carnegie Mellon University invested more than $25 million earned from its equity stake in the Web search engine Lycos into its computer science department, funding multiple endowed chairs and building state-of-the-art research facilities (Florida 1999).

Clearly, universities and academic science can benefit from the resources that flow to their proprietary activities. Indeed, in analyses conducted on a nearly twenty-year panel (1981–1998) of 89 research-intensive universities, I found an increasing pattern of positive linkages across patenting activity and the impact of published articles on campus. Several universities, including Emory, Rutgers, and Columbia, saw dramatic increases in their publication impact ratings that accompanied even more impressive growth in the size of their intellectual property portfolios (Owen-Smith 2003). While importing commercial mandates to universities offered some opportunities for mobility in the reputation-based world of academic science, such mobility is by no means assured, and attempts to leverage resources across academic and commercial efforts can carry dangers of their own.

Consider two well-known cases of such attempts: Boston University's controversial engagement with its spin-off firm, Seragen, and Columbia University's attempts to extend the life of their blockbuster patent on a technique for inserting gene fragments into a cell.[12] In 1987, Boston University invested $25 million of operating income to purchase a majority stake in Seragen, a biotechnology firm started by a BU faculty member. At the time, the purchase price represented a staggering 14 percent of the university's endowment. In subsequent years BU invested more money, eventually reaching a total of more than $85 million. In 1992, under pressure from the attorney general of Massachusetts, the university agreed to limit its investments in the company. By the end of 1997, BU purchased a unit of the flagging company outright and began providing the firm with manufacturing, clinical testing, and quality assurance services. In 1998, soon after the firm's flagship therapeutic, a treatment for lymphoma, was recommended for approval by an FDA advisory panel, the corporation was purchased by California-based Ligand Pharmaceuticals at a price that represented a more than 90 percent loss to the university. Shareholder losses were significant, and later that year a lawsuit filed against the university and its chancellor alleged self-dealing in their management of the firm (R. Rosenberg 1997; Barboza 1998).

Columbia University's patent extension case was similarly controversial. In 1983 a patent on what was to become one of the fundamental techniques underpinning drug development efforts in biotechnology was issued to the university. The "Axel patent," named for its senior inventor, covers a process known as co-transformation, which has been used to develop some of the most successful biotechnology drugs.[13] Columbia has broadly licensed the patent in return for a royalty of 1 percent of the sales of all drugs developed using the process. Royalty income on this patent has been estimated at $280 million dollars and has kept Columbia among the top universities in terms of licensing income for the last decade. But patents expire, and the loss of ownership rights also removes the royalty stream that Michael Crow so clearly associated with increased research capacity and flexibility.[14]

In an attempt to avoid this loss, Columbia officials persuaded U.S. senator Judd Gregg (R-NH) to sponsor an amendment to an agricultural spending bill that would extend the life of the patent.[15] The amendment drew upon and extended the logic of the 1984 Hatch-Waxman Act, which enables firms to apply for extensions to patented therapeutics as recompense for delays in the often-lengthy FDA approval process. An extension of Columbia's process patent would have significantly extended the reach of the 1984 act and thus met with significant resistance from industry lobbyists, patient groups, and senators. The attempt at extension eventually failed under a barrage of media criticism and attention (Babcock 2000; Pollack 2000; Marshall 2003).[16]

While there are clear benefits to parlaying commercial efforts into academic achievement, the Seragen case and the Axel patent offer cautionary notes while highlighting the extent to which grafting opens a limited window for innovation. Responses to that window by established players help a new system of stratification and constraint cohere. In one case, a university aggressively attempted to climb the public science hierarchy by gambling a significant portion of its endowment on the fortunes of a firm, in the process drawing attention from state regulators and undertaking some decidedly un-university-like actions such as leasing manufacturing services to a struggling corporation. At the other end of the continuum, an established, Ivy League university that had parlayed a blockbuster patent into increasing research capacity and flexibility faced the loss of that royalty stream. In response it pursued multiple controversial efforts to extend its ownership rights, prompting charges of unfair and illegitimate efforts to extend monopoly rights, a charge that on its face suggests distinctively un-university-like behavior.[17]

Both cases reflect the benefits and pitfalls of leveraging resources across institutional regimes. A failed gamble damaged the legitimacy of a striving organization whereas a commercial success helped another campus climb the status ladder at the cost of wedding success to continued income flows. This catch-22 placed the university in a position uncomfortably similar to its pharmaceutical and biotechnology clients whose uncertain longevity depends upon generating a series of risky blockbusters.

These cases are not isolated instances but instead represent multiple facets of university responses to a systematically changing institutional environment, an environment that is simultaneously constraining and a collective outcome of actions and approaches to technology licensing in the wake of the 1980 Bayh-Dole Act. In essence, importing commercial logics into the core mission of the academy opened a window for innovative action and increased possibilities for mobility before that very action crystallized a structure that constrained future opportunities and placed new limits on successful and striving universities alike. While collapsing once-separate regimes into a single organizational mission allows resources to be leveraged across realms, the practical fungibility of resources and strategies is temporally limited. The status order that structures opportunities and outcomes in the contemporary academy represents a hybrid that merges basic and proprietary logics, allowing advantage to cumulate within and across realms.

The process at work here is one of grafting. Institutional arrangements that were (at least titularly) separate were brought together. Their interface created contradictions that opened a space for novelty as participants struggled to react to their changing environment. That window opened only temporarily, though, because early entrants created a model that constrained their future moves and limited opportunities for innovation by latecomers who now faced a stratification system characterized by cumulative advantage across commercial and academic regimes. A grafting process transformed the characteristic and expected practices of universities. As a result, it is increasingly necessary to achieve in both commercial and academic activities in order to succeed at either, and attempts to accomplish results in both realms are routinely accompanied by broader and more explicit extensions from the university into other parts of society.[18] In an era when commercial successes and peccadilloes are trumpeted in the media and when many governors seek to parlay academic capacity into high-technology muscle, both intra-academy status competitions and transcampus constituencies feel the impacts of commercialization. In short, social constraints upon action that resulted from grafting also translate across extensive networks.

BLURRING BOUNDARIES AND UNIVERSITY NETWORKS

A language of extension and translation is central in stories of two universities struggling to achieve and maintain stature under an institutional order that blends basic and commercial science. The rules for inter-university competition have changed, and the networks that constitute and support ivory towers have shifted. Commercialization has changed the rules of the game, and those changes have altered the university's relationships to key constituents. Such alterations create a new sort of academic actor. In what follows I trace a few of the translations that allow commercial engagements to alter the very characteristics of contemporary universities.

Shifting Terrains for Academic R&D

Increasing academic commercialization has done more than shift the ways universities go about their traditional business. It has also transfigured relationships that reach across established divisions in society. Especially in arenas such as the life sciences, optics, nanotechnology, and computational chemistry, where technological advances, research requirements, and novel opportunities blur the academic and the commercial, university research has become a component of commercial application while industrial muscle now offers a key to fundamental discoveries. The shifting relationship between industrial and academic R&D efforts may lessen university scientists' ability to blaze new trails for industry (Woodhouse, this volume) while increasing the likelihood that commercial discoveries will shape academic research trajectories (N. Rosenberg 2000).

In a recent editorial in the journal *Science*, Donald Kennedy characterizes the contemporary world of life science research as a convergence of two distinct trends, the postwar engine of investigator-initiated and federally funded basic research and the development of "industrial strength basic science," which is "[d]one in mixed or commercial settings, usually by large teams with impressive infrastructure support" (Kennedy 2003).[19] The convergence of these two trends in the life sciences is most clearly visible in the recently completed Human Genome Project, which pitted a widely dispersed academic consortium spearheaded by the NIH's Francis Collins against a more focused for-profit team under the direction of Celera Genomics founder Craig Venter in a race to sequence the complete human genome. Arguably, the speed and technological innovations at the heart of the successful project owe much to the intense rivalry between these "teams." Although these trends are most pronounced in the life sciences, broad trends apparent in bibliometric analyses of patents and publications suggest that similar patterns appear across science and engineering fields.

In an era when both the frequency and size of scientific collaborations has increased dramatically, the center of gravity of the knowledge economy has shifted toward universities and nonprofit research centers and away from the corporate R&D laboratories that once dominated the U.S. R&D system (Hicks et al. 2001). Industrial R&D funds poured into U.S. research universities at unprecedented rates over the last decade, and the academy undertook an increasing share of American

R&D activity (National Science Board 2002). At the same time, patented science moved closer to academic concerns as evidenced by a more than doubled incidence of citations to academic articles in industrial patents (National Science Board 2002).

Universities provide an anchor for collaborative endeavors in an increasingly collaborative, multidisciplinary, and transsectoral research ecology. Consider Figure 1, created from the National Science Board's 2002 *Science and Engineering Indicators,* which describes the percentage of collaborative articles that involve organizations from multiple sectors. Figure 1 highlights patterns of inter-organizational collaboration in terms of the sectors to which collaborating organizations belong.[20] In the network image presented here, the size of a sector's node is proportional to the percentage of collaborative articles produced solely at organizations in the sector—for instance, a paper with authors from two or more firms would contribute to the size of the node labeled "industry" but not to any of the arrows. Arrows represent the proportion of collaborative pieces that involve other sectors. The size and grayscale tone of the arrows are proportional to the magnitude of the collaboration.[21] Figure 1's clear message is that the academy is both the most internally reliant sector (its node is the largest) and the central partner in a world defined by inter-sectoral collaboration. Across all fields, in 1999, the academy is the largest performer and the anchor for collaborative basic research.

Academe's role in the convergence of public and proprietary research may result from shifts in the character of science and from an institutional order that increasingly necessitates new strategies for simultaneously navigating both academic and commercial worlds. The university's centrality is comforting. Far from being destroyed by commercialization, the academy has become an obligatory passage point for research that spans basic and commercial concerns. This centrality makes the transformations at work in the academy issues of broad societal import while offering opportunities for a refigured academy to cement a new type of place in the pursuit of science.

Reaching beyond R&D

If theoretical work on enrollment and translation suggests anything, it is this: the cast of performers necessary to coax a new and stable entity or artifact into being is staggeringly diverse and often divisive. Similarly, the academy's research centrality rests on connections that extend into

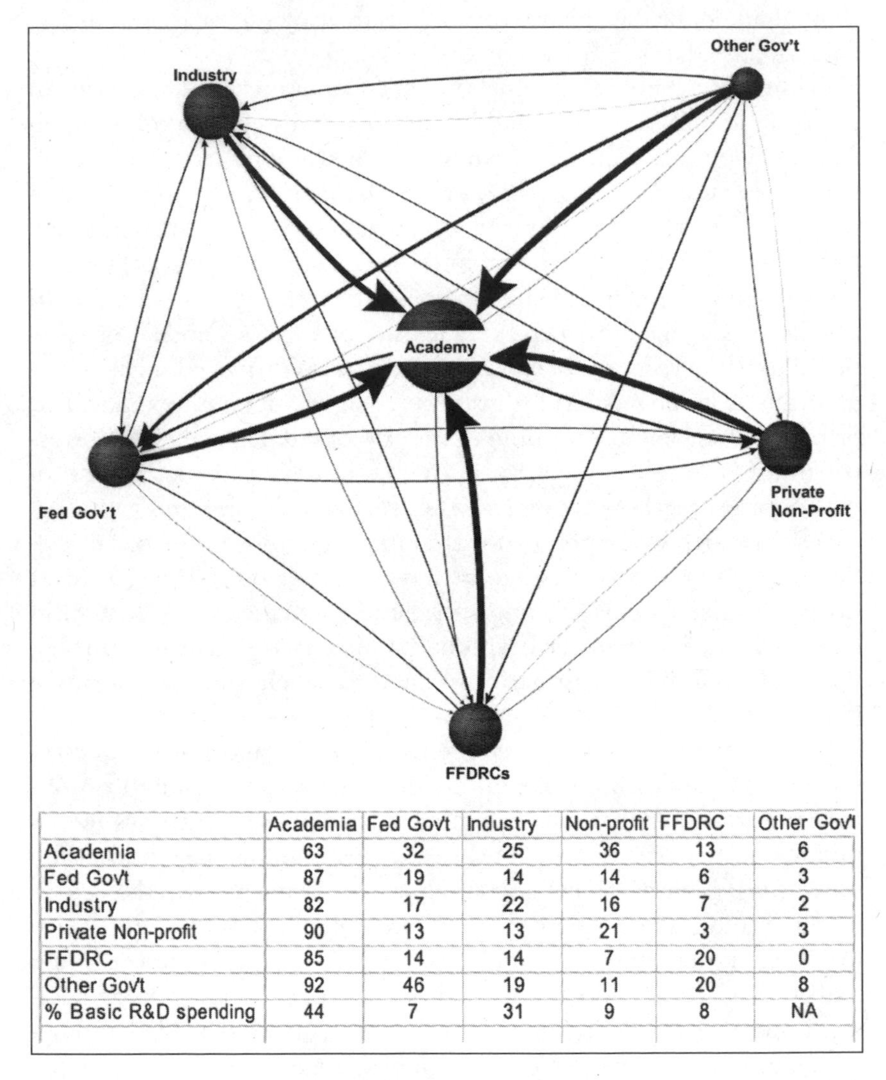

	Academia	Fed Gov't	Industry	Non-profit	FFDRC	Other Gov't
Academia	63	32	25	36	13	6
Fed Gov't	87	19	14	14	6	3
Industry	82	17	22	16	7	2
Private Non-profit	90	13	13	21	3	3
FFDRC	85	14	14	7	20	0
Other Gov't	92	46	19	11	20	8
% Basic R&D spending	44	7	31	9	8	NA

Figure 1. Percentage of collaborative articles involving other sectors, all fields, 1999. Initiators are listed in the left column. Collaborators are listed across the top. Percentages do not total one hundred because of multisector collaboration. Source: *Science and Engineering Indicators*, 2002.

the federal government, the international policy arena, activist groups, the news media, farther than the eye can see.[22] The emergence and constitution of an entrepreneurial academy and the associated convergence of two disparate models for R&D did not occur in a vacuum, and their effects have sparked tensions between the university and its industrial partners.[23]

Columbia's attempts to extend the life of its valuable patent and recent conflicts between the Wisconsin Alumni Research Foundation (WARF) and Geron, a California biotechnology firm, over the distribution of rights to recently isolated lines of human embryonic stem cells demonstrate why conflicts over access, exclusivity, and the right to profit from scientific discoveries generate concern about the university's ability to maintain its distinctive features in the face of increasing commercial involvement.[24] But the university is not a passive object buffered by stronger and savvier commercial partners. Patent fences give universities unprecedented leverage in the decisions of for-profit collaborators and clients. Using such leverage requires that academic institutions conceive of intellectual property (IP) as more than an alternative source of revenue. A broad conception of the ways that private rights can be turned to the service of public needs can and should inform policies and practices for the management of academic intellectual property.

Consider the case of anti-retroviral therapies for HIV/AIDS. These cocktails of drugs significantly increase survival rates among infected individuals. Universities have played a central role in the development of many of these therapeutics, with Yale, Minnesota, Emory, and Duke each holding patent rights to aspects of important anti-retroviral drugs (Kapczynski, Crone, and Merson 2003). The central role university science played in developing these drugs offers some support for the claim that real public benefits can accompany efforts to commercialize academic research. Such commercialization efforts also offer new power and possibilities to universities, who can use ownership strategically to ensure broad access to expensive but essential therapies.

Yale University holds a key patent on stavudine, a commonly used anti-retroviral drug exclusively licensed to the pharmaceutical firm Bristol-Myers Squibb (which markets it under the brand name Zerit). In early 2001, Doctors Without Borders—a nonprofit group dedicated to providing medical assistance to the populations of underdeveloped areas—requested that Yale allow them to distribute generic versions of the drug in South Africa, where the yearly cost of medication far exceeds the ability of most patients to pay. Yale refused, citing an existing contract with Bristol-Myers Squibb. John Soderstrom, managing director of Yale's Office of Cooperative Research, framed the issue in a letter: "Although Yale is indeed the patent holder, Yale has granted an exclusive license to Bristol-Myers Squibb, under the terms of which only that entity may respond to a request" (McNeil 2001). Alison Richard, then provost of the university, put the issue more colloquially in an interview with the *Yale Daily News:* "It's not ours to give away. We are hopeful that Bristol-Myers Squibb is going to do their best to make this work out" (Adrangi 2001).

Perhaps predictably, this interaction drew a huge amount of attention and sparked a storm of protest including an important campaign launched by a Yale law student and supported by the compound's inventor, then an emeritus professor (Borger 2001).[25] Among the responses was a letter from Doctors Without Borders suggesting that Yale's response was in violation of its own policy on intellectual property licensing. The relevant section of that policy reads as follows (my italics):

> *The objective of the University is to assure the development of its technology in furtherance of its own educational mission and for the benefit of society in general.* Therefore, as a general policy, the University will set the terms of its licenses so as to further the achievement of this objective. Exclusive licenses will be granted if it appears to the Office of Cooperative Research that this is the most effective way of ensuring development to the point that the public will benefit. *Any exclusive license agreement will be so drawn as to protect against failure of the licensee to carry out effective development and marketing* within a specified time period.[26]

After intensive talks between Yale and Bristol-Myers Squibb, the company announced that it would not enforce its patent rights in South Africa, paving the way for the manufacture and distribution of generic drug versions at a cost some thirty times less than the price of Zerit (Kapczynski et al. 2003).

This outcome was hailed as "historic" and largely resulted from the combination of internal and external pressures on Yale, pressures that may have been much less effective if applied directly against Bristol-Myers Squibb. Here an intellectual property owner formally and publicly committed to effective use and societal benefit offers a fulcrum for activists to successfully exert pressure on a corporate giant. That pressure depended both on Yale's concern with its public profile and its ownership of a fundamental and valuable piece of property. Somewhat paradoxically, Bristol-Myers Squibb's final decision may have hinged on the combination of Yale's open, academic character and its proprietary rights.

As noted by Toby Kasper, the head of the Doctors Without Borders unit responsible for the initial request, "A company has never given up its patent for a drug like this. Universities make a lot of money on these patents, so they're hesitant to give up their right. But Yale acted out of fear of public relations and fear of a student uprising. Besides, AIDS is a graveyard for corporate P.R., and it's an area that could cause potential harm to a university's P.R." (Lindsey 2001). Yale's brief press release in response to the announcement highlighted the university's role in the decision: "Yale worked diligently to remove any obstacles created by its license agreement with BMS [Bristol-Myers Squibb]. We are gratified that our efforts paved the way for the significant action announced by BMS today. We will continue to encourage all pharmaceutical companies to make their AIDS drugs affordable and widely available in Africa."[27] Perhaps more importantly, Yale has taken the opportunity to begin developing best practices for assuring broad access to essential medicines based on university innovations.

In what may turn out to be a prescient statement, a recent report from that effort notes that the upstream character of university research offers the possibility for academic institutions to exert "early leverage" in development and marketing. Another of that report's "key points" implicitly references blurring boundaries between academic research and commercial development and between academe and civil society: "The issue of access cannot be separated from that of innovation and universities must consider the impact of best strategies on both" (Merson 2002:1).

Clearly, Yale is not an unambiguous hero. The university's patent rights and licensing agreements could have prevented broad distribution of an important drug to patients in need. Yet those very rights may also have provided the necessary leverage to accomplish a transformation in corporate policy. This story turned out well. Its outcome has sparked a number of other attempts to use university patent rights to exert pressures for broader worldwide access to medication (Lindsey 2001). Here we see one way that the shifting focus of university competition and the convergence of multiple models of science might open new opportunities for academic institutions to exert influence on a broader field. An extensive hybrid, the entrepreneurial university anchors a science whose organizational division of labor spans diverse sectors. The contemporary research university is thus a constrained actor and a passage point, a position fraught with opportunity and danger.

Entrepreneurial Universities as Dual Hybrids

Changes at work in contemporary American universities have fundamentally altered the rules of the game for academic science and engineering. Opportunities for mobility along novel paths were opened through a process of grafting and then were recalcified into a new order that mixed commercial and academic logics. Through this process, one characteristic set of constraints was contingently transformed into another with significant consequences for striving and successful institutions alike.

Changes in the institutional order and stratification system of the academy have broader implications for contemporary society precisely because they cannot be understood in isolation. The emergence of a hybrid, entrepreneurial university is a relational accomplishment. The characteristics and attributes of contemporary research universities are being shaped by connections that reach across regimes and sectors. As universities adapt to the dictates of the new academic game they helped to create, such points of contact enable changes on campus to resonate through structures adjacent and distant, thus reconfiguring not only the university but also the outlines and features of contemporary society.

The two forms of hybridization I describe are inseparable. Grafting depends upon a stabilized and taken-for-granted system of relations that constitutes separate but potentially collapsible regimes. When such separated regimes collide, the very sets of relationships they depend upon are revealed to be contingent. Contradiction opens opportunities for once tightly constrained actors to forge novel connections in response to newly salient ambiguities.

Such moves reverberate and thus alter not only the distribution of resources and opportunities within a field but also the webs of translation and enrollment that are the wellsprings of stable social ontologies. The opportunities and dangers inherent in university research commercialization stem from the same source. Reordering and refiguring hybridizations jointly remove any possibility that the university's uniqueness and social efficacy require its separation from the world. Recognizing its embeddedness in and essential contribution to the making of a world beyond its grasp opens opportunities for new types of leverage and action on the part of universities. Careful attention to the systematic effects of actions and concern with the unintended consequences of ad hoc responses to contingent circumstances are necessary, but in this instance, I contend, collapsing distinctions and sticky imbroglios also offer opportunities for innovation and widespread reconfiguration of the order to which we have become accustomed.

Acknowledgments

This chapter benefited from thoughtful comments by the editors and by Elisabeth S. Clemens. Any remaining errors are my own.

End Notes

1. Rhetoric relating university research to national economic competitiveness is also commonplace in rationales offered for federal research funding (Slaughter and Rhoades 1996) and in academic analyses of national innovation systems (Mowery 1992).

2. Such discussions are often framed in terms of exemplary cases such as the contested tenure case of a Berkeley ecologist. Ignacio Chapela's controversial research on the spread of genetically modified corn strains into natural maize populations in Mexico and his outspoken resistance to the recently concluded deal linking Berkeley's prestigious plant science department to pharmaceutical giant Novartis have been offered as explanations for his unexpected denial (Dalton 2003).

3. It should come as no surprise that the entrepreneurial university represents neither the first nor the only academic hybrid. Among the more notable prior examples are the land grant universities that emerged around the turn of the last century (Geiger 1986) and postwar era research labs that supported both a dominant discipline and the policy nexus that generated the largely taken-for-granted ground for the figure of academic commercialization (Leslie 1993; Lowen 1997).

4. In a recent examination of an academic biology laboratory doing simultaneously academic and commercial research, Daniel Kleinman frames the issue as follows: "Although it is undoubtedly true that laboratories and scientists play a role in shaping the world, my aim has been to show what I see as a more important and significant trend: how the work done by scientists and the world they inhabit is shaped in significant ways by the larger (social) environment in

which the practice of science is undertaken" (Kleinman 2003:157). While I concur wholeheartedly with the sentiment expressed here, I am less interested in engaging with the ongoing discussion about the relative primacy and distinctiveness of the social and the technical. I contend that both are relational accomplishments that are naturalized through roughly analogous processes that obscure their contingent character.

5. Numerous historical analyses of the American university remind us that the now taken-for-granted character of academic research and development efforts was, indeed, an accomplishment (Guston and Keniston 1994; Kleinman 1995).

6. This section relies upon a recent article (Owen-Smith 2003).

7. This is not to say that articles convey all of the information necessary to actually accomplish a replication (let alone a novel extension) of a given finding.

8. It is precisely this transsectoral blurring that is the focus of the following section on hybridity through extension.

9. Reputation offers a particularly clear mechanism for the cumulative advantage in life and physical science fields where both article and granting agency peer review processes are single blind.

10. Commercial endeavors are by no means the only such loopholes. Indeed, academic earmarks—direct congressional appropriations for universities and colleges—have grown dramatically, from $11 million in 1980 to nearly $1.7 billion in 2001 (National Science Board 2002). These funds, which also bypass peer review (and thus have come under increasing fire), have been defended as a means for less successful universities to gain resources necessary for success in academic competitions. The chancellor of Boston University, John Silber, described such earmarks as a means to "force your way into" the "old boy network" that is the academic peer review system. The monies BU garnered through lobbying efforts, Silber contends, allowed the university to bootstrap its way into a higher position in the academic status order: "Our peer-reviewed grants and contracts have increased with every passing year. It is a result of having been able to put together the facilities to bring in the outstanding scientists who bring in those peer-reviewed grants" (Schlesinger 2001). While it is beyond the scope of this chapter, increases in university lobbying efforts and the antecedents and consequences of lobbying efforts offer fruitful areas of study for a political sociology of science.

11. Of the 142 U.S. universities surveyed by the Association for University Technology Managers, only 19 report gross licensing income in excess of $10 million. Almost half (70) report income less than $1 million, and none approach the $100 million mark. All told these institutions earned slightly more than $ 1.3 billion in royalties, slightly more than 7 percent of the $18+ billion they spent on federally funded R&D (Association of University Technology Managers 2000).

12. U.S. patent Number 4,399,216.

13. Among the more well known drugs are Amgen's anemia drug, Epogen, and Immunex's successful treatment for rheumatoid arthritis, Enbrel.

14. Issued in 1983, the Axel patent had a life of seventeen years and was set to expire in August 2000. More recently issued patents can provide protection for twenty years from the date of application.

15. The proposed extension of fourteen to eighteen months was estimated to be worth approximately $150 million in additional royalties.

16. The saga continued into 2003. Columbia filed a second patent on the Axel process in 1995, and that patent (USPTO # 6,455,275) was issued in September of 2002, offering the university the possibility of a further seventeen years of royalty income. Claiming that the new patent was not substantially different from the 1983 original and that it represented Columbia's "illegitimate effort to extend its patent monopoly," two Boston-area biotechnology firms, Genzyme and the Abbott Bioresearch Center, filed a lawsuit asking that the new patent be invalidated. Similar suits had already been filed in two other federal courts (a suit by Amgen was filed in Los Angeles and another by Genentech in San Francisco).

17. Patents, by definition, offer limited monopoly rights, and the question of how (and whether it is possible) to replace royalty streams from lucrative blockbuster patents is becoming an important one. In addition to the Axel patent, another fundamental and exceptionally valuable process patent (the Cohen-Boyer gene-splicing patent) recently expired, significantly reducing royalty streams at Stanford and the UCSF.

18. In part as a consequence of the actions described in this section, Columbia University and Boston University topped the list of spending on lobbyists in 2000 (data retrieved from opensecrets.org, www.opensecrets.org/lobbyists/indusclient.asp?code=W04&year=2000&txtSort=A, on January 6, 2003). Both universities rely largely on the services of Cassidy and Associates, the most prominent educational lobbying firm. Interestingly, Gerald Cassidy, the founder, chairman, and CEO of the lobbying firm, was recently named to Boston University's Board of Trustees (Deveney 2003).

19. The journal's editor and past provost of Stanford University.

20. At least those collaborations that result in published scientific articles in journals indexed by the Institute for Scientific Information.

21. Several caveats are in order. This image was created in Pajek using data presented in the table embedded in Figure 1. The data are asymmetric (meaning that industrial publications can more often draw on academic collaborators than vice versa), and the main diagonal represents the proportion of collaborative work undertaken within a sector. The largest node, representing the academy, reflects the fact that some 63 percent of all academic articles involving more than one organization involve only universities. FFDRCs are Federally Funded Research and Development Centers. Data

sum to more than 100 percent because of the counting methodology used to sum sector publications. Each institutional author was assigned a whole count of a publication.

22. Many of the arguments presented here are in accord with two other approaches to thinking about shifting networks constitutive of contemporary science and transformations underway in the academy (Gibbons et al. 1994; Etzkowitz and Leydesdorff 1998).

23. It should not be forgotten that a transformation in the academy necessarily resonates through associated fields in industry. Although this fact has been widely recognized, several scholars have suggested that it is industry that is more corrosive to distinctively academic approaches (Slaughter & Leslie 1997). While there is convergence across the worlds of the academy and industry, such transitions are "asymmetric" and favor firms at the potential expense of universities (Kleinman and Vallas 2001).

24. This challenge is becoming more pressing as a recent federal court decision (Madey v. Duke) essentially swept away the traditional concept of a research exemption for academic use of intellectual property on the grounds that research represents a central business interest of universities and the acknowledgment that those interests increasingly cross into the commercial realm (Eisenberg 2003). The Madey decision has accelerated concerns that widespread patenting of research tools will have a chilling effect on academic research as scientists (whether engaged in commercially valuable projects or not) are confronted by an increasing need to license tools and materials (at significant expense of effort and resources) from multiple sources.

25. Dr. William Prusoff expressed strong support for the student campaign to relax Yale's patent rights in South Africa: "I'd certainly join the students in that. I wish they would either supply the drug for free or allow India or Brazil to produce it cheaply for under-developed countries[*sic*] But the problem is, the big drug houses are not altruistic organizations. Their only purpose is to make money" (McNeil 2001). Interestingly, stavudine is not the first anti-retroviral developed by Prusoff. The first, idoxyuridine, he synthesized during the 1950s at Yale. That compound, the first retroviral to receive FDA approval, is still in use for treating eye infections resulting from the herpes simplex virus. Prusoff reminisces about a very different time in the history of university-industry relations: "My chairman at the time wanted to patent the drug, and Yale's lawyers wanted nothing to do with patents. It was beneath the dignity of the university to do something for money. So four or five different companies took the thing up and put it on the market—and there's no patent, anyone who wants to produce it can produce it. Those were the days" (Zuger 2001).

26. The policy was last revised in 1998 and so was in force at the time of the request. It was retrieved from www.yale.edu/ocr/invent_policies/patents.html on January 9, 2004.

27. Yale News Release, March 14, 2001. "Statement by Yale University Regarding Bristol-Myers Squibb Company's Program to Fight HIV/AIDS." www.yale.edu/opa/newsr/01-03-14-03.all.html (accessed January 9, 2004).

References

Adrangi, Sahm. 2001. "Doctor's Group Asks Yale to Relax Drug Patent." *Yale Daily News,* February 28.

Association of University Technology Managers. 1999. *Licensing Survey, FY 1999: Full Report.* Northbrook, IL: Association of University Technology Managers.

Babcock, C. R. 2000. "Senator Tries to Extend Alma Mater's Patent; Columbia Would Gain $100 Million a Year." *Washington Post,* May 19.

Barboza, David. 1998. "Loving a Stock, Not Wisely But Too Well." *New York Times,* September 20.

Bok, Derek. 2003. *Universities in the Marketplace: The Commercialization of Higher Education.* Princeton, NJ: Princeton University Press.

Borger, Julian, and Sarah Boseley. 2001. "Campus Revolt Challenges Yale over $40M AIDS Drug." *Guardian* (London), March 13.

Chubin, Daryl. 1994. "How Large an R&D Enterprise?" In *The Fragile Contract: University Science and the Federal Government,* ed. David H. Guston and Kenneth Kenniston, pp. 119–125. Cambridge, MA: MIT Press.

Clemens, Elisabeth S., and James M. Cook. 1999. "Politics and Institutionalism: Explaining Durability and Change." *Annual Review of Sociology* 25: 441–466.

Dalton, Rex. 2003. "Berkeley Accused of Biotech Bias as Ecologist Is Denied Tenure." *Nature* 426:591.

Dasgupta, Partha, and Paul David. 1987. "Information Disclosure and the Economics of Science and Technology." In *Arrow and the Ascent of Modern Economic Theory,* ed. George R. Feiwel, pp. 659–689. New York: New York University Press.

———. 1994. "Toward a New Economics of Science." *Research Policy* 23:487–521.

Deveney, Ann. 2003. "Boston University's Board of Trustees Names Three New Members." Boston University news release.

Eisenberg, Rebecca S. 2003. "Science and the Law—Patent Swords and Shields." *Science* 299(5609): 1018–1019.

Etzkowitz, Henry, and Loet Leydesdorff. 1998. "The Endless Transition: A 'Triple Helix' of University-Industry-Government Relations." *Minerva* 36: 203–208.

Etzkowitz, Henry, Andrew Webster, and Peter Healey. 1998. *Capitalizing Knowledge: New Intersections of Industry and Academia.* Albany: State University of New York Press.

Feller, Irwin. 1990. "Universities as Engines of R and D-Based Economic-Growth—They Think They Can." *Research Policy* 19:335–348.

Florida, Richard. 1999. "The Role of the University: Leveraging Talent, not Technology." *Issues in Science and Technology* 15:67–73.

Friedland, Roger, and R. Alford. 1991. "Bringing Society Back In: Symbols, Practices, and Institutional Contradictions." In *The New Institutionalism in Organizational Analysis,* ed. W. W. Powell and P. J. DiMaggio, pp. 232–266. Chicago: University of Chicago Press.

Geiger, Roger L. 1986. *To Advance Knowledge: The Growth of American Research Universities, 1900–1940.* New York: Oxford University Press.

Gibbons, Michael, Camille Limoges, Helga Nowotny, Simon Schwartzman, Peter Scott, and Martin Trow. 1994. *The New Production of Knowledge.* London: Sage.

Guston, David H., and Kenneth Keniston, eds. 1994. *The Fragile Contract: University Science and the Federal Government.* Cambridge, MA: MIT Press.

Hicks, Diana, Tony Breitzman, Dominic Olivastro, and Kimberly Hamilton. 2001. "The Changing Composition of Innovative Activity in the US—A Portrait Based on Patent Analysis." *Research Policy.* 30:681–703.

"Innovation's Golden Goose." 2002. *Economist,* December 13.

Kapczynski, Amy, E. Tyler Crone, and Michael Merson. 2003. "Global Health and University Patents." *Science* 301(5640): 1629.

Kennedy, Donald. 2003. "Industry and Academia in Transition." *Science* 302(5649): 1293–1293.

Kerr, Clark. 2002. "Shock Wave II: An Introduction to the Twenty-First Century." In *The Future of the City of Intellect,* ed. Steven G. Brint, pp. 1–19. Stanford, CA: Stanford University Press.

Kleinman, Daniel L. 1995. *Politics on the Endless Frontier: Postwar Research Policy in the United States.* Durham, NC: Duke University Press.

———. 2003. *Impure Cultures: University Biology and the World of Commerce.* Madison: University of Wisconsin Press.

Kleinman, Daniel L., and Steven P. Vallas. 2001. "Science, Capitalism, and the Rise of the 'Knowledge Worker': The Changing Structure of Knowledge Production in the United States." *Theory and Society* 30:451–492.

Krimsky, Sheldon. 2003. *Science in the Private Interest: Has the Lure of Profits Corrupted Biomedical Research?* Lanham, MA: Rowan & Littlefield.

Latour, Bruno. 1983. "Give Me a Laboratory and I Will Raise the World." In *Science Observed: Perspectives on the Sociology of Science,* ed. Karin Knorr-Cetina and Michael J. Mulkay, pp. 141–170. Los Angeles: Sage.

———. 1987. *Science in Action: How to Follow Scientists and Engineers through Society.* Cambridge, MA: Harvard University Press.

———. 1993. *We Have Never Been Modern.* New York: Harvester Wheatsheaf.

———. 1996. *Aramis, or, The Love of Technology.* Cambridge, MA: Harvard University Press.

——. 1999. *Pandora's Hope: Essays on the Reality of Science Studies*. Cambridge, MA: Harvard University Press.

Law, John. 1994. *Organizing Modernity*. Cambridge, MA: Blackwell.

Leslie, Stuart W. 1993. *The Cold War and American Science: The Military-Industrial-Academic Complex at MIT and Stanford*. New York: Columbia University Press.

Lindsey, Daryl. 2001. "Amy and Goliath." *Salon*, May 1.

Lowen, Rebecca S. 1997. *Creating the Cold War University: The Transformation of Stanford*. Berkeley: University of California Press.

March, James G., and Herbert A. Simon. 1958. *Organizations*. New York: Wiley.

Marshall, E. 2003. "Intellectual Property—Depth Charges Aimed at Columbia's 'Submarine Patent.'" *Science* 301(5632): 448.

McNeil, Donald G. 2001. "Yale Pressed to Help Cut Drug Costs in Africa." *New York Times*, March 12.

Merson, Michael. 2002. "Access to Essential Medicines and University Research: Building Best Practices." Workshop report, September 25. New Haven, CT: Yale University Center for Interdisciplinary Research on AIDS.

Merton, Robert K. 1968. "The Matthew Effect in Science." *Science* 159(3810): 56–62.

——. 1973[1942]. "The Normative Structure of Science." *The Sociology of Science*. Chicago: University of Chicago Press.

Meyer, John W., and Brian Rowan. 1977. "Institutionalized Organizations: Formal Structure as Myth and Ceremony." *American Journal of Sociology*. 83: 340–363.

Mol, Arthur, and John Law. 1994. "Regions, Networks and Fluids: Anemia and Social Topology." *Social Studies of Science* 24:641–671.

Mowery, David C. 1992. "The United States National Innovation System: Origins and Prospects for Change." *Research Policy* 21:125–144.

Mowery, David C., Richard R. Nelson, Bhavan N. Sampat, and Arvids A. Ziedonis. 2001. "The Growth of Patenting and Licensing by US Universities: An Assessment of the Effects of the Bayh-Dole Act of 1980." *Research Policy* 30:99–119.

National Science Board. 2002. *Science and Engineering Indicators*. Washington, DC, National Science Board.

Owen-Smith, Jason. 2003. "From Separate Systems to a Hybrid Order: Accumulative Advantage across Public and Private Science at Research One Universities." *Research Policy* 32:1081–1104.

Owen-Smith, Jason, and Walter W. Powell. 2001. "Careers and Contradictions: Faculty Responses to the Transformation of Knowledge and Its Uses in the Life Sciences." *Research in the Sociology of Work* 10:109–140.

Phillips, Damon J., and Ezra W. Zuckerman. 2001. "Middle-Status Conformity: Theoretical Restatement and Empirical Demonstration in Two Markets." *American Journal of Sociology* 107:379–429.

Pollack, Andrew. 2000. "Columbia Gets Help from Alumnus on Patent Extension." *New York Times,* May 21.

Press, Eyal, and Jennifer Washburn. 2000. "The Kept University." *Atlantic Monthly* 285:39–54.

Rip, Arie. 1986. "Mobilizing Resources through Texts." In *Mapping the Dynamics of Science and Technology: Sociology of Science in the Real World,* ed. Michel Callon, John Law, and Arie Rip. Houndmills, Basingstoke, UK: Macmillan Press.

Rosenberg, Nathan. 2000. *Schumpeter and the Endogeneity of Technology: Some American Perspectives.* New York: Routledge.

Rosenberg, Ronald. 1997. "Higher Learning, Higher Stakes: As BU Awaits a Big Seragen Payoff That May Never Come, Its Relationship Shifts." *Boston Globe,* April 13.

Schlesinger, Robert. 2001. "For BU, Lobbying Pays a Smart Return." *Boston Globe,* May 6.

Sewell, William H. 1992. "A Theory of Structure: Duality, Agency, and Transformation." *American Journal of Sociology* 98:1–29.

Slaughter, Shelia, and Larry L. Leslie. 1997. *Academic Capitalism: Politics, Policies, and the Entrepreneurial University.* Baltimore: Johns Hopkins University Press.

Slaughter, S. and G. Rhoades. 1996. "The Emergence of a Competitiveness Research and Development Policy Coalition and the Commercialization of Academic Science and Technology." *Science, Technology, & Human Values.* 21:303–339.

Wright, Erik O. 1984. *Classes.* New York: Verso.

Zucker, Lynne G. 1977. "The Role of Institutionalization in Cultural Persistence." *American Sociological Review.* 42:726–743.

Zuger, Abigail. 2001. "A Molecular Offspring, Off to Join the AIDS Wars." *New York Times,* March 20.

4

Commercial Restructuring of Collective Resources in Agrofood Systems of Innovation

STEVEN WOLF

Network organization has become widely recognized as important to research, policy, and practice. The network perspective, at its core, is relational. Explicit focus on relations between actors invites analysis of *institutions,* defined here as social, cognitive, and material structures that constrain and enable interactions and resource transfers among actors. Transfers of money, materials, know-how, and symbolic legitimacy underlie actors' acquisition of capabilities, and thus understanding ways in which institutions mediate such resource stocks and flows is important to an analysis of development processes, including scientific and technical practice. Attention to institutions will be required to structure innovation systems that bring us closer to the ideal of sustainable development in which investments in economic, social, and ecological objectives are complementary.

In this chapter I present an institutional analysis of knowledge production in agriculture. Drawing together ideas from science studies, innovation studies, and the economics of technical change, I derive the specific significance of network organization to scientific and technical practice through examination of *distributed innovation,* which is defined as interactive, collectively organized production of new technical knowledge by groups of heterogeneous actors operating in discrete organizations (Allaire and Wolf 2002; Allen 1983; von Hippel 1983). Orthodox economic approaches to analysis of innovation recognize incentives and mechanisms underlying both public and private engagement; on the one hand, knowledge is a public-good (i.e., accessible to all and nonrivalrous) legitimating public investment at the level of the nation-state, and on the other, individuals and private firms invest in innovation to produce, exploit, and keep secret their technical capabilities to create and sustain competitive advantage. Engagement of collective structures—an intermediate level of social organization—in learning has been considered much less thoroughly. Sociological analysis of institutions that mediate collective governance of knowledge production has much to contribute to economics and research policy, as paralleled by, or perhaps as part of, sensitization of economists and policymakers to the significance of social capital (Bowles and Gintis 2002).

This chapter focuses on changes in the institutional structures that support distributed innovation in agriculture. The pattern of institutional change in agricultural innovation systems parallels the global pattern (Kleinman and Vallas, this volume; Owen-Smith, this volume; Wolf and Zilberman 2001) in which the perceived traditional division of labor between public- and private-sector actors is blurring, the role of universities in production of public goods is increasingly ambiguous, and there is a tendency toward privatization and commercialization (Bok 2003). While distributed innovation occurs in networks, these networks are not necessarily collective or nonhierarchical in character. As I argue, network structures supporting distributed innovation are increasingly private and commercial. Applied to agriculture, this trajectory raises social concerns, including the power of farmers relative to nonfarm interests in agrofood systems and the capacity within systems of innovation to respond to demand for public goods such as improved ecological performance of agriculture.

For example, nonfarm agribusinesses are creating extremely large proprietary databases containing precisely geographically referenced, highly detailed records of how specific corn germplasm responded to management inputs such as fertilizer and pesticides. This resource allows firms to expand their knowledge base, refine the targeting of their research programs, increase the quality of the agronomic advice they provide to customers, and ultimately expand their market share. Under this structure, farmers' contribution to the production process is diminished, as know-how is incorporated into seeds, codified production prescriptions, and sales representatives of plant-breeding firms. At the same time, public capabilities to conduct research and shape technical systems are compromised as knowledge production is increasingly controlled by firms with insufficient incentives to invest in production of public goods such as water quality, biodiversity, and soil health.

On the basis of a general historical account of innovation in agriculture and case studies from previous research, I identify *professional structures* as critical elements of distributed innovation systems (Allaire and Wolf 2002). Professional structures coordinate interaction and facilitate production of collective goods—including new technical knowledge—among sets of interdependent actors who share a body of specialized expertise. The principal mechanism that supports this function is the production and maintenance of comparative references that allow practitioners to assess empirical observations. Agrofood systems are analyzed as a sector dependent on farmers' professional knowledge and collective organizational structures, more generally. The current tendency toward disinvestment and decline in collective organizational forms and the reciprocal movement toward private, corporate, and proprietary logic lead to a situation in which competitiveness requirements of firms, regions, and the sector as a whole are directly addressed, while the mechanisms for enhancing ecological and equity concerns are unspecified and may in fact be eroding. These findings point to two opportunities to sustain and enhance learning in agrofood systems. First, there is a need to recognize the essential contributions of collective organizational structures to processes of innovation.[1] Secondly, there is a need to complement commercial investments in knowledge creation capabilities with collective and public engagement.

The problem I address and the political economic dynamics I observe parallel those of Kleinman and Vallas (this volume). Maintenance of openness and accountability in knowledge production when the function of knowledge is increasingly strategic poses overarching social challenges. Recognition of collective structures as a third form of governance in knowledge systems may enrich the institutional repertoire of analysts and policymakers in useful ways (Williamson 1987; Menard 2002; Wolf 2004).

The remainder of the chapter is organized as follows. In the next section I examine the privileged status of innovation within the general problem of sustainable development. This brief evolutionary account of environmental management theory and policy orientations points to the ascendance of an institutional approach to ecological modernization. Turning to the specific case of agricultural sustainability, I identify a paradoxical element of the political economic critique of agricultural modernization represented by the farming systems research tradition that arose from the appropriate technology movement of the 1960s. While arguing explicitly for "bringing the farmer back in" to knowledge production and validation structures and routines, this literature tends to underestimate the historical and contemporary significance of contributions farm operators and farmers' professional structures make to production, refinement, and diffusion of agricultural technologies. Drawing explicitly on arguments presented in Allaire and Wolf (2002), I then elaborate the notion of distributed innovation in agriculture as a concrete illustration of the structure and function of network structures in knowledge production. My analysis is focused on collective organizational structures underlying knowledge production with specific reference to professional structures as defined above. Then, in order to highlight the power of commercialization and privatization logics within agricultural innovation systems, I present a pair of case studies that illustrate the contemporary institutional trajectory. In the final section of the chapter I conclude with a discussion of the policy implications of this pattern of restructuring of distributed innovation networks.

Innovation, Institutions, and Ecological Modernization

Analysis of innovation has taken center stage in ecological social science. At the broadest level, knowledge creation is a new focus political economy (Hollingworth and Boyer 1997). Within natural resource studies, ecosystem management is increasingly defined as a problem of adaptation in which learning is required to integrate local peoples' resource access claims with demand for public goods (Lee 1993; Walters and Holling 1990). Scholars of environmental sociology have developed ecological modernization theory, an analytic and normative framework for reconciling economic and ecological processes through technological and institutional change (Mol and Sonnenfeld 2000). Learning is also an overarching focus outside academic circles. Ecological health, social justice, and economic competitiveness, the three dominant performance criteria shaping sustainable development discourse and policy, are increasingly articulated in terms of an innovation dynamic. Individual firms, too, are focused on understanding and mobilizing knowledge creation capabilities to maximize performance.[2]

The conceptualization of innovation applied to sustainable development has changed over time. More specifically, the scope of the analysis has expanded, and continues to expand, in light of knowledge of constraints and strategies for the transformative greening of society. Early analyses of intervention strategies for environmental management focused on end-of-pipe mitigation and waste management technologies. The problem was framed in terms of engineering, including containing effluents, recycling, and creating cleaner versions of existing technologies. Classic examples include smokestack scrubbers, catalytic converters, and movement away from lead paint.

Recognition of cross-media environmental effects (for example, a change in a product to mitigate water pollution often negatively affected air quality) and the economic rationality of pollution prevention rather than remediation led to an expanded policy focus. Analysis and policy in this second era emphasized life-cycle analysis, full-cost accounting, process innovations, and system redesign according to the concepts of industrial ecology (Graedel and Allenby 1995). The problem was framed increasingly in economic terms, and the emphasis was squarely on creating incentives for firms and individuals to pursue eco-efficiency. Key examples include adoption of alternatives to bleaching in the pulp and paper industry and establishment of markets for trading SO_2 emissions. While this systemic approach provided an opportunity to pursue more comprehensive strategies and stimulated investment in R&D, the focus remained on problem solving rather than creation of new possibilities.

Based on interest in intervening at a higher level of organization in order to trigger more radical change, analysis of the technology-environment interface is moving to incorporate institutions. That is, attention is now being devoted to structures regulating access to and transfer of resources in order to spur innovation and create new material capabilities. The effort to alter the context of individuals' resource allocation decisions stems from recognition of limits of a static model in which innovation is viewed as a product of firms engaged in rational calculus in competitive markets (Porter and van der Linde 1995). Berkhout (2002:2) describes ecological modernization as seeking to understand reconciliation of economic development with ecological sustainability through "meso-level explanations. In particular, there has been a drive to include institutional contexts and processes in the picture, arguing that the correct focus should be on the co-evolution of technical and institutional innovations."

Within this increasingly ambitious analytic, programmatic, and developmental project, the significance of "getting the institutions right" has expanded.[3] Aligning private and public incentives addresses the need to change material flows according to the logic of eco-efficiency (i.e., economic competitiveness and natural resource conservation are complementary). And, simultaneously, institutional innovation is needed to change patterns of information generation and exchange so as to sustain and catalyze learning and technological change (Hemmelskamp, Rennings, and Leone 2000). In this framework, social and ecological coherence rest on identification of an institutional architecture capable of sustaining production and the application of new capabilities.

Local Knowledge Systems and the Linear Model

Turning to an examination of innovation processes in agriculture, I begin with identification of a paradox within the problem that structures the critical academic and development literature on agricultural and rural development. Actors engaged in efforts to diagnose past failures and initiate sustainable development focus heavily on criticizing the external imposition of industrial, or standardized, technologies (e.g., genetics, chemicals, and machines produced by agribusinesses) and legitimating local knowledge systems (e.g., know-how residing among groups of farmers in a specific location).

Participatory research and extension models in both industrial and developing settings are part of the farming systems research tradition (Chambers, Pacey, and Thrupp 1989) that emerged from a development studies and appropriate technology–inspired critique of the green revolution and the industrialization of agriculture. The criticism is aimed at reconfiguring the perceived Taylorist division of labor in which conception and design of work tasks are decoupled from their execution. At the levels of both policy and practice, the objective is to make practitioners (i.e., local people engaged in application of production tools in specific environments) active participants in innovation activities. These efforts to "bring the farmer back in" are articulated as a response to a perceived failure of the "technology transfer" model of development "in which research and extension systems are seen as the central source for generating and transferring information and technology" (Thrupp and Altieri 2001).

The critique of the technology transfer model holds that during the post-WWII period of agricultural modernization local actors adopted externally generated technical prescriptions composed of various combinations of a relatively small number of standardized industrial inputs (e.g., tractors, high yielding varieties, agrochemicals, plant and animal nutrition guides). These technologies are seen as products of Cartesian, universal science, or as Latour's "mobile immutables" (Kloppenburg 1991; Flora 2001). Discontinuities between standardized production tools and the local biophysical, socioeconomic, and cultural contexts in which implementation occurs give rise to dysfunction. On this basis, the problem is defined as one of liberating farmers from global (i.e., nonlocalized) technological packages created and supplied from outside and above. The reconstructive project is one of creating forums in which practitioners are active learners and meaningful participants in processes of shaping local production systems. "In these efforts, growers, researchers, extensionists, work within new organizational relations, learn together about principles of agroecology, blending knowledge and a variety of techniques, rather than being given fixed technological packages and inputs." (Thrupp and Altieri 2001:268)

Yet, it would appear that this critique is, in part, improperly derived. As Senker and Faulkner argue, "The linear model of innovation is now widely discredited. . . . It has been realized that technology often inspires science, and that many technological improvements have no connection with science. The linear model also overlooks the parallel and interactive activities which characterize innovation" (Senker and Faulkner 2001:207). In addition to providing a more accurate historical account of the connections between science and technology, the general field of innovation studies has enhanced our appreciation of localized aspects of learning (Lundvall 1992). Knowledge is no longer thought of as a freely circulating pure public good available to all off-the-shelf (Antonelli 1998). Individuals and organizations (e.g., family, firm, cooperative) must always confront the practical problem of accessing and decoding external generic information so as to create internally relevant knowledge. Premised as it is on the linear model of innovation, the technology transfer model insufficiently accounts for local action and local actors in processes of technological change. Although the technology transfer model served, and continues to serve, as a convenient and powerful representation for organizing modernization policy, investment, and narratives, it is a flawed description of the dynamics of technical change (Allaire and Wolf 2004).

Rosenberg (1982) has famously led an effort to look "inside the black box" of technological change. He concludes that to understand technical change from an economic and historical perspective, we must look beyond inventors and the revolutionary products associated with their names. Based on consideration of a series of empirical cases, Rosenberg argues that technological change "takes the form of a slow and often almost invisible accretion of individually small improvements in innovations" (62). More specifically, he identifies learning by using—"aspects of learning that are a function of [a product's] utilization by the final user" (122)—as a critical, if overlooked process.[4] Based on these insights, technological change should be seen as a product of combining knowledge assets of localized end users with those of researchers, product designers, and manufacturers engaged in R&D.

If we accept the decentralized, interactive model of innovation, we must revisit the basis of the political ecological critique of agricultural modernization (e.g., Chambers, Pacey, and Thrupp 1989; Kloppenburg 1991). If local practitioners' cumulative experience represents an essential component of technology generation and refinement processes, how should we interpret claims that the ecological and socioeconomic problems we observe are an outgrowth of an historical pattern of one-way, top-down flow of information that supports imposition of externally derived technologies?

Based on theoretical and empirical arguments below, I show technical change in agriculture to be an (asymmetrically) interactive process by means of theoretical and empirical demonstration of mechanisms through which end users' experiential knowledge is captured within the process of technological change. While insisting that knowledge production in agriculture is substantially dependent on inclusion of field- and farm-level processes, the analysis does not lead me to conclude that rural producers wield substantial power in the process of agricultural development by virtue of their centrality in technical knowledge generation networks. Farmers, as a class, do not participate as equals relative to agribusiness. Also, within the population of farmers, power and interests are unevenly distributed vis-à-vis targeting knowledge production capabilities.

Klein and Kleinman (2002) have argued for extension of the methods and mode of analysis of the social construction of technology (SCOT) approach through incorporation of political economic considerations. Based on a critique of the pluralist assumptions and failure to recognize power differentials within analysis of interactions among relevant social groups within SCOT, the authors demand analysts attend to structural considerations. While these authors' programmatic argument may be interpreted as quite orthodox from a social scientific perspective, the turn toward agency-centered social science in an age dominated by voluntaristic references makes it valuable. Kleinman defines *structures* as:

> specific formal and informal, explicit and implicit "rules of play" which establish distinctive resource distributions, capacities, and incapacities and define specific constraints and opportunities for actors depending on their structural location. Power and its operation are then understood within this structural context. The rules of play that define structures give certain actors advantages over others by endowing them with valued resources or indeed by serving as resources themselves (Kleinman 1998:289, as quoted in Klein and Kleinman 2002:35).

Endowments and access to valued resources are precisely the structural features that constrain farmers' participation in knowledge production, their ability to command a larger percentage of economic surplus in agrofood production systems (currently, farmers capture around $0.09 of each dollar spent on food in the United States), and perhaps their inability to act as stewards of the biophysical resources on which their livelihoods rest. The valued resources in question here include property rights over data, R&D capabilities, and the public legitimacy and state subsidies that support knowledge production within the agribusiness–U.S. Department of Agriculture–land grant university complex (Hightower 1973). Control of these strategic resources allows nonfarm interests to integrate farmers into knowledge production routines in ways that are in line with their political and economic interests.

Refutation of the linear model and adoption of an interactive, relational conception of innovation demand that we recognize farmers' participation in technoscientific networks. Acknowledging their participation in construction of technology does not, however, require that these networks be nonhierarchical. Although a network is defined by demarcation of an in-group and an out-group, power within networks can lead to differentiation among in-group members. How farmers are engaged in knowledge production and what political and economic resources they derive from engagement—as a class and as interest groups within this general class—is subject to structural considerations. The specific ways in which field- and farm-level technical practice has been integrated into science and technology are theorized in the next section.

Institutions and Learning: Collective Structures, Professional Knowledge, and Distributed Innovation

A system of innovation is composed of a collection of actors engaged in production, distribution, and application of knowledge, the resources available to them, and the institutions that shape interaction among these actors. As introduced above, this interactive conception of technical change stands in contrast to the linear model of innovation. Within this framework, innovation "is characterized by complicated feedback mechanisms and interactive relations involving science, technology, learning, production, policy and demand" (Edquist 1997). The organizational and institutional elements that support and constrain integration of knowledge from multiple sources are the key determinants of innovation and, ultimately, socioeconomic performance.[5] As argued in this chapter and others in this volume, knowledge production reshapes political and economic opportunities and the distribution of authority in social systems of production.

In considering systemic aspects of innovation, Allaire and Wolf (2002) have argued that technical change in agriculture, as is presumably the case in other extensive production systems, is a hands-on and decentralized process. Improvement of existing technologies and the development of new tools and techniques cannot be comprehensively performed in the laboratory through experimentation and simulation. Marked agroecological heterogeneity (i.e., biophysical variation) and idiosyncrasies in competencies of farm firms (i.e., socioeconomic variation) generate context-specificity in the economic and ecological performance of standardized technologies that characterize industrial agriculture.

> Agricultural technology differs from manufacturing, transportation or information technology in that it tends to be highly site-specific. Different crops have different requirements. . . . As a result, improvements have to be modified and adapted infinitely. A significant part of development cost is thus imposed on the user of an innovation, and the additional experimentation slows down the process. (Mokyr 1990:32)

Recognizing constraints on the diffusion of technical standards and the generally weak capacity of individual farms to undertake R&D to support adaptation of generic technologies, we identify two interrelated keys to innovation: 1) learning-by-using by farmers—experiential knowledge gained by individuals interacting with technical standards in more or less unique environments—must be purposively pursued as a knowledge resource, and 2) because learning based on individuals engaging in trial and error is inefficient and because their production environment is continually evolving, farmers must share experiential knowledge so as to continuously update their capabilities. These requirements point to value in *collective structures* that support *professional knowledge*. Allaire and Wolf (2002:1) define these terms as follows:

> Technical change is a process that occurs within, not to, a system of production. The experience of users of production tools is incorporated into the continual processes of research, design, and technical service support. At the local level, in addition to informal interaction, collective structures serve to aggregate, formalize, and transmit this experiential knowledge, thereby performing key functions within technical systems. Through circulation of information among and between users and suppliers of production tools, learning occurs and professional knowledge is created. Collective learning supports standardization, and permits individuals and groups to capture benefits from economies of scale.

Collective Structures for Learning in Technical Systems

In addition to the well-recognized contributions of market-coordinated private investment and publicly organized state action, collective structures powerfully support learning and innovation in technical systems.[6] Going beyond identification of informal mechanisms such as interpersonal networks that serve to convert distributed tacit knowledge to codified knowledge through processes of "socialization" and "internalization" as stressed in the organizational learning literature (Levitt and March 1988; Nonaka 1994), we emphasize formal collective structures within systems of innovation.

My arguments are in dialogue with the increasingly well-established critique of a conceptualization of knowledge as a pure public good. For me, knowledge is a localized, specific resource premised on site-specific adaptation by firms, each of which is distinguished by more or less idiosyncratic organization and competencies (e.g., Antonelli 1998). In the absence of mechanisms to overcome the "stickiness" arising from the tacit component of knowledge—friction stemming from the difficult-to-communicate properties of know-how—knowledge cannot circulate effectively among heterogeneous actors. As a result, cumulativeness of learning and the pace of innovation are constrained. To mitigate these costs efficiently, an array of collective intermediaries, including professions, facilitates inter-organizational communication and standardization. Standardization as the term is used here refers to diffusion of technical and cognitive conventions such that barriers to effective communication and coordination are relaxed. Such harmonization enhances the capacity for collections of actors to capture economies of scale in production and consumption.

Collective structures and the professional knowledge they produce also support adaptive management at the scale of both individual enterprises and sectors, and the capacity to reallocate resources across organizational boundaries in response to constraints and opportunities generated by new knowledge and external changes. For example, American agriculture simultaneously suffers from overproduction of commodities such as milk and underproduction of services such as food safety, ecological stewardship, and agro-tourism. The capacity to adapt to change—to shift resources out of overly mature markets into emerging markets—is collective in the sense that a broad range of firms and organizations within the sector are implicated. Efficiency of adaptation within the sector is enhanced by collective structures that support investment in infrastructure (re)education and (re)training of workers, refocusing of knowledge inputs from specialized organizations such as universities, refocusing of state regulation, and coordination of other activities that individual firms are poorly positioned to pursue.

Incentives at several levels support investment in collectively structured professional knowledge in innovation systems. Farmers have an incentive to participate actively in sharing experiential knowledge in order to enhance their learning capabilities and the competitive position of their firm, territory, or sector. Farm input suppliers and farm product buyers have an incentive to support localized collective structures because they enhance farmers' ability to learn efficiently so as to keep up with the technological frontier, thereby remaining valued customers and suppliers. The public has an incentive to support such professional knowledge in order to create efficiency at the sectoral and societal level—for example, the ability to respond to food safety and other service demands. Despite recognition of value in collective competencies, firms' inability to capture returns to such investments perfectly leads to a general problem of under-investment.

Professional competencies are essential resources for entrepreneurship and development. As Laurent Thevnot has commented, "one cannot innovate alone" (Thevnot 1998). Material change in practices and production systems rests, in part, on institutions and resources that connect interdependent actors and enable their collective production of new technical knowledge.

PROFESSIONS AS NETWORK ORGANIZATIONS FOR PRODUCTION AND MANAGEMENT OF TECHNICAL KNOWLEDGE

Professionals are workers operating in uncertain and contingent settings who must solve new problems in the course of production. They must creatively apply the tools of their trade and develop new production routines. This mode of production is different from that of semi- and non-professionals who rely on variously formalized "jigs," pre-prepared routines that allow them to produce in cookie-cutter fashion (Braverman 1974). Such standardization of work is possible only when the context of production and the quality of inputs are stable or predictable. In such settings, monitoring of workers' performance and output (quality measurement) is straightforward, and we expect to observe hierarchical organization. In contrast, in nonstandardized production environments, we observe higher levels of worker autonomy and more segmented production systems (i.e., lower levels of vertical integration) (Robertson 1999).[7]

Control of specialized technical knowledge is a key defining feature of professions (Elliot 1972; Freidson 1986). The "market enclosure" (i.e., limited monopoly) granted by the state and society to professions rests on responsible use and maintenance of this knowledge. Given that professionals' power and status stem from knowledge, professional structures serve to coordinate members in order to facilitate learning.

> As a working definition of professional networks, we can think of them as communities of independent practitioners who share a core competence, and who form strategic alliances across ownership boundaries. Professional networks identify core competencies, build capabilities, share them across the membership, and internalize knowledge flows

> without integrating ownership. Each professional's decisions and abilities are constrained by the capabilities of the network as a whole, as well as by other institutions. . . .
>
> While individual practitioners remain independent, they make long-term commitment of their substantial human capital to a 'hubless', indeed a bossless, network. . . . Without the exchange of cash payments, members willingly exchange information and technology and collaborate in production—that is, share routines—without authoritarian supervision, and without integrating external management functions into their day-to-day operations. In fact, network members remain competitors across many dimensions, attempting to take advantage of their capabilities more quickly and ably than others. So professions operate a complicated production strategy, furthering the interests of individual members as well as the interests of the network as a whole. (Savage and Robertson 1997:158–159).

While the quotation from Savage and Robertson identifies the tension between self-interest and group interest, the collective nature of learning serves to sustain information pooling. Professions are collective networks by virtue of their distributed structure (i.e., composed of references contributed by individual practitioners), access regime (i.e., those who contribute have access), and self-governance (i.e., rules of membership and conduct are made and enforced by participants, generally under state authority).

REFERENCE NETWORKS

Professional knowledge is based on a variety of collective goods including reference networks (Allaire and Wolf 2002). References are normalized empirical observations drawn from production. Reference networks are composed of sets of individuals engaged in pooling references. An individual reference is a record of performance (i.e., yield, quality, and cost) in a specified environmental and managerial setting—for example, bushels of corn per acre given varietal selection, soil type, tillage system, fertilizer and pest management inputs, and any other information collectively regarded as sufficiently important to warrant inclusion. These references represent a resource that supports farmers' capacity to successfully select an appropriate corn seed from a catalogue and grow a profitable crop. And over a longer time frame, such references allow plant breeders to adapt varieties to emerging localized constraints and opportunities.[8]

As a means to aggregate, compare, and synthesize cross-sectional and longitudinal observations, reference networks support collective learning across organizational boundaries, what we have defined earlier as distributed innovation. Reference networks play an important role in transforming tacit knowledge into codified knowledge (Nonaka 1994). This codified knowledge can be archived and circulated. Thus, reference networks represent a key communication channel, horizontally across sites of primary production and vertically between buyers and suppliers. These resources allow producers of goods and services to adapt evolving generic technology according to evolving local resource conditions. Local practitioners are able to evaluate and selectively incorporate new production tools and techniques. In this way, collective structures contribute to short-run technical coherence through enhancing productivity of inputs. Also, they contribute to long-run competitiveness through reducing learning costs and supporting system-wide capacity to adapt.

In the absence of opportunities to make comparisons, identification of differences (in the sense of Bateson 1980) is impossible, and information is scarce. As a result, learning is tremendously slow, particularly in a setting in which production outcomes are multivariate functions (i.e., hard-to-identify causality) and there are only a limited number of occasions to observe outcomes (i.e., annual harvests, long gestation times). To interpret empirical observations made in the course of production and more or less formalized experimentation (e.g., is 150 bushels/acre yield high or low given my approach to weed control?), references are required. References also support decisions as to how to allocate search capabilities across a vast array of potentially profitable projects (what is the likelihood of increasing my yield by changing the timing of nitrogen fertilization?). Through access to others' production experience, a farmer can compress the time required to learn and increase returns on experimental trials, most of which will result in dead ends. Reference networks enhance the efficiency of localized learning. Of course, experience of other farm operators is not perfectly translatable due to the idiosyncratic nature of resource endowments and operational routines, yet what is learned by using is partly transferable. Given the alternatives of relying solely on one's own internal record of trial and error or investing in proprietary R&D, there is value in the experiences of others.

To aggregate experiential knowledge of individuals and make it accessible requires coordination. First, creating a powerful body of references capable of context-specific decision support requires that all contributors adhere to standardized data collection and reporting protocols. Second, these protocols assume the existence of a shared technical vocabulary and a degree of agreement as to which technical questions and methodologies merit attention. Of course, these elements of group identity do not exist in a state of nature. While self-organization plays a role in communities of practice, such conventions must be nurtured, if not manufactured, and always there are maintenance considerations. Third, large numbers of observations representing suitable cross-sectional variability in inputs and production settings must be represented in the database. At the same time, the network must be bounded so as to remain coherent. Operators with production systems so different so as to make their experience non-aggregateable and noncomparable must be discouraged from participation as their records will be sources of noise. Fourth, specifically trained and equipped people must organize and analyze the data in line with a set of theoretical or operational questions and must format output in useful ways. And, fifth, references must be continuously updated to remain a productive resource, particularly in evolving settings such as agriculture, where, for example, pest resistance, introduction of new production tools, new regulatory constraints, and shifting consumer demand erode the value of existing knowledge.

These coordination requirements suggest a need for a more or less formal structure. Often, professional groups provide leadership and administrative services to create and sustain reference networks. As groups defined by their control of specialized technical knowledge (Freidson 1986), professions are well suited to perform such coordinating functions.

Empirical Description of Institutional Change in Agricultural Innovation Networks

Beyond illustrating distributed aspects of innovation, analysis of learning structures in these technical domains points to an important contemporary political economic dynamic in the organization of agricultural innovation systems. Increasingly, reference networks in farming systems are controlled, and in some cases constructed, by private commercial entities engaged in vertical relationships with farmers. In line with Castells's (2000) notion of "informationalism" in which knowledge production has become the focus of organizational and strategic logic, we observe downstream agrofood processors and upstream input suppliers positioning themselves as "data hubs" in production systems. These developments prompt us to ask, how are these data collected and who has access? Why are agribusinesses investing in field- and farm-level data recording? What are the implications of this change in the institutional configuration of reference networks and the larger systems of innovation? These questions structure presentation of empirical details presented below.

I argue that contemporary reorganization of learning structures constitutes heightened emphasis on commercial relations and a diminished collective character, and represents an accelerated period within a more-than-one-century-old pattern of erosion of farmers' position as participants in systems of innovation. Marcus (1985) identifies an early and decisive moment in this trajectory of decline. In studying several decades of debate preceding the creation of the state agricultural experiment station system in the United States in the last half of the nineteenth century, Marcus finds that the experiment station concept ultimately chartered by Congress and implemented by the individual states was the result of the defeat of a decentralized, systems approach based on integration of informed empirical observations of a legion of professional "experimental farmers." The victors were a cadre of professional agricultural scientists dedicated to reductionist principles and experimental design. Their work was to be coordinated by the newly created post of secretary of agriculture.

Importantly, the emergence of a class of professional agricultural scientists and centralized research administration diminished, but did not eliminate, farmers' status within processes of knowledge production. Although there is substantial evidence that modern agricultural research does not serve farmers' interests (Friedland, Barton, and Thomas 1981), farmers are incorporated into research processes. A variety of mechanisms link empirical field- and farm-level observations to technoscientific activities in centers of R&D.

Field- and farm-level observations are collected and circulated in various ways including farmer participation on administrative committees and advisory boards. Statistical censuses are important sources of empirical information on local activity and conditions. Production settings are (imperfectly) simulated at experiment stations. Extension and other intermediaries such as sales representatives of input manufacturers conduct on-farm field trials, enlist farmers in demonstrating techniques, and provide a conduit for information to circulate between

R&D facilities and farms. And lastly in what surely is not a comprehensive list, reference networks function to sustain iterative feedback between users and producers of production tools. This connectivity is essential because knowledge production cannot be completely removed from the field to an experimental facility. As stated above, biophysical variance and differences in farm firms' competencies require intensive and continuous dialogue between the laboratory and the sites of production. The restructuring of this dialogue is the story of reorganization of reference networks.

Allaire and Wolf (2002) focused on a theoretical specification of reference networks and professional knowledge within the context of an analysis of the collective dimension of innovation. I extend this analysis through presentation of two empirical examples of reference networks—hog production management and agronomic consulting. Information presented here was collected during interviews of farmers, agribusinesses, representatives of commodity associations (i.e., professional trade groups), environmental organizations, regulators, and researchers employed by both commercial firms and universities.

Case 1: Hog Management

Pig Improvement Company (PIC) is the world's largest supplier of hog breeding stock and a vendor of a popular software-based, farm management tool, PigCare. Participating growers keep detailed daily records on individual animals and route their data to PIC, which pools and analyzes the data. PIC provides a periodic assessment to each grower, ranking their performance against that of other comparable farms. Such benchmarking allows farm managers to analyze their performance relative to peers. Also, because the software synthesizes various information and produces easy to interpret output, farmers are more easily able to assess their own performance over time. Are my live births per sow above or below that of peers? Are my veterinary costs per animal rising and are they in line with those of other similar operations? Did my investment in construction of new rearing facilities result in increased average daily weight gain per animal? Such references allow the individual farmer to interpret data collected on the farm and to evaluate strategic choices.

Apart from using the PigCare service as a sales tool to market swine genetics and to support the competitiveness of the hog industry, PIC uses the data they receive from individual growers to evaluate proprietary genetic lines as well as their corporate strategy. PIC can statistically analyze how various combinations of genetics and farm management such as enterprise type, building type, farm size, feed and medication inputs, and managers' experience correlate with performance measures such as feed conversion ratio and rearing costs. While at this time farm sales and carcass data are not systematically incorporated in the database, analysts foresee this evolution. When such data are available, performance measures such as leanness of meat, size of cuts such as chops and hams, and pH could be analyzed as a function of genetics and husbandry. The database supports PIC's ability to assess the performance of certain breeding lines and to identify how to allocate their experimental and financial resources in line with data on changing preferences of pork consumers and cost constraints at the farm level.

We note that data useful for assessment and potential remediation of pressing environmental problems such as hog waste management and odor control are not collected. The largest hog farms produce as much sewage as a large city. Hog manure management is a major environmental and political concern in states where production is concentrated, such as Iowa, North Carolina, and Missouri. For example, a spill from a manure lagoon released over two million gallons of waste into a creek in North Carolina in 1999. Similarly, this database is not being used to analyze the costs and efficacy of alternatives to prophylactic antibiotics to manage animal health. The spread of resistant strains of bacteria and the erosion of the usefulness of antibiotics in human health care have emerged on policymakers' agenda. Despite demand for a response from the industry to ecological and human health concerns, the capabilities of these proprietary databases remain focused on short-term, private concerns.

Hog producers who choose to share their data with PIC receive the PigCare software and data analysis services at a discounted price. Growers who want access to the software but who do not want to release data on their operations must purchase the software. For example, Murphy Family Farms, the second largest producer of hogs in the United States ($650 million in sales last year), chooses to retain their data. As a result they do not receive comparable performance data from other growers. Evidently such firms perceive that there is more to lose than gain through disclosure. In effect, they are so large they are able to use the diversity within their own operation to identify management opportunities.

PigCare, however, is not the industry leader. PigChamp, developed by the University of Minnesota, represents the most widely used software package in the United States and the largest database. The university recently awarded an exclusive marketing license for PigChamp to a private entity, but the university retains ownership. PigChamp data remain available to university researchers free of charge, and contracts are negotiated with commercial firms seeking access to the data. Additionally, the National Pork Producers Council (NPPC), the association that ostensibly represents U.S. hog farmers, has launched a database construction project. The NPPC initiative is a cooperative effort by farmers to capture value in data as well as an attempt to guard against internalization (privatization) of production data by nonfarm agribusinesses.

This collective project comes at a time when there is talk of formation of "institutes"—organizations that would facilitate data pooling by a set of mega-farms (e.g., Murphy Family Farms, Premium Standard). In line with the highly developed division of labor and extensive resources of its members, the database within such a members-only club would be more detailed and more highly customized than the management tools described above. The incentive to form and join such an enterprise for Murphy and others who presently do not pool data would be to compare their performance to that of like competitors and to access innovative thinking (and maybe genetics). Limited to their private databases, firms can only access what information and diversity they have in-house. The motivation to "open it up" and engage in cross-fertilization is, of course, in tension with the competitive advantage they enjoy from retaining sole access to their records.

The emergence of these rather similar, competing database projects highlights incentives underlying various actors' construction and control of reference networks. The range of alternatives represented—Who is a member? Who has access to the data? Who decides which aspects of technical systems merit scrutiny?—highlight the salience of institutional changes underway that will structure how distributed innovation processes are organized and what roles farmers, agribusiness, universities, professional associations, and consumers will occupy.

Case 2: Agronomic Consulting

In the 1990s, commercial agronomic service markets in the United States exploded. During this period, consulting agronomists who provided varied decision-support services to farmers became increasingly important in several respects. At the level of the individual farm, technicians working for pesticide and fertilizer retail dealers assumed heightened levels of responsibility for influencing or making production decisions on a growing percentage of crop acres. The partial outsourcing of production management and decision making by farmers—particularly the collection, management, and interpretation of data used to guide decisions as to which crop varieties and how much of which kinds of pesticides and fertilizers to apply where, when, and how—raises important questions about agro-environmental management and agro-industrial organization.

A variety of factors have contributed to the rise of commercial service markets and the general pattern of outsourcing by farmers. At this time, technologies geared toward site-specific management (e.g., information technologies, precision farming, irrigation technologies, pest scouting, soil and plant tissue testing) were refined and increasingly incorporated into the conventional agricultural production model. These sophisticated and integrated sets of technologies generally require specialized equipment and management skills, investment in which often makes sense for a local input vendor rather than an individual farmer. The cost-price squeeze on farmers tightened due to globalization and trade liberalization, causing them to seek to manage inputs more effectively. And environmental pressures, particularly stemming from water quality concerns, raised the stakes of agrochemical management. In this context, many farmers looked to input suppliers for assistance in managing data and enhancing the use of information in managing their farms. Not coincidentally, at this same time, publicly operated extension services were being commercialized and privatized in dozens of countries worldwide (Feder et al. 2001).

At the level of the input supply sector, the emergence of a knowledge-intensive service orientation—the self-perception and production of a public image in which their role is to support farm management in addition to simply delivering inputs—created new forms of competition and required acquisition of new competencies. In order to diversify and create a profit center in services, input vendors had to be able to add value to increasingly sophisticated farmers' operations by providing detailed, timely, site-specific management support. Opportunities to market analytic and diagnostic services for a fee were hindered because for many years farmers had purchased chemicals and in the bargain received casual over-the-counter or in-field advice for free.

In response to the need to enhance the service capabilities of people and firms in the industry, and to maintain legitimacy in the face of mounting environmental criticism as to the external costs of high-input, petrochemical-based agriculture, a national professional certification program was initiated, the Certified Crop Advisor (CCA) program (Wolf 1998). The concept of certification of private-sector agronomists arose out of meetings between the Fertilizer Institute (an industry consortium) and the USDA–Natural Resource Conservation Service. Input supply dealers were prohibited from participating in a federally funded program to promote and subsidize alternative agrochemical management practices to protect water quality. These commercial actors were excluded due to a perceived conflict of interest in that their principal business is agrochemical sales.

The program consists of a national- and state-level written examination, continuing education requirements, letters of recommendation, and an ethics pledge. There are now over 14,000 CCAs in the United States and Canada. By virtue of their professional competencies, agronomists working for input vendors are positioned to synthesize local observations, thereby providing informed advice to farmers. The farmer benefits from the consultation services of an individual who specializes in specific aspects of production (e.g., nutrient management, weed management, insecticide treatments). In theory, a CCA visits several farms a day in a local area during the growing season and checks in by phone or at the sales counter with many more farmers. At the same time, the contract services provided by dealers to farmers and the records used to perform these agronomic functions (e.g., digital soil and crop yield maps and records of chemical management) represent additional data of value. In this sense, the agronomist and the input dealership they represent are the hub in the local reference network. By virtue of their privileged position, they will learn efficiently and become increasingly powerful actors in the local production system.

This dynamic will accelerate the existing trend in agricultural production, which is outsourcing. Increasingly, dealers are doing far more than selling advice coupled with inputs. They are now engaged in making farm management decisions and providing custom services. In a farming community in Illinois, I was told corn farmers plant the crop and harvest it, but everything in between is left to the local input dealer. In Florida, a local dealership was selling a comprehensive weed management service to farmers (payment of a one-time per acre fee buys the farmer guaranteed weed control for the season). In Mississippi, cotton farmers told me that they did not know what insecticides are used on their land because they pay consulting entomologists to make spraying decisions and hire aerial applicators for them. In order to provide such services efficiently and ensure their economic survival in a shrinking and rapidly consolidating sector, dealers are aggressively motivated to be efficient learners.

The pattern of data acquisition and management under the commercial agronomic consulting model discussed here represents a further acceleration of the pattern of erosion of farmers' capabilities to be active learners. As discussed earlier in reference to the critique represented within the farming systems research tradition, public agencies and universities in the United States have been criticized heavily for functioning to promote industrial technologies rather than build local capabilities. The emerging organizational model in which localized references are controlled by and embodied within commercial input dealers similarly fails to contribute to farmers' autonomy and scope of their contributions to agrofood innovation systems. The nature of farmers' dependence and subordination may have changed, but there appears to be no real mechanism in place to improve their status. Erosion of public agencies' engagement in data acquisition, technology generation, and oversight of agricultural production raises important questions regarding who is monitoring and educating the relevant decision makers. Accountability applied to environmental management and natural resource conservation is not assured when both farmers and public agencies have heavily outsourced information management functions to the commercial sector.

Analysis: Implications of Reorganization of Collective Learning Structures

In terms of the institutional basis of coordination, the commercial, proprietary character of these reference networks represents a departure from previously localized, informal or publicly facilitated collective structures. The vertical-orientation (i.e., buyer-supplier) of information transmission differs from previous emphases on horizontal communication among localized sets of farmers. The local professional group, a more or less formally delimited collection of farmers with similar enterprises committed to pooling various resources, including knowledge, is of apparently decreasing socioeconomic importance. And in terms of function, these private, corporate knowledge structures do not serve to develop, maintain, and advance farmers' professional knowledge. In fact, the structure of these commercial reference networks ensures that new capabilities and capacity to learn accrue in nonfarm segments of agrofood systems.

These data-pooling networks support refinement of technological standards, including management routines, but they do not contribute to professional workers' capability to exercise judgment and create new problem-solving routines in production. As a result, the dynamism and power of farmers as a class is expected to become further compromised in the increasingly knowledge-intensive and innovation-driven agrofood sector.[9] Additionally, incentives and mechanisms to pursue knowledge production to enhance the ecological performance of agrofood production through system redesign appear to be absent. Waste-minimization—optimization of input use—will be effectively achieved through this approach, but radical innovations are unlikely to emerge through these learning routines. The institutional changes discussed in the first section of the chapter with reference to strategies for ecological modernization are not evident within the specific context of U.S. agriculture.

Reflecting on the relationship between knowledge and authority, Savage and Robertson (1997) note that performance of distributed production systems improves when knowledge is possessed by the relevant decision makers. To achieve performance-enhancing alignment, two avenues are available: "One is by moving the knowledge to those with the decision rights; the other is by moving the decision rights to those with the knowledge." (Jensen and Meckling 1992:253, quoted in Savage and Robertson 1997:157). They go on to argue that to the extent that property rights and know-how are transferable among organizations, knowledge will accumulate to those who can put it to best and highest use. Thus, in somewhat of a functionalist (apolitical) vein, prevailing patterns of industrial organization and social relations of production can be said to reflect the distribution of know-how. As corporate, off-farm interests assume increasing control of reference networks and knowledge production apparatus, we can imagine that their power will grow relative to farmers and the state.

Enhanced integration of reference networks into farmers' commercial relationships with buyers and suppliers reflects a general pattern of agrofood system restructuring and specifically vertical coordination of commodity chains (Heffernan 1999). As Savage and Robertson (1997) suggest, we observe parallel concentration of market power and knowledge-generating capability in nonfarm agribusinesses. While more empirical analysis is required, we can suggest that the observed restructuring of reference networks does not bode well for farmers' professional knowledge and development of their capabilities. Informing field- and farm-level management decisions remains an important output of these new reference networks. The emphasis is on how to use the sponsoring firm's inputs effectively and achieve qualities specified by downstream buyers. Commercial objectives of the actor operating at the center of the network take precedence.

In taking a dynamic perspective and seeking to study co-evolution of technology and institutions, we see that the phrase "knowledge is power" is only partially correct. Because knowledge can lose significant value over time—particularly in settings like agriculture where the production environment is unstable (e.g., changes in pest dynamics and resistance, dietary patterns, and environmental regulations), and there is potential for radical innovation (e.g., biotech)—it is capacity to produce and absorb knowledge (that is to say, to learn) that represents the basis of power and dynamism.

Reference networks enhance two essential management functions: short-term input allocation efficiency and longer-term dynamic adjustment (adaptive capacity). But references are useful only within a fairly narrow technical spectrum. They cannot reliably support inference outside of their empirical breadth. They serve to refine management quantitatively (incrementally), not qualitatively. As such, the technological application or technical routine under construction will likely retain its central features over time. In this sense, corporate control of reference networks raises the probability of lock-in, or what is referred to as closure and stabilization with SCOT (social construction of technology). Lock-in here refers both to production tools and techniques and the distribution of authority within the production system. The relative power of incumbents controlling reference networks will increase, potentially at an accelerating rate, as they learn (and learn to learn) more efficiently than others. Thus, reference networks should be seen as resources for continual improvement within pre-existing technological regimes and not as triggers of radical innovation.

The commercialization of reference networks further erodes traditional roles and justifications for public agency investments in agricultural development. One could ask, now that private firms are willing to coordinate reference networks, what should be the public-sector role and on what basis does the social contract between farmers and the public rest? Commercialization also complicates public agencies' access to data and potentially compromises the capacity of public researchers to engage in projects to address ecological problems stemming from production. The public sector's capacity to pursue system redesign—radical innovation—is weakened as a result of the internalization of data that accompanies commercialization of reference networks. Within such a configuration, mechanisms for production of public goods–type knowledge are unclear.

Conclusion: Institutional Hybridity

In making the claim that the governance, function, and form of reference networks are changing and that these changes imply an increasingly subordinate role for farmers, a qualification is required. I have not presented substantial historical evidence that in a previous era agricultural reference networks were noncommercial, strictly oriented to public goods production, and served only to reinforce farmers' autonomy as professionals (however, see Marcus 1985 for a claim not too distant). Such a pure, collective organizational form would be difficult if not impossible to identify. There have always been private investments in agricultural science and private benefits associated with the outputs of public agencies engaged in agricultural research and education. The financing and legal basis of the USDA/land grant university system reflect the explicit blending of public and commercial interests and modes of functioning (National Research Council 1995).

As is the case today, in earlier times organizational structures in agricultural knowledge systems were hybrid. They were composites sustained by an admixture of privately motivated self-interest, collective action within variously scaled communities, and public action for the common good. This hybridity implies diversity of justifications, sources of legitimacy, and resources for coordination, and this diversity is the basis of organizational strength, adaptability, and, ultimately, persistence over time. In adopting a framework of hybridity, no definitive historical rupture in the organization of reference networks can be identified. However, the composition of the hybrids, and the relative importance of the institutional modes of coordination does change across time, place, and project.

Thus, in presenting contemporary reference networks and contrasting them to earlier structures, I seek to avoid the reifying effect of the labels *public* and *private*. Also, I want to avoid an overly narrow specification of their function within systems of innovation, as there is significant diversity and opportunity for complementarity among private, collective, and public modes of production. For example, information on how nitrogen fertilization affects corn yields can, in theory, benefit farmers' production efficiency, the corn sector's competitiveness versus other feed grains in the local market, ammonia manufacturers' efforts to market fertilizer, the region's water quality, and consumers' nutritional status. Commercial restructuring of reference networks does not imply elimination of such complementarities, but the function of knowledge production apparatuses changes as their organizational and institutional composition changes.

Recognition of the historical contributions that farmers have made to development, refinement, and diffusion leads us to identify important structural features of innovation systems and contemporary patterns of change. As argued in this chapter, the mechanisms through which farm-level actors participate in knowledge creation is changing in such a way as to erode the professional knowledge of local actors and the status of the state (i.e., public sphere) in systems of innovation. These changes raise difficult questions about how to support innovation efforts targeted at ecological and social costs of production. In seeking to address these challenges, this chapter points to existing elements of connectivity in systems of agrofood innovation. The challenge then is to identify and complement existing collective structures so as to maintain commercial incentives for investment in knowledge production, professional capabilities of farmers, and appropriate levels of public-agency involvement. Obviously, redundancy considerations must be addressed within efforts to identify and enhance the division of labor that supports coordinated engagement. Through engaging a broader array of actors in knowledge systems and a broader set of objectives, diversity can be maintained and the complete range of economic, social, and ecological requirements of social systems can potentially be advanced.

Acknowledgments

Gille Allaire's essential contributions to this paper are gratefully acknowledged. Also, I appreciate the constructive comments of the editors and Elisabeth S. Clemens.

Endnotes

1. See Allaire and Wolf (2004) for an analysis of the rational myths that sustain ignorance of the essentially collective nature of innovation.

2. In the aftermath of 9/11, I note that (national) security has assumed a status that supercedes this traditional troika.

3. The institutional problematic has in some ways eclipsed the ecological economics problematic of "getting the prices right" (Daly and Cobb 1989).

4. My treatment of learning by using will depart significantly from that of Rosenberg (1982), who focused on users learning to interact more efficiently with an unchanging capital good (i.e., a piece of manufacturing machinery) over the course of its productive life. I focus on adaptation by users and adaptation of the technologies with which they interact. In other words, for me, the attributes of technologies co-evolve with users' knowledge and skills.

5. See Edquist (1997) for a useful review of the systems of innovation literature and its relationship to evolutionary economics. The institutional orientation of this literature and its explicit focus on knowledge production suggest direct links with themes developed in this volume.

6. This section draws directly on Allaire and Wolf (2002).

7. Self-employment, however, is now the exception rather than the rule for professionals. Nonetheless, by virtue of engaging in production of technical services, entry into employment contracts does not correlate directly with loss of control over the labor process. Freidson (1986) argues that self-employment is most often a problem, not a luxury, for professionals.

8. Such a database can also be used to identify economically rational limits on fertilizer and pesticide applications, so-called best management practices (BMPs).

9. The argument of obtaining new skills is also potentially relevant here. Farmers' professional competencies as managers of biological resources may erode as a function of reorganization of reference networks. It is conceivable, however, that in the case of large farms, outsourcing technical responsibilities (including production) may be part of a larger enterprise strategy in which the

"farmer" is defined as a specialized, professional agribusiness manager, not a natural resource manager.

References

Allaire, Gille, and Steven Wolf. 2002. "Collective Goods and Accountability in Technical Information Systems: Innovation and Conflict in Agriculture." Unpublished paper, INRA, Toulouse, France, and Cornell University, Ithaca, NY.

———. 2004. "Cognitive Models and Institutional Hybridity in Agrofood Innovation." *Science, Technology, & Human Values* 29:431–458.

Allen, Robert C. 1983. "Collective Invention." *Journal of Economic Behavior and Organization* 4:1–24.

Antonelli, Cristiano. 1998. *The Microdynamics of Technological Change.* London: Routledge.

Bateson, Gregory. 1980. *Mind and Nature: A Necessary Unity.* Toronto: Bantam.

Berkhout, Frans. 2002. "Technological Regimes, Path Dependency and the Environment." *Global Environmental Change* 12:1–4.

Bok, Derek. 2003. *Universities in the Marketplace: The Commercialization of Higher Education.* Princeton, NJ: Princeton University Press.

Bowles, Samuel, and Herbert Gintis. 2002. "Social Capital and Community Governance." *The Economic Journal* 112:419–436.

Braverman, Harry. 1974. *Labor and Monopoly Capital: The Degradation of Work in the Twentieth Century.* New York: Monthly Review Press.

Castells, Manuel. 2000. *The Rise of the Network Society.* Oxford: Blackwell.

Chambers, Robert, Arnold Pacey, and Lori Ann Thrupp, eds. 1989. *Farmer First: Farmer Innovation and Agricultural Research.* New York: Bootstrap Press.

Daly, Herman, and John Cobb. 1989. *For the Common Good: Redirecting the Economy toward Community, the Environment, and a Sustainable Future.* Boston: Beacon Press.

Edquist, Charles. 1997. *Systems of Innovation: Technologies, Institutions and Organizations.* London: Pinter.

Elliott. Philip. 1972. *The Sociology of the Professions.* New York: Herder and Herder.

Feder, Gershon, Anthony Willett, and Willem Zijp. 2001. "Challenges for Public Agricultural Research and Extension in a World of Proprietary Science and Technology." In *Knowledge Generation and Technical Change: Institutional Innovation in Agriculture,* ed. Steven Wolf and David Zilberman. Norwell, MA: Kluwer Academic.

Flora, Cornelia B. 2001. "Agricultural Knowledge Systems: Issues of Accountability." In *Knowledge Generation and Technical Change: Institutional Innovation in Agriculture,* ed. Steven Wolf and David Zilberman, pp. 111–124. Norwell, MA: Kluwer Academic.

Freidson, Eliot. 1986. *Professional Powers: A Study of the Institutionalization of Knowledge.* Chicago: University of Chicago Press.

Friedland, William, Amy E. Barton, and Robert J. Thomas. 1981. *Manufacturing Green Gold.* New York: Cambridge University Press.

Graedel, Thomas E., and Braden R. Allenby. 1995. *Industrial Ecology.* Englewood Cliffs, NJ: Prentice Hall.

Heffernan, William. 1999. *Consolidation in the Food and Agricultural Industry.* Report to the National Farmers Union. http://home.hiwaay.net/~becraft/NFU FarmCrisis.htm (accessed June 8, 2005).

Hemmelskamp, Jens, Klaus Rennings, and Fabio Leone, eds. 2000. *Innovation-Oriented Environmental Regulation.* ZEW Economic Studies 10. Heidelberg: Physica-Verlag.

Hightower, Jim. 1973. *Hard Tomatoes, Hard Times: A Report of the Agribusiness Accountability Project on the Failure of America's Land Grant College Complex.* Cambridge: Schenkman.

Hollingsworth, J. Rogers, and Robert Boyer, eds. 1997. *Contemporary Capitalism: The Embeddedness of Institutions.* New York: Cambridge University Press.

Klein, Hans, and Daniel L. Kleinman. 2002. "The Social Construction of Technology: Structural Considerations." *Science, Technology & Human Values* 27:28–52.

Kleinman, Daniel L. 1998. "Untangling Context: Understanding a University Laboratory in a Commercial World." *Science, Technology & Human Values* 23: 285–314.

Kloppenburg, Jack, Jr. 1991. "Social Theory and the De/reconstruction of Agricultural Science: Local Knowledge for Sustainable Agriculture." *Rural Sociology* 56:519–548.

Lee, Kai. N. 1993. *Compass and Gyroscope : Integrating Science and Politics for the Environment.* Washington, DC: Island Press.

Levitt, Barbara, and James G. March. 1988. "Organizational Learning." *Annual Review of Sociology* 14:319–340.

Lundvall, Bergt-Åke. 1992. *National Systems of Innovation.* London: Pinter.

Marcus, Alan. 1985. *Agricultural Science and the Quest for Legitimacy: Farmers, Agricultural Colleges, and Experiment Stations, 1870–1890.* Ames: Iowa State University Press.

Menard, Claude. 2002. "The Economics of Hybrid Organizations." Presidential address, International Society for New Institutional Economics, Paris, September 29.

Mokyr, Joel. 1990. *The Lever of Riches: Technological Creativity and Economic Progress.* New York: Oxford University Press.

Mol, Arthur P. J., and David A. Sonnenfeld. 2000. *Ecological Modernisation around the World: Perspectives and Critical Debates.* London: Frank Cass.

National Research Council. 1995. *Colleges of Agriculture at the Land Grant Universities.* Washington, D.C.: National Academy Press.

Nonaka, Ikujiro. 1994. "A Dynamic Theory of Organizational Knowledge Creation." *Organization Science* 51 (February): 14–37.

Porter, Michael, and Claas van der Linde. 1995. "Toward a New Conception of the Environment-Competitiveness Relationship." *Journal of Economic Perspectives* 9:97–118.

Reinhardt, Nola, and Peggy Bartlett. 1989. "The Persistence of Family Farms in U.S. Agriculture." *Sociologia Ruralis* 29:203–225.

Robertson, Paul L., ed. 1999. *Authority and Control in Modern Industry: Theoretical and Empirical Perspectives.* London: Routledge.

Rosenberg, Nathan. 1982. *Inside the Black Box: Technology and Economics.* New York: Cambridge University Press.

Savage, Deborah A., and Paul L. Robertson. 1997. "The Maintenance of Professional Authority: The Case of Physicians and Hospitals in the United States." Social Science Research Network. http://papers.ssrn.com/sol3/papers.cfm?abstract_id=86528 (accessed May 6, 2005).

Senker, Jacqueline, and Wendy Faulkner. 2001. "Origins of Public-Private Knowledge Flows and Current State-of-the-Art: Can Agriculture Learn from Industry?" In *Knowledge Generation and Technical Change: Institutional Innovation in Agriculture,* ed. Steven Wolf and David Zilberman, pp. 203–232. Norwell, MA: Kluwer Academic.

Thrupp, Lori Ann, and Miguel A. Altieri. 2001. "Innovative Models of Technology Generation and Transfer: Lessons Learned from the South." In *Knowledge Generation and Technical Change: Institutional Innovation in Agriculture,* ed. S. Wolf and D. Zilberman, pp. 267–290. Norwell, MA: Kluwer Academic.

Von Hippel, Eric, 1983. "Cooperation between Rivals: Informal Know-how Trading." In *Industrial Dynamics: Technological, Organizational, and Structural Changes in Industries and Firms,* ed. Bo Carlsson. Boston: Kluwer Academic.

Walters, Carl J, and Crawford S. Holling. 1990. "Large-Scale Management Experiments and Learning by Doing." *Ecology* 71:2060–2068.

Williamson, Oliver. 1987. *The Economic Institutions of Capitalism: Firms, Markets, Relational Contracting.* London: Collier Macmillan.

Wolf, Steven. 2004. "Community Governance and Natural Resources." Manuscript submitted for publication.

———, ed. 1998. *Privatization of Information and Agricultural Industrialization.* Boca Raton, FL: CRC Press.

Wolf, Steven, and David Zilberman, eds. 2001. *Knowledge Generation and Technical Change: Institutional Innovation in Agriculture.* Norwell, MA: Kluwer Academic.

5

Antiangiogenesis Research and the Dynamics of Scientific Fields

Historical and Institutional Perspectives in the Sociology of Science

DAVID J. HESS

In 1958 a Harvard Medical School student named Judah Folkman worked with an MIT engineer to develop an implantable pacemaker. Because the medical school did not seek patents at that time, the two researchers published the results and left the product in the public domain for firms to commercialize (Cooke 2001:37, Folkman and Watkins 1957). In the intervening years Folkman developed a theory of cancer based on angiogenesis (that is, the idea that tumors required the growth of blood vessels), and he also helped pioneer a new level of university-firm collaboration. By 1998 his work was generating increasing attention among researchers, biotechnology firms, drug companies, and the major media. In contrast, in 1957 a young associate professor at Columbia Presbyterian Medical Center named John Prudden found that bovine cartilage could accelerate the healing of wounds and reduce inflammation in rats (Prudden, Nishikara, and Baker 1957). Over the decades he developed a cartilage-based therapy for cancer, but when he died in 1998, his approach was largely lost to mainstream medicine. The story of the growth of antiangiogenesis drugs, considered in comparison with the parallel story of the stunted development of cartilage research and related natural products, provides an opportunity for the sociology of scientific knowledge to consider theoretical frameworks that examine institutional factors such as changing regulatory policy, commercialization, and social movements.

Theoretical Background

What is at stake in revived attention to institutional factors such as states, markets, and social movements in science and technology studies? In *Politics on the Endless Frontier* Kleinman suggests that the issue involves the conflict over democratic participation in science and technology policy, which in the United States dates back at least to the state funding order that emerged after World War II. Whereas Vannevar Bush, a former vice-president of MIT and the head of the Office of Scientific Research and Development during World War II, advocated a large degree of autonomy for the scientific community, the New Deal senator Harley Kilgore advocated a funding model that included participation by representatives of farmers, labor unions, and the public. The Bush model eventually triumphed, and the institution of science was cloaked in the policy of "exceptionalism," that is, the view that high levels of autonomy are socially valuable.

The autonomy assumption was defended not only by scientists but also by sociologists and philosophers of science (Daniels 1967; Mulkay 1976; Fuller 2000). The embrace of the autonomy assumption in science studies was evident in various research traditions, including Merton's (1973) depictions of science as a self-regulating system and Kuhn's (1970) account of paradigm change as governed by epistemic relations internal to the scientific field. To some degree the subsequent generation of studies in the sociology of scientific knowledge (e.g., Knorr-Cetina and Mulkay 1983)—which emphasized the microsociology of laboratories, discourse, networks, and controversies—also represented a continuation of the autonomy assumption (Hess 2001a:39–42). Although those studies tended to emphasize the agency of scientists, their networks, and related microsociological units of analysis, they provided glimpses of the causal shaping role of states, firms, and social movements in the making of scientific knowledge. Attention to institutional factors was also evident in other science and technology studies (STS) traditions, especially the Marxist literature (e.g., Bernal 1969; Hessen 1971) and the interests analyses of the late 1970s (MacKenzie and Barnes 1979), but also Merton's more Weberian work (e.g., Merton 1970). Likewise, the work of anthropologists and feminists in the 1980s and 1990s drew attention to macrosociological categories of analysis, social problems, culture and power, and interactions with lay groups and social movements (Hess 2001b). In significant ways their work was parallel to renewed attention to structure and external institutions in the sociology of scientific knowledge (e.g., Kleinman 2003).

Drawing on the two strands of the post–laboratory studies literature, this chapter contributes to the renewed attention to institutional factors such as states, markets, and social movements by arguing for the value of a more deeply historicized sociology of scientific knowledge. Neither the Mertonian nor the constructivist research traditions emphasized the questions raised in a historical sociology of modernity, yet those questions are often very close to the surface of the new studies of regulatory politics and expertise, commercialization and privatization, civil society, social movements, public understanding of science, and public participation in science (Misa, Brey, and Feenberg 2003). The new problem areas often involve reference to contemporary historical change, which is framed under various rubrics (e.g., late capitalism, postmodernity, reflexive modernization, and globalization). My own view is that modernity needs to be conceptualized not as an event but as an ongoing process that has taken on specific forms during the last decades of the twentieth century, but those forms are largely continuous with long-term historical developments since at least the sixteenth century. As a result, early- and mid-twentieth-century social theories of modernity continue to be of value, even if they are in need of revision.

This chapter hypothesizes specific tendencies in the contemporary historical development of science as a field of action. The term is borrowed from Bourdieu (e.g., 2001) but is located in a more historical sociological perspective based on the following four processes.

1. *Expansion of scale.* In laboratory sciences the cost and scale of research has increased and outpaced the ability of public institutions to fund it. As scale increases, new arrangements with the private sector have become necessary, and ongoing negotiations over central control and local autonomy occur. The Bush/Kilgore debate is merely one example of an ongoing negotiation of the relative autonomy of scientific fields, which continues today in debates over commercialization and the university.
2. *Differentiation of institutions.* Human and organizational actors in scientific fields increasingly face conflicts of coordination and

alignment of roles and organizational goals with those of other fields of action. For example, scientists and universities develop increasingly complex goals as they negotiate their roles in education, research, fundraising, management, policymaking, citizenship, community development, and entrepreneurship. New boundary roles and organizations emerge to negotiate the increased complexity, which in turn generates further differentiation of fields of action (Frickel 2004; Guston 2001; Moore 1996).

3. *Universalization of values.* The cultures of scientific fields tend to become increasingly universalistic in the sense of developing increasingly formalized methodologies and methods of dispute resolution among competing research networks. In applied fields, such as clinical medicine, regulatory policies encode the universalism through mandated standards that determine the translation of laboratory findings into clinical applications. The formalization of methods and standards for acceptance of both facts and artifacts creates conflicts over access to the means of knowledge production and clashes between expert and lay positions on knowledge-making priorities.
4. *Denaturalization of the material world.* Both research technologies and the technologies/products generated by research tend to become increasingly synthetic or distanced from living entities over time. In scientific fields research innovation is driven in part by the problem of diminishing returns of research efforts to a given method, which results in efforts to find new research methods and technologies (Rescher 1978). Patent law and the commercialization of research also drive an increasing emphasis on invention, innovation, and synthesis. However, technological innovation also generates new hazards, side effects, and risks (Beck 1992), and as a result it drives an ongoing negotiation between new technologies and their societal and environmental implications. The safety and environmental concerns raised by civil society organizations create an ongoing negotiation of innovation oriented toward profits versus societal and environmental amelioration.

The Case Studies: Background

The empirical research presented here develops a comparative analysis of two American research and therapy fields for cancer, one successful and the other unsuccessful. The dual case studies provide a good example of why the sociology of scientific knowledge needs to take into account historical changes such as commercialization, new regulations, and civil society participation. For example, the development of the two research programs takes place within a rapidly changing context of commercialization of medical research. The Bayh-Dole Act of 1980, which facilitated patenting and licensing for universities, is generally considered the watershed moment in commercialization of university-based research in the United States, but in the case examined here some of the partnerships predated the act and suggest that commercialization was a much-longer-term process. The commercial appeal of patented drugs made it easier for one pathway of scientific research, drug-based angiogenesis research, to overcome significant opposition from scientific and medical elites, whereas an alternative pathway, cartilage-based research, remained underfunded and enveloped in controversy.

With respect to the regulatory function of states, another watershed moment was the passage in 1962 of the Kefauver-Harris amendments to the Food, Drug, and Cosmetics Act of 1938. The amendments, which were in response to the safety concerns raised by thalidomide, created new standards of efficacy and higher standards of safety for drug approval, but they significantly increased the cost of bringing a new drug to market. By the late 1990s the research and development cost of bringing a drug to market has been estimated to be as high as $800 million, although some studies indicate that after-tax research costs are only a tenth of that figure (Young and Surrusco 2001). Still, even the lower end of the estimate range represents a significant investment that drives the preferences of capital-bearing private-sector firms to favor the financial security of drug-based patents, in contrast with the uncertainties of the intellectual property rights associated with food-based products. However, the food/drug distinction has also undergone change; under the Dietary Supplement, Health, and Education Act (DSHEA) of 1994, the regulatory system in the United States grappled with an emergent category between food and drug—the nutritional supplement or nutraceutical—and granted wide over-the-counter access while restricting manufacturers from making claims about treating disease.

A third area of historical change has been the development of civil society organizations. Behind the appeal of both antiangiogenesis drugs and cartilage-based products is a patient-driven reform movement that has demanded changes in the therapeutic regimes of chemotherapy and radiation therapy. The movement existed before the 1960s, but it coalesced into a mass social movement to protest the suppression of laetrile in the 1970s, and by the mid 1980s it had diversified into a broad-based alternative cancer therapy movement (Hess 2003). Alongside the patients stand many doctors who are frustrated by the high toxicity and low efficacy of the conventional cancer therapies, but there is also a history of opportunists who have developed proprietary products, made excessive health claims, and preyed on the vulnerabilities of cancer patients. As a result, the leaders of the patient advocacy movement are very cautious about leading patients toward products that lack efficacy, and they sometimes put the brakes on overstated claims, including those associated with shark cartilage.

The changes in regulatory policies, intellectual property regimes, and social movements intersected with many other historical changes that can only be flagged here. For example, the medical profession itself was losing autonomy due to the rise of countervailing powers such as health maintenance organizations and patient advocacy movements. The resulting decline in autonomy helped open the door to the proliferation of complementary and alternative cancer therapies. Likewise, research methods were undergoing shifts that both favored drug development (through the increasing emphasis on clinical trials as the standard of evaluation) and allowed spaces for evidence-based medical claims for the complementary and alternative therapies (through the development of retrospective methods and databases with historical controls).

The case studies presented here draw on primary and secondary sources, and they are part of a broader research project that has involved over a decade of ethnographic observation, semi-structured interviewing, and archival research in the United States and other countries. The case study method is widely used in the STS field, and it is modified here in two ways that are consistent with the theoretical framework. First, the use of closely related but inverted "twin" cases of success and failure is used to facilitate the development of a non-autonomous analysis that includes regulatory, private-sector, and social movement factors (see also the similar comparative strategy adopted by Woodhouse, this volume). Second, the historical scope of the cases is long term (that is, decades) rather than short term. The longer temporal perspective facilitates a more deeply historical analysis.

THE MAKING OF A RESEARCH FIELD AND INDUSTRY

Angiogenesis is no newcomer to science; the surgeon John Hunter used the term to describe blood vessel growth in 1787, and as early as 1907 researchers had observed tumor vascularization (Angiogenesis Foundation

2003; Goldman 1907). In 1941 a medical researcher reported that tumors implanted into guinea pigs' eyes would grow and develop blood vessels (vascularization), but in cases where vascularization did not occur, the tumors also did not grow (Greene 1941). A few years later, cancer researchers published the hypothesis that blood vessels grow toward tumors (Algire and Chalkley 1945). In the 1960s Judah Folkman and colleagues observed the same process in transplanted animal tumors, and the young surgeon went on to play the central role in the development of the research field in the United States. Because of his central role, this section focuses on the shifting position of his work in the fields of cancer research and, eventually, cancer therapy.

As a surgeon, Folkman entered the field of oncology research as an outsider, but he had credibility within the broader medical field because of his reputation as a stellar medical student, his rapid rise to prominence as a professor of surgery at Harvard Medical School, and his position as chief of surgery at Boston Children's Hospital. However, during the first decades of his work on angiogenesis inhibitors, his position in the medical field was due to his work as a surgeon. His laboratory work was tolerated as a voluntary activity, and his first publications on angiogenesis and cancer were largely ignored, even when in top journals (e.g., Folkman 1971). Skeptical scientists argued that the growth of blood vessels toward tumors was due to inflammation, and clinicians remained uninterested because applications seemed remote (Cooke 2001:100). When the renowned Boston oncologist Sidney Farber encouraged the public relations coordinator of the American Cancer Society to feature Folkman in its annual press seminar, the ensuing media attention that he garnered only increased his isolation among cancer researchers (116–119).

During the 1960s and 1970s cancer research in the United States was dominated by a network that had first pursued viral oncology and then shifted to oncogene research (Chubin 1984; Fujimura 1996). A few other researchers were studying the problem of angiogenesis (e.g., Greenblatt and Shubik 1968), but the emergent research field was both small in size and marginal to the emergent molecular frameworks for cancer research. Even into the 1970s the field of angiogenesis research was producing only about three papers per year (Birmingham 2002). One factor that helped shift the position of this marginal research field was the finding that the problem of vascularization in tumors was related to the problem of endothelial cell growth. As a result, research on tumor angiogenesis could be connected to another, somewhat larger research field, and Folkman's work became part of a network called the "blood vessel club," which was attempting to isolate endothelial cell growth factors (Cooke 2001:131). By 1974 Folkman's laboratory and another lab had reported the successful cultivation of endothelial cells in culture (Gimbrone et al. 1973; Jaffe et al. 1973). With the new success behind him, Folkman attempted to get a major grant from the National Cancer Institute, but the reviewers demanded that he first team up with a biochemist (Cooke 2001:134).

Up to this point the story is largely one that can be told within a perspective limited to the position of a "challenger" research program within a broader research field. However, a new actor now enters the stage. With the grant now approved, Folkman's new partner, a professor of biochemistry named Bert Vallee, argued that they needed to scale up significantly in order to produce the tumor angiogenic factor that Folkman was now able to isolate. Vallee had a consulting arrangement with Monsanto, and as a result he was able to facilitate arrangements that led to a new form of university–medical school partnership (Cooke 2001:136–148). The Harvard-Monsanto agreement is now a classic case in the history of technology transfer and private-sector partnerships. It provided the Folkman and Vallee laboratories with $22 million over twelve years and granted them the right to publish their work in return for Monsanto's right to patent products coming out of their laboratories (Culliton 1977). The agreement necessitated a sea change in Harvard's intellectual property policies and provided a model for emerging policies at other medical schools (Cooke 2001:145–187). In the world before the Bayh-Dole Act, Harvard did not seek patents on health agents, and the university had no patent attorney.

In addition to the historic importance of the Monsanto agreement for the commercialization of biomedical research in the United States, it also moved Folkman's research program a step closer to institutionalization by providing a funding base. However, winning a secure funding base was only part of the picture; Folkman and his colleagues also needed to win acceptance by the scientific community, a process that would take more time. In fact, backlash against the Monsanto agreement was tremendous, both within Harvard and within broader scientific research communities. Folkman found his next NIH grant proposals turned down and his research program dismissed as quixotic. When he took the podium at one conference, he watched as a hundred people walked out of the room, and he heard postdocs tell of how they were advised to avoid his lab (Cooke 2001:145–187). A skeptical article in *Science* (Culliton 1977) led to a negative external committee review of Folkman's work. On top of this the administrators at Children's Hospital asked him to choose between surgery and research, and in 1981 he reached the key decision to step down as chief of surgery (Cooke 2001:198).

Notwithstanding the professional setbacks, the laboratory was slowly accumulating a successful track record. In 1976 Robert Langer, a chemical engineer who at the time was a postdoctoral researcher in Folkman's laboratory, processed huge amounts of bovine cartilage, then shark cartilage, and found that he could isolate a substance that inhibited angiogenesis (Langer et al. 1976). In 1979 a microbiologist in the laboratory succeeded in getting a special kind of endothelial cells, those from the capillary, to grow in culture (Folkman et al. 1979). The achievement led to international recognition, and Folkman's lab began training researchers in the technique (Cooke 2001:194). Subsequently the laboratories of Folkman and Vallee isolated angiogenic growth factors (Shing et al. 1984; Fett et al. 1985). As a result, by the mid 1980s, research on angiogenesis had become part of the burgeoning field of growth factor research, which was attracting increasing attention from industry. A researcher at Genentech found that a factor they had identified, vascular endothelial growth factor, was identical to the tumor angiogenesis factor of Folkman's laboratory (Leung et al. 1989). Although Monsanto decided to focus on agricultural biotechnology and consequently did not renew its agreement with Harvard, Folkman soon had new support of one million dollars per year from a Japanese company that wanted to enter the market, Takeda Chemical Industries (Cooke 2001:209–217). The ability to find support from firms was crucial to keeping a laboratory alive that was challenging some of the dominant assumptions of cancer research and opening up the doors to a therapeutic approach that did not rely on cytotoxic chemotherapy.

With the development of clinical applications, the position of the research field underwent another level of transformation. In 1989 alpha-interferon became the first antiangionesis agent used clinically, and in 1992 Takeda's TNP-470 became the first antiangiogenesis drug to enter into a clinical trial (Folkman 1996:150). Angiogenesis research spread rapidly to many laboratories, and research on leukemia and angiogenesis became a burgeoning field (Cooke 2001:236–237). Competition among postdoctoral candidates for a position in the once-spurned laboratory became intense (245) However, even at this point the National Cancer Institute turned down a major grant proposal from Folkman's laboratory, and the major pharmaceutical companies remained interested only in research that would result in rapid clinical applications. Consequently, in 1992 Folkman worked out an arrangement that brought in support from EntreMed, a biotechnology start-up company (248–250). The collaboration led to the development of angiostatin, an angiogenesis inhibitor that blocked metastases in murine models (O'Reilly et al. 1994). By 1996 seven antiangiogenesis drugs were in clinical trials (Folkman 1996:154). With the transition of the field into drug development, the size of the research field grew to hundreds of papers per year (Cooke 2001:260).

In May 1998, *New York Times* journalist Gina Kolata reported on the excitement in a front-page article that had international repercussions. She quoted Nobel Prize laureate Francis Crick as saying, "Judah is going to cure cancer in two years." Although Crick denied the quote, the controversial story set off a storm of international media attention, and EntreMed's stock prices soared. The article also set the stage for subsequent critical coverage of angiogenesis research when one of the drugs encountered some difficulties in replication attempts (e.g., King 1998). EntreMed also suffered some setbacks, including a lawsuit from Abbot, which had an agreement with Takeda. Meanwhile, other companies initiated clinical trials, and old drugs such as thalidomide were reintroduced for their antiangiogenic properties.

By the first decade of the twenty-first century angiogenesis and antiangiogenesis research had become mainstream. As Folkman noted in an interview in 2002, the field of angiogenesis research was growing at a rate of forty papers per week, that is, over two thousand papers per year (Birmingham 2002). Previously disconnected diseases such as cancer, cardiovascular disease, arthritis, diabetes, and macular degeneration were now connected through the common thread of angiogenesis. A whole industry of drugs designed to enhance angiogenesis in some cases, such as cardiovascular disease, and to inhibit it in others, such as cancer, had emerged. According to the Angiogenesis Foundation (n.d.), by 1999 there was a "massive wave" of both angiogenic and antiangiogenic drugs undergoing clinical trials for cancer, macular degeneration, diabetic retinopathy, psoriasis, coronary artery disease, peripheral vascular disease, stroke, and wound healing. By 2002 there were three hundred companies worldwide involved in angiogenesis research, embracing seventy-one agents, 10,000 patients, and $4 billion dollars of research (Angiogenesis Foundation 2002).

It might be tempting to describe the ascendancy of angiogenesis research and therapies by using the concept of a paradigm shift or scientific revolution, or even the framework of the rise of one network to dominance over another network. However, those interpretive frameworks miss some of the complexity of the transition. To date, molecular approaches to basic research and chemotherapeutic approaches to clinical applications remain dominant in the cancer field. Although traditional cancer chemotherapy drugs are now recognized to exhibit antiangiogenic effects at low doses, the clinical trials of antiangiogenic drugs have tended to position the drugs as additions to the traditional cancer chemotherapy armamentarium. A similar process is occurring with monoclonal antibodies, which are being tested for antiangiogenic properties. Rather than viewing antiangiogenic drugs as replacing existing research programs and drug cocktails, it seems more accurate to describe their development as being integrated into a diversifying therapeutic field. Angiogenesis research and drugs have not so much replaced existing frameworks and research programs as grown into them.

FOOD, CARTILAGE, AND ANGIOGENESIS

In contrast to angiogenesis research, the story of cartilage research for cancer represents a case of what I have called "undone science" (Hess 2001a; Woodhouse et al. 2002). The field of cartilage-based therapies for cancer in the United States was developed by John F. Prudden, whose career was in some ways similar to that of Folkman, but with a much less positive outcome. Prudden graduated from Harvard Medical School somewhat earlier than Folkman, in 1945, and then received a doctorate in medical science from Columbia University (Moss 1993).[1] After a stint in the army, he practiced as a surgeon at Columbia Presbyterian Hospital and, during the late 1960s and early 1970s, was an associate professor of clinical surgery at Columbia. In the 1950s he found that placing pieces of cartilage in wounds accelerated their healing (Prudden, Nishikara, and Baker 1957). Although he built his reputation for research on the enzyme lysozyme, he remained intrigued by the therapeutic potential of cartilage and soon obtained an investigational new drug permit from the Food and Drug Administration to treat cancer patients with bovine cartilage. He began treating cancer patients with subcutaneous injections of bovine cartilage in 1972, but the chair of the surgery department did not like the research, and Prudden eventually left Columbia to develop affiliations with other hospitals. In 1985 he published a review of thirty-one patients, which concluded that the drug was so safe that no upper limit of toxicity was reached. Furthermore, in a subset of patients for whom the treatment was applied consistently, all of whom were late-stage patients who had failed conventional therapy, he claimed to have 61 percent complete responders (Prudden 1985). In an interview in 1993, Prudden claimed that in a subsequent study with renal cell carcinoma, a very lethal form of cancer, the cartilage drug had a 25 percent complete or partial response rate (Moss 1993). He attributed the failure of cancer researchers and clinicians to follow up on the research as due to their dislike of natural products. He died in 1998, unable at that point to have brought bovine cartilage into the mainstream of cancer treatment.

Prudden's research was eclipsed not only by the growing attention to antiangiogenesis drugs but also by the growing attention to shark cartilage. The leading advocate of shark cartilage, I. William Lane, did not have a medical degree and lacked a university position, but he did have significant credentials relevant to the use of shark cartilage as a nutritional supplement. He received a master's degree in nutrition from Cornell University and a doctorate in agricultural biochemistry and nutrition from Rutgers University, and he had also served as the vice-president of the Marine Resources Division of W. R. Grace and Co. [2] During subsequent consulting work in the 1970s, he became interested in shark fishing. He learned about bovine cartilage from a business associate, met with Prudden in 1981, and tried Prudden's cartilage pills for his back pain. Finding that the pills helped not only his own back pain but also the severe arthritic symptoms suffered by the wife of a colleague, Lane became very interested in the therapeutic potential of cartilage. A few months later he met with Langer, the chemical engineer who had worked with Folkman, and he became even more convinced of the therapeutic potential of cartilage. At Lane's urging, the Institute Jules Bordet in Brussels conducted toxicity and dose-response studies in rats as well as human arthritis patients. According to Lane, the results were all positive, but they apparently went unpublished, and Lane was unable to interest the National Institutes of Health, whose representatives told Lane that they did not want to research natural products.

Thwarted in the United States, he pursued partnerships with clinicians in Panama, Mexico, Costa Rica, and Cuba. In 1992 Lane published the book *Sharks Don't Get Cancer,* and in 1993 the CBS newsmagazine *60 Minutes* covered the Cuban trial, for which Lane claimed that 40 percent of the 18 patients showed significant improvement.

By the mid-1990s cartilage products were one of the leading over-the-counter supplements products, and shark cartilage had displaced bovine cartilage. Retail sales for shark cartilage in the United States at that time were $50–60 million per year, and Lane estimated in an interview that 25,000 people were using shark cartilage products (Flint and Lerner 1996). Under the DSHEA regulations, cartilage products could be sold in stores as food supplements without requiring a prescription, provided that manufacturers made only structure and function claims (e.g., they promote healthy joints and bones). If manufacturers were to make disease claims (i.e., they can successfully treat cancer), the supplements would become classified as drugs and would be required to go through the expensive approval process using clinical trials. In other words, it is not the "naturalness" of the product that determines its legal status but the health claims that are associated with it.

Although the DSHEA regulations created a loophole through which over-the-counter supplements could be made available for off-book therapeutic uses, Lane went the official route and in 1994 obtained an investigational new drug permit from the Food and Drug Administration (Lane and Comac 1996). He described the Food and Drug Administration at that time as cooperative, but by the late 1990s the agency came to believe that shark cartilage products were being used in unapproved ways. In 1999 it filed a lawsuit against Lane Labs USA to limit distribution of products unless they were for approved clinical trials (Angiogenesis Foundation 1999). The Federal Trade Commission also intervened to stop the marketing of shark cartilage products by various firms that were making claims related to cancer treatment. In the case of Lane Labs, the settlement reached in 2000 mandated that the company fund a Phase III study of their shark cartilage product (Health Supplement Retailer 2000). In my review of U.S. Web sites for cartilage products in late 2003, the claims were carefully restricted to the legally allowable categories of health structure and function.

In addition to the regulatory and evidential problems, during the 1990s advocates of shark and bovine cartilage became caught up in their own controversies, including differences between the shark and bovine cartilage advocates. One debate involved the mechanism of action: Prudden believed that the therapeutic effect involved activation of the immune system via mucopolysaccharides (carbohydrates), whereas Lane believed that it was via antiangiogenesis factors (proteins). Environmentalists were also raising concerns with overfishing due to the growth of the shark cartilage industry. Although Lane responded that the overfishing problem was more related to Asian demand for shark fins and nonsustainable harvesting practices, both of which were problems that needed government regulation (Lane and Comac 1996:70–72), there was no parallel problem for bovine cartilage. To my knowledge, the growing concerns about "mad cow disease" have not yet been utilized in the shark/bovine controversy, but they could add yet another chapter to the ongoing conflict.

A more general controversy emerged around the question of absorption of any cartilage product when it is delivered orally or rectally, rather than by injection. Folkman, who injected cartilage rather than administering it orally, claimed that the pharmacologically active substances in cartilage are unlikely to be absorbed by the gut, and that a cancer patient would have to eat hundreds of pounds of cartilage daily to derive a therapeutic benefit (Beardsley 1993). Although an independent review of the issue indicated that gut absorption was possible, it also raised concern about the high doses of cartilage needed and the potential risk of excess calcium from oral cartilage (Flint and Lerner 1996).

Cartilage research was additionally weakened as the leaders of the CAM (complementary and alternative medicine) cancer therapy movement shifted from optimism to more cautious or even critical statements. Initial reports by leaders of the CAM cancer therapy movement, such as Ralph Moss (1991, 1993) and Ross Pelton (Pelton and Overholser 1994), as well as other CAM leaders (e.g., Williams 1993) were optimistic, but by the late 1990s the leaders were more skeptical. A key study led by Michael Lerner, head of the patient support organization Commonweal and a moderate voice in the CAM cancer therapy movement, concluded that the therapy remained unproven (Flint and Lerner 1996). Patrick McGrady Jr., the founder of a patient-oriented cancer information-providing service called CanHelp, told me in the late 1990s that he was very skeptical of both bovine and shark cartilage products (Hess 1999:35). Likewise, Ross Pelton, who in 1994 published a major book on CAM cancer therapies that had given shark cartilage relatively positive coverage, told me half a decade later that he preferred fermented soy products, which also had antiangiogenic properties (151). Robert Houston, a journalist who had been a consultant to *60 Minutes* for the Cuban story and was widely recognized as a pre-eminent scholar of CAM cancer therapies, confirmed that Lane's analysis of the Belgian data was essentially correct, but he remained unconvinced that shark cartilage was dramatically effective in humans (141). Ralph Moss, in many ways the "dean" of the CAM cancer therapy movement in the United States, subsequently added a comment to his 1991 article stating that the "jury is still out" (Moss 1991) and, in 1997, described himself as speaking in "measured tones" about the product (Moss 1998).[3] In summary, although there are clinicians in the United States, Mexico, and other countries who continue to use cartilage products and claim to see some benefit at the bedside, by the late 1990s several U.S. patient advocacy leaders were cautious about the claims for therapeutic efficacy, even though they continued to support the need for increased public funding for evaluation of natural products with antiangiogenic effects.

The lack of support from CAM-oriented patient advocacy leaders may appear counterintuitive. One might expect from them an uncritical embrace of all alternative cancer therapies. However, the patient advocacy leaders today are generally well-educated and quite sophisticated both methodologically and politically. Several hold doctorates in the social sciences and humanities, so they understand how to do research and how to interrogate both its methodology and politics. They understand that overhyped claims can come from CAM clinicians or innovators as easily as from oncologists and pharmaceutical companies. They are particularly critical of some CAM advocates who assume that products are safe and efficacious because they are natural. Instead, they tend to keep their eye on the bottom line of decreasing toxic side effects and increasing efficacy, notwithstanding the "naturalness" of the product. If a new class of drugs is proving to have few side effects and potentially high efficacy, such as the antiangiogeneis drugs at the current stage of their historical development, the CAM advocates could end up preferring the nontoxic drugs to a natural product that is bogged down in a variety of yet unresolved controversies. Furthermore, the patient advocacy leaders tend to warn patients not to chase after a single therapy (whether it is an experimental drug or a new food supplement), just as newcomers to investing may select one favorite stock. Instead, the patient advocacy leaders tend to support diversified, individualized therapeutic portfolios of surgical, nutritional, immunological, and mind-body protocols that are offered under the guidance of qualified clinicians. Although the advocates disagree on many specific issues, including the value of some or any concomitant chemotherapy and radiation therapy, they generally agree that more funding is needed to evaluate CAM cancer therapies, and they are skeptical of magic bullets.

In the past it took mass mobilization from the patient advocacy groups to pressure the federal government to fund clinical trials of controversial substances such as laetrile. By the mid-1990s an integration process was well underway (Hess 2003), and some federal funding was available for cartilage-based research. By 2003 the National Center for Complementary and Alternative Medicine (2003) listed two cartilage trials that it had funded, and the National Cancer Institute (2003b) had also funded two clinical trials for genistein, a bioflavonoid found in soy that has antiangiogenic properties. Likewise, other food components that may have antiangiogenic properties were being explored, such as thiol compounds (found in garlic) and vitamin A analogs (Boik 1996:29–30). Notwithstanding the availability of limited government funding as well as funding from supplements companies and clinicians from their own income streams, research on cartilage was progressing at a snail's pace in comparison with that on antiangiogenesis drugs. According to the National Cancer Institute listing (2003a), between the 1970s and 2003 there were eight clinical trials and one case series of cartilage products, three of which were for a bovine product, four for a shark product, and two for a purified cartilage-based drug called Neovastat. None of the listed trials was at a Phase III level. Why?

Leaders of the CAM movement have frequently noted that food-based or other "natural" products become the orphans of clinical research because private-sector firms are unwilling to invest the capital in a product that cannot be patented. Because patentability is a precondition for the heavy private-sector investment that is needed to bring most drugs to market, there is an indirect relationship between "naturalness" or proximity of a supplement to food and animal products and the status of the product as a drug. It is true that the distinction is increasingly murky because of the various ways in which intellectual property rights are becoming associated with food and food supplements. For example, the emerging nutraceutical industry can acquire intellectual property rights for food substances through trademarks, just as it can patent processes used to derive a purified form of the food product. Furthermore, it is possible to develop patents for therapeutic use of natural products. Indeed, in 1991 Lane obtained a patent on the use of cartilage as an angiogenesis inhibitor at a dose of twenty grams, and Prudden held a more general patent on the therapeutic use of any type of cartilage for cancer (Flint and Lerner 1996). However, just as trademarks or process patents are relatively weak forms of intellectual property, so the patent rights that Prudden and Lane held were weak because they covered mechanism or use rather than the substance itself. The Lane patent was particularly vulnerable because it was limited to a specific dosage (Flint and Lerner 1996). As a result, investment in developing drug status for a food-derived product can run the risks of creating free riders who can subsequently enter the market with similar products and benefit from a market leader's investment costs. Unless the public sector steps in to pick up the tab, the research field is condemned to developing products at a very slow pace, or it must market its products as supplements that lack disease-curing legal status and run the risk of regulatory intervention when off-book uses become too prominent. To keep pace with the angiogenesis industry, the public investment in natural products with purported antiangiogenic properties would need to be on the order of hundreds of millions, if not billions, of dollars.

Conclusion

By the first decade of the twenty-first century, antiangiogenesis drugs were attracting increasing excitement among mainstream researchers and clinicians, as well as patients and some patient advocates, whereas cartilage-based research remained enveloped in various circles of controversy. Arguably, the situation was not optimal from the point of view of cancer patients. In other words, investing more public resources in food-based angiogenesis products, such as cartilage or genistein, might have been a wise use of public funds. If successful, food-based drugs would be less expensive and more readily available, particularly to segments of the world's population that are off the medical grid of health insurance and pharmaceutical products. If not, then the thousands of users of those products would have good critical information that might steer them away from inefficacious products.

The National Cancer Institute and National Center for Complementary and Alternative Medicine have funded some relevant research, but advocates in the CAM community, and some of their supporters in

Congress, have argued that funding is tiny in comparison with total health-related research expenditures and disproportionately small when contrasted with the large number of patients who are using such products or who could benefit from them. As a result, the current confluence of private- and public-sector resources in favor of drug-based research creates a situation of a rapidly changing world of drug-based research and a very slowly developing world of research for food-based therapies. Whereas in many ways a consensus shift occurred during the 1990s around the value of angiogenesis research and antiangiogenesis drugs, it may take decades for a similar shift to occur around the therapeutic value of food-based interventions such as cartilage and soy products.

Understanding the current situation, where two research fields and associated therapies have developed radically different levels of credibility and research funding, requires a sociology of knowledge that is attendant to industrial priorities, regulatory policies, and social movement politics. However, the argument here goes beyond the problem of bringing markets, states, and social movements back into the study of scientific change. The point is also to raise the historical sociological question of the ways in which scientific and technological fields are themselves undergoing change. In returning to the four processes outlined at the start, a few elaborations are now possible.

Clearly, the issue of the increasing scale of institutional structures is evident. In my earlier historical research on another nondominant but nonetheless biological approach to cancer research (the networks of researchers who studied bacterial etiologies and the clinicians who employed bacterial vaccines; Hess 1997), the costs of doing animal-based research and developing vaccines were relatively small during the middle decades of the twentieth century. The costs could be internalized by clinicians or microbiologists on a part-time basis, somewhat akin to Folkman's work at the earliest stages of his career. In contrast, as the antiangiogenesis research program developed, it soon grew into a complex series of related problem areas that required collaboration with biochemists, molecular biologists, and microbiologists. Purification of the antiangiogenesis factors was prohibitively expensive, and the need to scale up drove the collaboration with Monsanto. Furthermore, the translation of such research into a legally approved drug has become extremely expensive in comparison with the relatively open and unregulated clinical testing environment of prior decades, when the first bacterial vaccines and sera were being tested. During the earlier period, a potential scientific or therapeutic "revolution" in the biomedical field only needed a low-tech laboratory, some mice, a vaccine or serum, and a clinical setting for small-scale testing. The costs and size of network that were needed to develop a therapeutic product and bring it into a clinical setting were smaller. The research and therapy programs related to cartilage have lacked the level of capital infusion found in drug development, and as a result they have had to rely on self-capitalization from sales of cartilage-based supplements products or meager government funding resources. Whereas the strategy of self-capitalization might have worked fifty years earlier, before the tighter regulatory environment engendered by the Kefauver-Harris amendments, by the late twentieth century the strategy created a mismatch between the funding and the scale of the projects needed for success in a competitive world of cancer drugs.

A second major historical change has involved the ongoing differentiation of institutions and roles. A scientist such as Folkman juggled conflicts among his roles as medical school instructor, manager of a laboratory, research scientist, clinician, public spokesperson, fundraiser, and party to contracts with private-sector firms. At some points the roles spilled over in uncomfortable ways, such as in the "backfire" (Jansen and Martin 2003) that occurred in the wake of his media attention or private-sector contracts. At one point he even hired a public relations person to handle his relations with the press (Cooke 2001). The increasingly complex set of roles that scientists must juggle accompanies a parallel growth of new organizations that have emerged in the interstices of previously separated organizational fields: the medical school technology transfer office (between the university and private sector), the supporting foundation (among researchers, clinicians, patients, and donors), and the biotechnology start-up company (among researchers, investors, and the pharmaceutical industry). The level of role conflict and negotiation, coupled with the formalization of requirements for role specificity, creates the conditions for actual or apparent conflicts of interest and subsequent crises of credibility. However, the crises of credibility have been greater for Lane than for Folkman. Rather than attempt to reduce the difference to a psychology of personal integrity, a sociological perspective would point to how a scientist needs conviction to stay with a research program and battle for its success, but an entrepreneur with equal conviction can run into legal constraints on issues such as health claims rules. The battle for the acceptance of a research program that is linked to a new therapy hinges on maintaining the separation of roles between researcher and entrepreneur, but the processes of commercialization make it increasingly difficult for the roles to remain separate, particularly for small-scale defenders of natural products.

A third change is in the culture of biomedical research and its clinical applications. On the research side, there is an increasing concern with mechanism, with understanding causal pathways at the molecular level of growth factors and gene expression. The black-boxing of therapeutic agents that occur in foods, herbs, cartilage, and other naturally occurring products is anathema to a research culture that is focused on mechanisms. Although the reason why there is so much focus on mechanism is beyond the scope of the study, the hypothesis that drug-based research priorities drive such a concern would be worthy of study. As a result, there is ongoing resistance from establishment research communities to the empiricism of food-based research when it is accompanied by weakly understood mechanisms (see also the article by Woodhouse in this volume on various types of scientific momentum). On the clinical side, there is an increasing formalization of the hurdles required for clinical approval. While in theory the three phases of clinical trials required for drug approval in the United States constitute a level playing field, in practice it is a pseudo-universalism similar to the American criminal justice system. As I have sometimes heard in CAM-oriented cancer conferences, the idea that the randomized clinical trial represents the "gold standard" of research is well named because those who have the gold set the standards. The older model of clinicians who tinker with therapies, introduce them to patients, and present case study series has been rejected, at least in the United States and other wealthy countries (less so in Mexico, which is home to many of the rejected American cancer therapies).

Regarding the technological and natural world, the cancer therapy field is characterized by increasing recognition of the failure of conventional therapies and the emergence of movement for less toxic cancer therapies. Ralph Moss (1992) even made "toxicity" the central issue in a survey of CAM cancer therapies. Concern with the negative side effects of radiation therapy and chemotherapy, and with their inability to cause remission or prevent recurrence at desirable levels, has spurred a general movement among cancer patients and some clinicians to reject those therapies or, at the minimum, to seek nutritional interventions that mitigate the toxicities of conventional therapies (Hess 1999). Yet this "greening" of cancer therapy is accompanied by a denaturalization process; in other words, the older generation of high-dose chemotherapy with its undesirable side effects is being replaced not so much by natural products and nutritional interventions as by a new, less toxic wave of biological therapies, such as antiangiogenesis drugs. Even when traditional chemotherapy continues to be used, often its mode of delivery has been modified to reduce toxicity, such as by emphasizing low doses and slow infusion over a long period of time instead of short-term blasts followed by a recovery period. In fact, antiangiogenesis research suggests that chemotherapy used in this manner may have antiangiogenic properties.

The theoretical frameworks developed in this chapter and the others in this volume urge research on science, technology, and society to pay more attention not only to factors such as commercialization, regulatory policy, and civil society participation but also, as I would argue, to the patterns of historical change that characterize the recent development of science and technology. The new theoretical frameworks promise to provide social scientists and historians with a helpful lens for understanding change in science, technology, and society, and they may also be helpful for reform movements in science, industry, and society that are strategizing efforts for political and technological change.

Acknowledgments

I wish to thank Joerg Albrecht, Arthur Daemmrich, Scott Frickel, Daniel Kleinman, and Kelly Moore for helpful comments on an earlier draft, as well as comments from faculty and students at the University of Pennsylvania, University of Sydney, and various conferences.

Endnotes

1. The biographical information in the remainder of the paragraph is based on an interview between Ralph Moss and John Prudden in 1993 (Moss 1993).

2. The biographical material in this paragraph is based on Lane and Comac (1993).

3. However, after the approval of Avastin in 2004, Moss was also critical of the costs, low efficacy, and side effects of the Avastin-chemotherapy protocol, and he continued to support the need for more funding for evaluation of low-cost natural products that have antiangiogenic effects (Moss 2004).

References

Algire, Glen, and Harold Chalkley. 1945. "Vascular Reactions of Normal and Malignant Tissues in Vivo: 1, Vascular Reactions of Mice to Wounds and to Normal and Neoplastic Transplants." *Journal of the National Cancer Institute* 6:73–85.

Angiogenesis Foundation. n.d. "Historical Highlights of the Angiogenesis Field." www.angio.org/researcher/library/highlight.html (accessed October 23, 2003).

———. 1999. "FDA Demands Halt to Sale of Lane Lab's BeneFin Citing Unproven Claims." www.angio.org/newsandviews/archive1999/dec_17_1999.html (accessed October 22, 2003).

———. 1999a. "Historical Highlights of the Angiogenesis Field." www.angio.org/researcher/library/library.html (accessed October 28, 2003).

———. 2002. "Angiogenesis State of the Art Highlighted at Florida Summit." www.angio.org/newsandviews/archive2002/June02.html (accessed October 22, 2003).

Beardsley, Tim. 1993. "Sharks Do Get Cancer." *Scientific American,* October: 24–25.

Beck, Ullrich. 1992. *The Risk Society*. Newbury Park, CA.: Sage.

Bernal, John. 1969. *Science in History*. Cambridge, MA.: MIT Press.

Birmingham, Karen. 2002. "Judah Folkman." *Nature Medicine* 8:1052.

Boik, John. 1996. *Cancer and Natural Medicine*. Princeton, MN: Oregon Medical Press.

Bourdieu, Pierre. 2001. *Science de la science et réflexivité*. Paris: Raisons d'agir.

Chubin, Daryl. 1984. "Research Mission and the Public: Over-Selling and Buying on the U.S. War on Cancer." *Citizen Participation in Science Policy,* ed. James Petersen, pp. 109–129. Amherst, MA.: University of Massachusetts Press.

Cooke, Robert. 2001. *Dr. Folkman's War: Angiogenesis and the Struggle to Defeat Cancer*. New York: Random House.

Culliton, Barbara. 1977. "Harvard and Monsanto: The $23 Million Alliance." *Science* 195:759–763.

Daniels, George. 1976. "The Pure-Science Ideal and Democratic Culture." *Science* 156:1699–1705.

Fett, James, D.J. Strydom, R. R. Lobb, E. M. Alderman, J. L. Bethune, J. F. Riordan, and B. L. Vallee. 1985. "Isolation and Characterization of Angiogenin, an Angiogenic Protein from Human Carcinoma Cells." *Biochemistry* 24:5480–5486.

Flint, Vivekan, and Michael Lerner. 1996. *Does Cartilage Cure Cancer?* Bolinas, CA: Commonweal.

Folkman, Judah. 1971. "Tumor Angiogenesis: Therapeutic Implications." *New England Journal of Medicine* 285:1182–1186.

———. 1996. "Fighting Cancer by Attacking Its Blood Supply." *Scientific American* 275:150–154.

Folkman, Judah, Christian Haudenschild, and Bruce Zetter. 1979. "Long-Term Cultivation of Capillary Endothelial Cells." *Proceedings of the National Academy of Sciences* 76:5217–5221.

Folkman, Judah, and Elton Watkins. 1957. "An Artificial Conduction System for the Management of Experimental Complete Heart Block." *Surgical Forum* 8:331–334.

Frickel, Scott. 2004. "Just Science? Organizing Scientist Activism in the U.S. Environmental Justice Movement." *Science as Culture* 13:459–469.

Fujimura, Joan. 1996. *Crafting Science.* Cambridge, MA.: Harvard University Press.

Fuller, Steve. 2000. *Thomas Kuhn: A Philosophical History for Our Times.* Chicago: University of Chicago Press.

Gimbrone, Michael, Ramzi S. Cotran, Christian Haudenschild, and Judah Folkman. 1973. "Growth and Ultrastructure of Human Vascular Endothelial and Smooth-Muscle Cells in Culture." *Journal of Cell Biology* 59:A109.

Goldman, E. 1907. "The Growth of Malignant Disease in Man and the Lower Animals with Special Reference to the Vascular System." *Lancet* 2: 1236–1240.

Greenblatt, Melvin, and Philippe Shubik. 1968. "Tumor Angiogenesis." *Journal of the National Cancer Institute* 41:111–116.

Greene, Harry S. N. 1941. "Heterologous Transplantation of Mammalian Tumors: 1, The Transfer of Rabbit Tumors to Alien Species. 2, The Transfer of Human Tumors to Alien Species." *Journal of Experimental Medicine* 73: 461–486.

Guston, David. 2001. "Boundary Organizations in Environmental Policy and Science: An Introduction." *Science, Technology, & Human Values* 26:399–408.

Health Supplement Retailer. 2000. "FTC: Firms Can No Longer Make Shark Cartilage/Cancer Claims." Government Watch column, September. www.hsrmagazine.com/articles/091govwa.html (accessed October 22, 2003).

Hess, David J. 1997. *Can Bacteria Cause Cancer? Alternative Medicine Confronts Big Science.* New York: New York University Press.

———, ed. 1999. *Evaluating Alternative Cancer Therapies: A Guide to the Science and Politics of an Emerging Medical Field.* New Brunswick, NJ: Rutgers University Press.

———. 2001a. *Alternative Pathways in Globalization.* Niskayuna, NY: Letters and Sciences. http://home.earthlink.net/~davidhesshomepage (accessed October 22, 2003).

———. 2001b. "Ethnography and the Development of STS." In *Handbook of Ethnography,* ed. Paul Atkinson, Amanda Coffey, Sara Delmont, John Lofland, and Lyn Lofland, pp. 234–245. Thousand Oaks, CA.: Sage.

———. 2003. "CAM Cancer Therapies in Twentieth-Century North America: Examining Continuities and Change." In *The Politics of Healing,* ed. Robert Johnston, pp. 231–243. New York: Routledge.

Hessen, Boris. 1971. *The Social and Economic Roots of Newton's Principia.* New York: Howard Fertig.

Jaffe, Eric, Ralph Nachman, Carl Becker, and C. Richard Minick. 1973. "Culture of Human Endothelial Cells Derived from Uumbilical Veins." *Journal of Clinical Investigation* 52:2745–2756.

Jansen, Sue Curry, and Brian Martin. 2003. "Making Censorship Backfire." *Counterpoise* 7(3): 5–15. www.uow.edu.au/arts/sts/bmartin/pubs/03counterpoise .html (accessed December 3, 2003).

King, Ralph. 1998. "Laboratory Hitch: Novel Cancer Approach from Noted Scientist Hits Stumbling Block." *Wall Street Journal,* November 12, A1.

Kleinman, Daniel Lee. 1995. *Politics on the Endless Frontier.* Durham, NC: Duke University Press.

———. 2003. *Impure Cultures: University Biology and the World of Commerce.* Madison: University of Wisconsin Press.

Knorr-Cetina, Karin, and Michael Mulkay. 1983. *Science Observed.* Beverly Hills, CA: Sage.

Kolata, Gina. 1998. "Hope in the Lab: A Special Report." *New York Times,* May 3, 1:1.

Kuhn, Thomas. 1970. *The Structure of Scientific Revolutions.* 2nd ed. Chicago: University of Chicago Press.

Lane, I. William, and Linda Comac. 1992. *Sharks Don't Get Cancer.* Garden City Park, NY: Avery.

———. 1996. *Sharks Still Don't Get Cancer.* Garden City Park, NY: Avery.

Langer, Robert, Henry Brem, Kenneth Falterman, Michael Klein, and Judah Folkman. 1976. "Isolation of a Cartilage Factor That Inhibits Tumor Neovascularization." *Science* 193:70–72.

Leung, D. W., G. Cachianes, W. J. Kuang, D. V. Goeddel, and N. Ferrara. 1989. "Vascular Endothelial Growth-Factor Is a Secreted Angiogenic Mitogen." *Science* 246:1306–1309.

MacKenzie, Donald, and Barry Barnes. 1979. "Scientific Judgment: The Biometry-Mendelism Controversy." In *Natural Order,* ed. Barry Barnes and Steve Shapin, pp. 191–210. Beverly Hills, CA: Sage.

Merton, Robert. 1970. *Science, Technology, and Society in Seventeenth-Century England.* New York: Howard Fertig.

———. 1973. *The Sociology of Science.* Chicago: University of Chicago Press.

Misa, Thomas, Philip Brey, and Andrew Feenberg, eds. 2003. *Modernity and Technology*. Cambridge, MA: MIT Press.

Moore, Kelly. 1996. "Organizing Integrity: American Science and the Creation of Public Interest Science Organizations, 1955–1975." *American Journal of Sociology* 101:1592–1627.

Moss, Ralph. 1991. "Sharks May Take a Bite out of Cancer." *Cancer Chronicles*, no. 10, n.p. www.ralphmoss.com/html/shark.shtml (accessed November 18, 2003).

———. 1992. *Cancer Therapy*. Brooklyn, NY: Equinox Press.

———. 1993. "A Potent Normalization." *Cancer Chronicles* no.16: n.p. www.ralphmoss.com.html/bovine.shtml (accessed October 21, 2003).

———. 1998. "Visit to Toronto." *Cancer Chronicles*, Winter 1997–1998, n.p. www.ralphmoss.com/html/toronto1.shtml (accessed November 26, 2003).

———. 2004. "FDA Approves Avastin." *Townsend Letters for Doctors and Patients*, May, 30–31.

Mulkay, Michael. 1976. "Norms and Ideology in Science." *Social Science Information* 15:637–656.

National Center for Complementary and Alternative Medicine. 2003. "Shark Cartilage Trials." www.nccam.nih.gov/clinicaltrials/sharkcartilage.htm/ (accessed November 26, 2003).

National Cancer Institute. 2003a. "Cartilage (Bovine and Shark)." Last modified July 9, 2003. www.cancer.gov/cancerinfo/pdq/cam/cartilage (accessed November 26, 2003).

———. 2003b. "Phase II Randomized Study of Genistein in Patients with Localized Prostate Cancer Treated with Radical Prostectomy; Phase II Randomized Study of Soy Isoflavone in Patients with Breast Cancer." http://cancer.gov/search/clinical_trials/results_clinicaltrialsadvanced.aspx?protocolsearchid=412785 (accessed October 22, 2003).

O'Reilly, Michael, Lars Holmgren, Yuen Shing, Catherine Chen, Rosalind Rosenthal, Marsha Moses, William Lane, Yihai Cao, E. Helene Sage, and Judah Folkman. 1994. "Angiostatin: A Novel Angiogenesis Inhibitor That Mediates the Suppression of Metastases by a Lewis Lung-Carcinoma." *Cell* 79:315–328.

Pelton, Ross, and Lee Overholser. 1994. *Alternatives in Cancer Therapy*. New York: Simon and Schuster.

Prudden, John. 1985. "The Treatment of Human Cancer with Agents Prepared from Bovine Cartilage." *Journal of Biological Response Modifiers* 4:551–584.

Prudden, John, Gentaro Nishikara, and Lester Baker. 1957. "The Acceleration of Wound Healing with Cartilage-1." *Surgery, Gynecology, and Obstetrics* 105: 283–286.

Rescher, Nicholas. 1978. *Scientific Progress*. Pittsburgh: University of Pittsburgh Press.

Shing, Y., J. Folkman, R. Sullivan, C. Butterfield, J. Murray, and M. Klagsbrun. 1984. "Heparin Affinity: Purification of a Tumor-Derived Capillary Endothelial Growth Factor." *Science* 223:1296–1299.

Williams, David. 1993. *The Amazing New Anti-Cancer Secret That's About to Take the World by Storm.* Ingram, TX: Mountain Home.

Woodhouse, Edward, David Hess, Steve Breyman, and Brian Martin. 2002. "Science Studies and Activism: Possibilities and Problems for Reconstructivist Agendas." *Social Studies of Science* 32:297–319

Young, Robert, and Michael Surrusco. 2001. "Rx R&D Myths: The Case against the Drug Industry's R&D 'Scare Card.'" Washington, DC: Public Citizen. www.citizen.org/documents/ACFDC.PDF (accessed October 23, 2003).

6

Nanoscience, Green Chemistry, and the Privileged Position of Science

EDWARD J. WOODHOUSE

This chapter compares two cases of forefront science. The first, nanoscience, is a classic hot research arena: scientists rush into each niche as soon as it opens; conferences and professional publications buzz with the latest results; pundits offer glowing predictions of benefits to environment, world hunger, and medicine; government officials generously dole out taxpayers' money; and voices even arise to counsel the need for prudent foresight.[1] A very different profile characterizes the second case, that of "green chemistry," which aims at redesigning molecules and chemical production processes to make them more benign: research is coming nearly a century later than it could; conferences are few in number; funding is niggardly; public attention is slight; and little fame accrues to participants.[2] What can we learn from these polar opposite cases about the influence relations in and around contemporary science?

The first section of this chapter describes nanoscience and technology R&D and the social forces impelling the activities, both within the scientific community and more broadly. Section two explains some of what is and is not occurring in green chemistry, traces the factors that have caused it to lag, and discusses a recent surge that may ultimately bring green chemistry onto mainstream chemical agendas. The remainder of the chapter analyzes implications of the two cases for our understanding of scientists as participants in a system of power, including their roles as allies of business, and my analysis is intended as a contribution to reinvigorating the interests tradition in the sociology of knowledge. As the title of the chapter suggests, however, I do not see forefront technoscientists primarily as pawns of political-economic elites but argue that they also use these connections to enjoy a structurally privileged position in contemporary social life—exercising considerable discretion over matters of great public consequence and themselves being among the foremost beneficiaries of science funding and technological innovation.

Although I start from the constructivist assumption that social forces have shaped the nanoscience-technology juggernaut and the green chemistry laggard, and although I inquire into what those forces have been, I aim to move beyond social *construction* of science and technology toward *re*constructivist scholarship aimed at clarifying alternative possibilities, both substantive and procedural.[3] I think there are lessons in the juxtaposition of the two cases for those who seek to reconstruct scientific research and technological practice along fairer, wiser, more democratic, or otherwise "better" lines. If technoscience in some respects constitutes a form of legislation that reshapes the everyday lives of billions of persons, is it not about time to develop procedures capable of holding the scientists and technologists doing the legislating a good deal more accountable for their actions?[4]

Nanoscience and Nanotechnology

Nanoscience and nanotechnology are "the art and science of building complex, practical devices with atomic precision," with components measured in nanometers, billionths of a meter.[5] This is not a typical scientific field inasmuch as researchers do not pursue common substantive knowledge: "smallness" is the unifying attribute, so it may be more appropriate to term research and development at the nanoscale as an approach rather than a field. Indeed, in private some scientists go so far as to suggest that "nano" functions more as a label to legitimize receiving grant monies than as a coherent set of research activities.

Nobelist Richard Feynman is generally credited with calling attention to the possibility of working at the atomic level on nonradioactive materials in a 1959 lecture at Cal Tech titled "There's Plenty of Room at the Bottom,"[6] and he no doubt was a source of inspiration for at least some of the nanoscience that gradually began to develop. Most such research actually being conducted is relatively mundane, whereas the hype and concern about nanotechnology are due more to the dramatic notions first presented in then-MIT graduate student K. Eric Drexler's *Engines of Creation: The Coming Era of Nanotechnology*.[7] This visionary/fictional 1986 account for nontechnical readers sketched a manufacturing technology that would construct usable items from scratch by placing individual atoms precisely where the designers wanted. This he contrasted with contemporary manufacturing, which starts with large, preformed chunks of raw materials and then rather crudely combines, molds, cuts, and otherwise works them into products. The current approach uses far more energy than "molecular manufacturing" (MNT) would require, while leaving enormous quantities of waste products that molecular manufacturing would not.

Moreover, molecular manufacturing conceivably could become self-sustaining, with tiny factories building tiny factories to build tiny machines. However, because some of these might escape their designers' control, Drexler warned from early on that special controls would be needed: "Assembler based replicators could beat the most advanced modern organisms. . . . Tough, omnivorous 'bacteria' could out-compete real bacteria: they could spread like blowing pollen, replicate swiftly, and reduce the biosphere to dust in a matter of days."[8] This warning was reiterated to a larger audience in Bill Joy's April 2000 *Wired* magazine article "Why the Future Doesn't Need Us," describing a world of self-replicating, exponentially proliferating "nanobots" that could drown the planet in an uncontrollable "gray goo."[9] Michael Crichton gave the warning a more explicitly sci-fi spin in his 2002 novel *Prey*, featuring swarms of intelligent, predatory, nearly unstoppable nanobots.[10] Crichton, of course, could be dismissed, but it was impossible to brand Joy a Luddite, given his role at the time as chief scientist at Sun Microsystems, and his standing as an architect of the world's information infrastructure. Nevertheless, the research and technology communities quickly mobilized like antibodies to neutralize him.

More ordinary, but still potentially transformative, aspects of nanoscience include work projected to accelerate present trends toward faster computing by aiding in miniaturization and by providing new ways to store information at the atomic level, as by facilitating "quantum" computing.[11] One effort concerns developing microchip-like functionality from single molecules, enabling tiny, inexpensive computers with thousands of times more computing capacity than current machines, perhaps introducing a second computer revolution.[12] New materials include carbon nanotubes and other very strong and very light advanced materials. Nanomix Corporation is working to develop "nanostructured materials to store solid-state hydrogen for automotive and portable power applications" for what is being touted as the coming "hydrogen economy."[13]

Other endeavors attempt to replicate biological functions with synthetic ones, such as designing and synthesizing organic molecules and supramolecular arrays that can mimic green plants' photosynthetic processes—perhaps opening the way for solar energy in a more fundamental sense than what the term so far has meant.[14] Most of the research presently is at a precommercial stage, although nanoparticles are beginning to come onto the market (e.g., titanium dioxide in sunscreens), and health and environmental concerns around nanoparticles may come to be the first specific point of contention among environmental groups, business, and government regulatory bodies.[15]

Along with tangible investments and research trajectories comes a good deal of hype of the sort that commonly shows up in the early years of new cycles of innovation. "Imagine highly specialized machines you ingest, systems for security smaller than a piece of dust and collectively intelligent household appliances and cars. The implications for defense, public safety and health are astounding."[16] Even normally staid government reports burst with promotional fervor,[17] and a university web page says that

> Our world is riddled with flaws and limitations. Metals that rust. Plastics that break. Semiconductors that can't conduct any faster. . . . Nanotechnology can make it all better—literally—by re-engineering the fundamental building blocks of matter. It is one of the most exciting research areas on the planet, and it may lead to the greatest advances of this century.[18]

Another R&D pathway is nanotechnology applied to biotechnology, or nano-bio for short. A mundane example is the use of nanoscale bumps on artificial joints in order to better mimic natural bone and thereby trick the body into accepting transplants. Pharmaceutical manufacturing increasingly will rely on nanoscale techniques, according to some observers. More generally, as one advocacy organization puts it, "Recent developments in nanotechnology are transforming the fields of biosensors, medical devices, diagnostics, high-throughput screening and drug delivery—and this is only the beginning!"[19]

Although relatively routine at present, some of the potential consequences are profound, especially the joining of nanotechnology with biotechnology to blur the dividing line between living and nonliving matter. For example, neural implants could make machine intelligence directly available biologically, and tiny machines may live (if that is the right word) in the body either just as sensors or also to treat incipient illness. An example in this category is the development of sensors that could "detect minute quantities of biological and chemical hazards" and devices as small as "a hypodermic needle's tip" with "the ability to detect thousands of diseases."[20] These innovations may continue the trend of increasing costs, widening the divide between medical haves and have-nots, and accentuating tendencies to substitute medicine for a healthy lifestyle.

Nanotechnology has been embraced enthusiastically by government officials. The Japanese Ministry of International Trade and Industry in 1992 launched the first major nanoscience initiative funded at what seemed a generous amount—$185 million over ten years. That has now been dwarfed, with U.S. government support for civilian nanoscience and technology research at approximately $900 million annually under the 21st Century Nanotechnology Research and Development Act, which in 2003 formalized a variety of programs and activities undertaken by the Clinton administration's National Nanotechnology Initiative (NNI). European and Japanese funding likewise is increasing rapidly.

As occurred with the so-called War on Cancer launched by the Nixon administration, researchers have repackaged their work to jump on the nanotech-funding bandwagon. Many are actually revising their research interests as they become involved in nanoscience, and new graduate students and postdoctoral researchers are able to move into the hot new areas of study without much loss of accumulated intellectual capital. Is this simply a response to ready availability of funding, or is there more to it?

We lack good methodologies for sorting out such matters, and even participant observation and interviewing are in the early stages. However, I think anyone who has been around nanoscientists can attest to the genuine enthusiasm prevalent among them. Conferences already abound devoted specifically to nanoscience and technology, and large international meetings of chemists, physicists, and others include increasing numbers of papers with "nano" themes and methods. A sampling: Nanotechnology Growth Opportunities for the Biotech and Medical Device Sectors, Irvine, California, April 2004; Nanomechanics: Sensors and Activators, Reno, Nevada, May 2004; Second International Conference of Microchannels and Minichannels, Rochester, New York, June 2004; European Micro and Nano Systems Conference, Paris, October 2004.

Nanotechnology relationships are tight among university, industry, and government, continuing the triple-helix trend of the past generation. Several of the above conferences have obvious commercial themes, and dozens of other events are even more targeted toward nano-oriented businesses. NanoSIG, part of NASA/Ames and sponsored in part by the business sector, focuses on accelerating the commercial development of nanotechnology. The organization's nanoBio Forum works "at the interface of biotechnology and nanotechnology," sponsoring discussions on topics such as nanoparticles in proteomics, genomics, and cellomics. In addition, the Silicon Valley Organization provides advice on how to start a nanobiotechnology company, get government funding, and work with local government agencies.[21]

IBM and Xerox are among an increasing number of large corporations engaged in nanotechnology R&D, and start-up firms hoping to mimic the explosive success of Silicon Valley are racing to get products onto the market. Carbon Nanotechnologies, Inc., for example, claims to be a "world leading producer of single-wall carbon nanotubes . . . the stiffest, strongest, and toughest fibers known." Their most advanced product, "BuckyPlus Fluorinated Single-wall Carbon Nanotubes," was selling for $900 per gram in 2004, many times the price of gold.[22] Altogether, by 2005 there were at least a thousand companies world wide claiming to be in some facet of the nanotechnology business.

In sum, nanoscience and technology R&D consists of myriad minor, relatively useful, and harmless trajectories combined just about inextricably with some fascinating, potentially helpful, and potentially disastrous radical innovations. Although interest is high among business executives and elected officials, it would be a mistake to overlook their dependence on the technologists. As Kleinman and Vallas say in chapter 2, scientists mobilize lines of action that can modify, mediate, or contest aspects of the emerging knowledge regime," a theme to which I return later in the chapter.

Brown Chemistry versus Green Chemistry

Twentieth-century chemists, chemical engineers, and chemical industry executives made fundamental choices that substantially shaped humanity's experiences with chemicals. Most educated people know the main facts about some of those choices, such as DDT, PCBs, and chemical waste dumps. However, hardly anyone yet understands the deeper story behind the social construction of chemicals: it turns out that there was far greater technological malleability than almost anyone appreciated, and that chemicals as we know them by no means constitute the only path that chemistry and the chemical industry could have taken.[23]

One of the choices was that of relying on petroleum-based feedstocks instead of on fatty/oily or woody plant materials from which chemicals also can be made. Research on lipid chemistry and carbohydrate chemistry has a long history, with a number of journals devoted entirely to it.[24] But the bulk of chemists' attention went first to coal-tar derivatives and later to natural gas.

A second choice was that of opting for wet chemistry. Most contemporary chemical reactions occur in solution, and getting chemicals into solution requires solvents. There is a minority tradition of dry synthesis, but it received little attention in the past century. The result is that solvents played a huge role in the chemical industry, and many of those solvents such as benzene and toluene are extremely toxic.

A third choice was to emphasize stoichiometric processes, where two or more chemicals are combined to produce an output, which goes on to interact with another chemical in a subsequent stage. Eventually, after as many as thirty steps in the "synthesis pathway," a final product emerges. Along the way, byproducts are produced at each step—outputs not directly usable in further steps toward a desired final product. Some byproduct chemicals can be used elsewhere, but some are just wastes—hazardous wastes. For example, formaldehyde and cyanide were among the byproducts that used to be created during production of the painkiller ibuprofen. Since the mid-1970s, chemical companies have had to track these wastes from cradle to grave, paying huge sums for paperwork, trucking, incineration, deep well injection, and other methods of disposal. Prior to that date, the hazardous wastes were treated more haphazardly, with some finding their way into water supplies at places such as Hooker Chemical Company's former site, Love Canal, and at Woburn, Massachusetts.[25] Estimates vary, but it is clear that brown chemistry has produced millions of tons of hazardous wastes.

A fourth distinguishing feature of the twentieth-century chemical industry was rapid scale-up from original synthesis through pilot plants to full-scale production—a process that rarely required as much as a decade. The result was megaton quantities of pesticides, plastics, finishes, and myriad other chemical products and processes, and few people hesitated to release the synthetic organic chemicals into ecosystems and human environments. Many of those involved did not know better, one assumes, and yet not knowing depended on selective attention, perception, or recall, because warning signs began to accumulate early on. For example, when orchardists sprayed fruit trees, massive die-offs of bees occurred until those doing the spraying learned to wait until well after flowering had finished.[26] Eggshell thinning and other more subtle signs took longer to observe, but there were enough early warnings and enough farmers with old-fashioned distrust of new-fangled inventions that chemists and others on the lookout for warnings would have found them. Instead, industrial chemists and their bosses and customers skipped the gradual scale-up that would have allowed lower-cost learning from experience.

Following Rachel Carson's cogent criticisms and the rise of the environmental movement, of course everything began to change in chemistry and in chemical engineering. . . . Well, no, actually it did not. Indeed, even after passage of bookshelves of environmental legislation, more than 100,000 chemical researchers worldwide continue to collaborate with purveyors and purchasers of chemicals within the brown chemistry paradigm. Although the synthetic organic chemical industry is now in its second century, with many tens of thousands of chemicals in commerce and considerable opportunity to learn from the bad experiences, some of the world's brightest and most highly trained experts continue to poison their fellow humans and the ecosystem without fundamental reconsideration of whether there is a better way to do things. More strangely yet, although environmental organizations have worked hard to persuade governments to reduce toxic emissions in air and water, most environmentalists have not yet realized that it might make sense to change the basic dynamics creating toxins in the first place.

There have been selected exceptions, such as a push for less persistent pesticides, a reduction of volatile organic compounds in paints and coatings, and efforts in a few countries to phase out a handful of the worst chemicals. But the basics of brown chemistry remain intact, not only in chemical practice but in human thinking. The reason is pretty simple: almost everyone assumes that there is no realistic alternative. Faced with a choice between better living through chemistry and back-to-the-cave living, postmoderns may bitch about the uncaring executives in the chemical industry, but few of us seriously entertain the notion of doing without our plastics, chrome plating, leather seating, gasoline additives, lawn and garden chemicals, inexpensive food raised with pesticides, and other products produced via chlorinated and other synthetic chemicals.

It turns out we have been wrong, however, to assume that chemicals are chemicals, and there's nothing much to be done about it. Very, very slowly, over enormous cognitive, institutional, and economic momentum, an alternative, green-chemistry paradigm has begun to emerge:

- Design each new molecule so as to accelerate both excretion from living organisms and biodegradation in ecosystems.
- Create the chemical from a carbohydrate (sugar/starch/cellulose) or oleic (oily/fatty) feedstock.
- Rely on a catalyst, often biological, in a small-scale process that uses no solvents or benign ones.
- Create little or no hazardous waste byproducts.
- Initially manufacture only small quantities of the new chemical for exhaustive toxicology and other testing.
- If preliminary results are favorable, follow up with very gradual scale-up and learning by doing.

The chemical research community belatedly has begun to experiment with aspects of this second formula. Sometimes called "sustainable chemistry" or "benign by design," the organizational heart of the enterprise has been located not at major universities but in a tiny program in the Pollution Prevention and Toxics branch of the U.S. Environmental Protection Agency. Rather than focusing on cleanup and correction of problems already existing, the emphasis there since 1994 has been on prevention of problems before they occur—partly by redesigning chemical production processes and products at the molecular level to make them radically less dangerous.

There are very few historians of twentieth-century chemistry, and most of these work on medicinal chemistry or on the World War I ("chemists' war") era, with none I know of studying the origins of green chemistry or barriers to it. Hence, we have no scholarship based on lab notes, internal industry memos, or other archival sources. Nevertheless, it is apparent that several small tributaries of chemical investigation came together in the 1990s to create a rivulet of activities with enough coherence to be labeled "green chemistry."

One contributing element was research on supercritical fluids (SCFs).[27] Discovered more than a century ago was the fact that certain substances "exist in a hybrid state between liquid and gas above a critical temperature and pressure and as such have some bizarre and very useful solvent properties."[28] Supercritical carbon dioxide ($scCO_2$), for example, is nonflammable and nontoxic, has extremely low viscosity—and is very inexpensive. Supercritical decaffeination of coffee began in the 1960s, but only since the mid 1990s have chemists begun taking advantage of SCF properties for chemical synthesis of ordinary industrial chemicals. Industry sometimes can actually save money; instead of becoming contaminated by the chemicals being processed as do ordinary solvents (think of used paint thinner), $scCO_2$ can be reused repeatedly because it simply boils off for collection once the pressure is reduced. For similar reasons, it is being championed as an alternative to the dangerous solvent perchloroethylene, now widely used for professional dry cleaning, for industrial degreasing operations, and for microelectronics fabrication facilities.

Opinions differ on why the delay in making use of SCFs. Some point to the complexity of the equipment, but four thousand pounds per square inch and temperatures ranging up to a few hundred degrees certainly are within the range often found in industrial practice. Other observers nominate maintenance difficulties and costs of the equipment as the culprit. Still others argue that safety is harder to assure when dealing with pressurized systems. No doubt there is some validity in these claims, but given the life cycle costs and environmental-social costs of many petroleum-based solvents, it seems to me clear that a huge number of chemists in and out of industry have for decades not been paying appropriate attention to the potential advantages of $scCO_2$ and other supercritical fluids. As Kleinman and Vallas put it in chapter 2, "deeply entrenched norms and practices . . . constrain actors in their efforts to remake" their fields.

A university professor working on SCF in the United Kingdom suggests that the explanation rests partly with the faddish way new techniques sometimes are approached. There have been recurrent cycles, he says, in attention to SCF potentials. Typically the potential is oversold as enthusiasts propose and try out fancy schemes far beyond the existing state of knowledge; when these fail, interpretations hold that SCF has not worked out, and attention turns to some other hot topic. An obvious alternative would be patient, steady exploration of whatever fundamental questions remain, coupled with modest chemical and other engineering innovations designed to apply relatively simple SCF technology in relatively simple manufacturing and other processes. For example, one entrepreneur is now utilizing SCFs to make building materials out of fly ash from a coal-burning power plant—a low-level application of no interest to most forefront researchers.

Nevertheless, through conferences and professional networks, the idea is spreading of using SCFs for the relatively mundane task of replacing dangerous solvents such as benzene and toluene, of which millions of pounds are used annually. Environmentally conscious chemists and engineers had already been worrying about solvents because of their obvious health and environmental risks—for example, the paint industry has greatly reduced volatile organic compounds in paints, thinners, varnishes, and other coatings, partly by building on the techniques used in latex, water-based paints. The confluence of these other solvent activities with the new push in supercritical fluids stimulated a miniboom in the late 1990s. A credible, but not peer-reviewed, plan emerged at a University of Massachusetts workshop predicting that within two decades it should be possible to replace all solvents and acid-based catalysts that have adverse environmental effects with solids, water-based substitutes, or other green alternatives.[29] Among other approaches, the solvent replacers began turning toward dry chemistry—probing the technical feasibility of skipping altogether one of the first and messiest steps in the brown chemistry formula: that of dissolving the compounds prior to reacting them. No dissolution, no need for solvents.

Another progenitor of green chemistry was Stanford chemistry professor Barry Trost, who first proposed the concept of "atom economy" in 1973. Rather than judging a chemical process successful if it produced a usable product at a satisfactory cost—the mainstream standard—Trost argued for elegant efficiency, for using the highest possible percentage of input atoms in the usable output, ideally leaving zero waste. This originally seemed a utopian concept, but an increasing number of biocatalytic and other chemical processes now are being explored that achieve exactly such an outcome.[30] What Trost contributed, I think, was not a concrete method like supercriticality, nor a goal such as solvent replacement, but a vision that the would-be green chemists could use to construct a counternarrative to the dominant approach—which now could be labeled as messy, inefficient, inelegant, and otherwise outdated.[31]

A fourth contributing factor was the work of a few renegades who continued to probe alternatives to the use of fossil fuels as chemical feedstocks. These researchers lacked funding or standing to make a real difference in their day, but they kept the flame going by training a few students and by winning at least brief mentions in the field's texts and in chemists' peripheral consciousness. The research took place under two main headings—carbohydrate chemistry (i.e., cellulose and other plant materials) and oleic chemistry (fatty and oily substances, mostly also from plants). A vivid illustration of how this type of research differs from conventional organic chemistry is that it actually is possible to obtain the necessary research materials from an ordinary grocery store! This minority tradition gained a bit of new life during the recurrent energy crises late in the last century, and won a measure of vindication in 2002 when Cargill Dow opened the world's first commercial-scale plant to make polyaspartic acid for plastics out of corn.

Why didn't these and other sources for green chemistry come together sooner and more powerfully? One answer is that the brown chemistry formula was technically just too sweet, economically too attractive, and institutionally too convenient. Fossil feedstocks became plentiful, reliable, and cheap at just about the same time that chemists were discovering lots of ways to combine chlorine molecules with carbon and hydrogen to create chlorinated hydrocarbons such as DDT, PCBs, dieldrin, polyvinyl chloride, and others. These new products had the great virtue of using up the surplus chlorine that industry executives wanted to get rid of, chlorine generated by the chlor-alkali process that produced one of the basic industrial chemicals on which the industry was built.[32] Coinciding, or lagging slightly, "demand" for pesticides and other organic chemical products escalated.

Also slowing the emergence of green chemistry has been the market milieu. More than perhaps any other science, chemistry has been captive to industry, although geology comes close. Industrial "needs" determine chemistry curricula to a far greater extent than is true, say, of physics curricula. Likewise, the American Chemical Society has had a closer connection with the chemical industry than biologists have had with medicine, though the rise of biotechnology is changing this. Asked why there has been relatively little change in chemical engineering curricula in the face of environmental pressures, one department chair I interviewed actually dragged out the old cliché "There just isn't room in the curriculum." A further indicator of the backwardness: whereas about half of PhD-granting chemistry departments in the United States still require students to pass a test in a foreign language, none requires an equivalent test in toxicology, according to green chemist John Warner of the University of Massachusetts–Boston.[33]

Moreover, professional licensing tests for chemical engineers showed very little change between 1980 and 2000 in terms of requiring environmental competence as part of being a certified chemical engineer. Professional licensing in the United States is conducted under auspices of the American Institute of Chemical Engineers, and rather than drawing in young turks with state-of-the-art knowledge, AIChE relies primarily on retired chemical engineers to devise the tests. There is an irony in this: credible estimates suggest that greener processes may actually cut in half the development time to produce new polymers, partly by simplifying manufacturing requirements and partly by reducing environmental compliance transactions.[34] Hence, it could be said that industry actually needs university researchers to point the way, but the captive relationship undercuts this leadership potential, a point worth reflecting on in the context of Owen-Smith's analysis of the "dual hybrid" university in chapter 3.

More generally, government officials in the European Union, Japan, and the United States alike have little concept of how green chemistry might be used to modify traditional approaches to environmental regulation. The top levels at EPA are not very knowledgeable about forefront science, perhaps any science, and among environmental scientists the general reputation of the executive office of the president is low. Several European nations do better on both counts, with Sweden generally regarded as being in the forefront via their Chemical Inspectorate's planned phaseout of the most dangerous chemicals. But this move is timid compared with the bold possibilities foreseen by a handful of green chemistry visionaries who argue that it is technically and economically feasible to eliminate virtually all chlorinated chemicals.[35] Social scientists need not take sides among the technical disputants in order to recognize that except for a brief period in the mid-1990s, there has been little public dispute regarding the possibility of a broad phaseout of brown chemicals.[36] Even major environmental interest groups have hardly any PhD chemists on staff, which may be part of the reason many continue to focus on land preservation, endangered species, and other obviously worthy issues but fail to call attention to the potential malleability of the chemical universe to attack toxics problems at the source by moving toward a vision of benign by design.

On the other hand, green chemistry is becoming institutionalized within the research community, as indicated by the annual Gordon Conference on Green Chemistry. The Organisation for Economic Co-operation and Development (OECD) has been organizing periodic workshops on sustainable chemistry, as has the International Union for Pure and Applied Chemistry; and meetings of the American Chemical Society and equivalent organizations now include panels on green chemistry. *Chemical and Engineering News* has begun to feature articles on the subject. The "Reinventing Government" initiative of the Clinton-Gore administration created an annual Presidential Green Chemistry Challenge award competition starting in 1995, and the National Academy of Engineering in conjunction with the EPA Office of Research and Development in 2005 launched a program for fifty teams of college students who will "research, develop and design sustainable solutions to environmental challenges."[37] A Green Chemistry Institute within the American Chemical Society has several dozen affiliate organizations throughout the world, and ACS staff and volunteers have undertaken significant efforts to revise chemistry textbooks and precollegiate curricula to include green chemistry. The Royal Society of Chemistry has launched the journal *Green Chemistry*.[38] Finally, the U.S. House of Representatives Committee on Science has held hearings on the subject, leading to a proposed 2004 Green Chemistry Research and Development Act, which easily passed in the House despite opposition by the Bush administration, but the measure died in the Senate.

Government, Business, and Science

What makes one research area so compelling while other areas are ignored? Why the belated push on supercritical fluids and solvent replacement, yet almost no movement on the use of medicinal chemistry principles for industrial chemicals? Why nano-bio, nano computing, and many other facets of nanoscale inquiry, yet very little support for what seems to have the most important potential, molecular manufacturing? I do not presently have answers to these questions that are as good as I would like, partly because of the shortage of social science and historical scholarship together with the absence of a perspective on contemporary events that time eventually will lend. Some of the story is hard to miss, however.

The commercial possibilities for nanotechnology seem uppermost in the minds of government officials. They manifest "an almost religious belief . . . in the powers of science-based technology. . . . [O]nly scientific and technological supremacy over the rest of the world will allow the country to prosper economically."[39] The 21st Century Act mandates ongoing reporting and priority setting on what needs to be done to keep the United States "competitive" in nanotechnology commerce. As a first approximation, I believe that major technoscientific initiatives fostering the interests of political-economic elites are more likely to succeed than those that do not. As Hess puts it in chapter 5, understanding "where two research fields . . . have developed radically different levels of credibility and research funding, requires a sociology of knowledge that is attendant to industrial priorities, regulatory policies, and social movement politics" (or the lack thereof). The solicitous attention from elected officials comes about partly because of the privileged position that the business sector enjoys in what are known as market-oriented societies, but that might equally well be termed "business-oriented" societies.

Business executives occupy a role unlike that of any other social interest, in part because they are structurally located to make key economic decisions including creating jobs, choosing industrial plant and equipment, and deciding which new products to develop and market. Many direct and indirect supports, interferences, and other partly reciprocal connections between science and business arise during this process. Sometimes referred to incorrectly as the "private sector," business actually performs many tasks that are public in the sense that they matter greatly to almost everyone. Even if no business executive or lobbyist ever interacted in any way with government officials, business would be highly political in the sense of exercising influence over key public choices. However, business also is privileged in a second sense, in that executives have unrivaled funds, access, organization, and expertise to deploy in efforts to influence government officials.[40]

Although the connection with business and government in a system of political-economic power is crucial to understanding the new political sociology of science, just as Hess finds the story of cancer research and treatment more complex than the standard interests perspective alone can handle, so it is with nanoscience and green chemistry. Government officials' willingness to splurge on nano is due in part to the fact that the five myths of science analyzed by Daniel Sarewitz are alive and well in the halls of Congress and in government corridors throughout the world. These include "The myth of infinite benefit: More science and more technology will lead to more public good," "The myth of accountability: Peer review, reproducibility of results, and other controls on the quality of scientific research embody the principal ethical responsibilities of the research system," and "The myth of the endless frontier: New knowledge generated at the frontiers of science is autonomous from its moral and practical consequences in society."[41]

There has been considerable analysis of changes in university-industry relations in recent decades, including a number of chapters in this volume, with talk of triple helixes and new forms of intellectual property and so forth.[42] Sufficient for present purposes, however, is an utterly obvious and simple fact underlying the sophisticated analyses: more chemists work with and for industry than for any other social institution, and the same is true, or soon will be true, of nanoscientists and technologists. Both the great successes and the horrible failings of the past century's chemistry can be traced in part to the close relationship between science and business, and the same probably will prove true of nano capacities. A higher percentage of chemistry majors work for industry than is true of any other science major. Many are direct employees whose pay and career literally depend on the continuing good will of their bosses, but even researchers in universities bend curricula to perceived corporate "needs," earn extra money by consulting, and seek grants and contracts from businesses. Moreover, NSF and other government programs explicitly tout the economic payoff.[43]

Another consideration that must be included in a well-rounded understanding of science and power is that those with authority in business and government are likely to make use of scientists to the extent they find it convenient, profitable, or otherwise in line with their own aspirations. Sometimes this takes a straightforward route, with "members of a relatively small, like-minded elite . . . justifying their power over important scientific and technological choices by referring to the need for an efficient response to both commercial and military threats."[44] Getting a jump on foreign competition is surely in the minds of some legislators voting to support nanoscience, and hesitancy on green chemistry has something to do with excess capacity in the world chemical industry, enormous sunk costs in plant and equipment, and declining fortunes of the U.S. chemical industry (due partly to domestic prices for the main feedstock, natural gas, that are higher than most competitors pay in other nations). Conversely, there is no question but that brown chemistry was extremely useful to businesses, and those deploying the chemicals had little motivation to worry about long-term health and ecosystem effects.

Although a full replay of the twentieth-century cowboy economy is unlikely given heightened environmental awareness, businesses exploring nanoscale manufacturing or new products utilizing nanoscience are likely to find that near- and middle-term profits may be compatible with longer-term problems for society. For example, diagnosis of ill health made possible by tiny ingestible sensors may increase medical costs and rates of iatrogenic illness by inducing physicians and patients to intervene where they previously would not, but this need not impact profitability of the company manufacturing the new sensor. If nanoscale surveillance makes privacy even less protected than is now the case, the problem will be borne largely by persons other than executives of the relevant companies. Hence, the emerging technical potentials can be useful to business and government even if in some larger, longer-term sense they cause more problems than they solve (a cost-benefit calculation with results no one can know in advance).

The Privileged Position of Science

That the myths concerning science are widely shared is part of the basis for a privileged position for science that grants forefront researchers considerable autonomy, and that calls on social scientists to produce a nuanced understanding of science and social power. Science is not as directly central to daily life as is business, of course, and scientists lack the monetary inducements deployed by business executives; nevertheless, there are some interesting and important parallels with the privileged position of business. Thus, just as public well-being depends on what is called a "healthy" business sector, so has technological civilization come to rely on scientists, engineers, and other technical specialists to conduct inquiries into matters beyond most people's competence, train future generations of technical specialists, and perform functions at the interface between scientific knowledge and technological innovation. As scientists link with business, they obtain certain privileges.

The linkages with business and government obviously magnify the influence of scientists on the social construction of everyday life, but it would be a serious mistake to suppose that the scientists are merely the handmaidens in this relationship. Many technoscientific researchers originate and undertake their inquiry to a far greater extent than that inquiry is foisted upon them: It was not consumers or workers or business executives or government officials who chose not to pursue green chemistry—none of them had an inkling of how molecules are put together. If anyone chose, it was the chemists—although unpacking their "choice" could occupy a good number of historians, sociologists, philosophers, and others, and I doubt whether "choosing" is the best term to describe the sociotechnical processes that led to brown chemistry or any other complex technological trajectory.

The same goes for nanotechnology: "There's plenty of room at the bottom," Eisenhower said in his farewell speech in 1961? No, of course not—it was Richard Feynman, the Nobel Prize–winning physicist, who first suggested that there might be a future in working at the atomic level with nonradioactive materials.[45] And every other contributor to the discussion likewise has been a trained scientist or engineer, except for public figures touting and voting to fund the new initiatives (which virtually none of them much understand). As mentioned above, mainstream nanotechnology leaders at NSF and elsewhere have worked to downplay the potential for the more radical innovations associated with molecular self-assembly. Thus, a semifinal draft of the 21st Century Nanotechnology Act called for the National Research Council to submit to Congress in 2005 a review of the technical feasibility of molecular self-assembly for the manufacture of materials and devices at the molecular scale, and to assess the need for standards and strategies for ensuring "responsible development" of self-replicating nanoscale devices. However, this was substantially watered down in the final version of the bill, and Christine Peterson of the Foresight Institute blamed the deletion on "entrenched interests." "That's sad," Peterson said, "immense payoffs for medicine, the environment and national security are being delayed by politics."[46] The politics to which she refers is occurring largely *within* the technoscientific community.

Nobel Prize–winner Richard Smalley dismisses both the promises and the problems associated with molecular manufacturing: "My advice is, don't worry about self-replication nanobots. . . . It's not real now and will never be in the future."[47] Smalley, developer of carbon nanotubes and a major force in the U.S. National Nanotechnology Initiative, says the necessary chemistry simply will never be available. His argument, presented in a number of highly visible publications including a cover story of *Chemical and Engineering News,* holds that there is no way to place atoms or molecules sufficiently precisely. Disagreeing with him point by point in a series of published letters is Eric Drexler, a dominant force within the Foresight Institute, which aims to help prepare humanity for the nanotechnology era he believes is coming.[48]

Those that count within the nanoscience community tend to side with Smalley in dismissing Drexler's view. U.S. nanotechnology czar Mihail Roco might have reason to downplay the more radical potentials, for fear of provoking the kind of resistance that agricultural biotechnology has encountered in Europe. Although Congress ultimately defeated a requirement to set aside for ethical, legal, and social issues 5 percent of the $3.6 billion nanotechnology authorization, the House Science Committee inserted a requirement for public consultation in the 2003 nanotechnology legislation precisely to try to head off a GMO (genetically modified organism) like fate.[49] An undersecretary for technology at the U.S. Commerce Department explicitly noted that rationale at an NSF-sponsored symposium on societal implications of nanotechnology in December 2003. More generally, government funding and industry interest both depend in part on public quiescence, on treating nano capacities as ordinary, boring science and technology, rather than as an issue similar to GMOs or nuclear power perhaps deserving intense, widespread scrutiny.

However, it is not as clear why so many others are willing to go along in dismissing molecular manufacturing, with even Greenpeace issuing a pretty tame report on nanotechnology as a public issue.[50] ETC Group, the nongovernmental organization (NGO) that did a great deal to bring the GMO issue onto public agendas, is calling for a moratorium on certain nanotechnology research and diffusion, but they are more worried about the environmental and health effects of potentially ingestible nanoparticles such as carbon nanotubes.[51] The ETC proposal so far has not won much of a following. Within the NSF sphere, Roco and his allies as of this writing were managing to keep the controversy over molecular manufacturing entirely off the table in public meetings sponsored by the NNI. Not only presenters but even audience members asking questions at such meetings somehow get the quiet message that retaining one's credibility requires not discussing molecular manufacturing. I have asked participants and other observers how that message is telegraphed, and the answers are not very illuminating. Everyone "just knows" not to talk about it.[52]

As a decision theorist, I have no professional opinion about which of the technoscientists is correct. I do notice that the anti-Drexlerian arguments shift quite a bit over time as if earlier arguments were found wanting. And I notice that Smalley, Whitehead, and Roco utilize assertion and flamboyant language, in a way that reminds me of other realms of politics. When they do mount an argument, metaphors such as "slippery fingers" play a larger role than equations. That may be unavoidable in debating futuristic potentials, one must acknowledge; still, there is a somewhat eerie resonance with many previous arguments about what technical innovation cannot do—from flying across the Atlantic to open-heart surgery. Hence, it may be premature to dismiss the possibility of molecular manufacturing, for good or ill.

Whoever may be winning in the scientific community's internal battles, if scientists are exerting broader social power in the nanoscience and brown/green chemistry cases, what is the nature of the influence? I do not refer to ordinary interpersonal relations in the laboratory or battles for influence within subfields, but to the world-shaping influence exercised by the development and deployment of new knowledge—or the failure to develop and deploy it.[53]

Influence of this sort does not normally take the form that A has power over B when B does what A chooses. Such a formulation probably is too simple for understanding influence relations in any complex social situation, but is especially questionable when it comes to the "cascading series of unintended consequences" that followed in part from chemists' inquiries into organic chemistry, and that may follow from nanoscientists' current enthusiasms.[54] Social scientists need an approach to scientific influence that is consonant with the bull-in-the-china-shop or sorcerer's-apprentice character of the havoc that science-based practices sometimes wreak. This also must allow room for the nearly magical and extremely useful outcomes sometimes catalyzed by scientific knowledge, as well as for the fact that a high percentage of scientific work reaches and benefits or harms no one at all outside of a few researchers in a subfield.

Suppose we were to suspend certain assumptions about the inevitability and rightness and naturalness of science as now conducted, backing far enough away to look afresh at the whole setup. Isn't there a sense in which it's a bit weird to have quietly allowed twentieth-century chemists and chemical engineers to go along their merry ways synthesizing tens of thousands of new substances that the world had never before experienced and helping produce and distribute billions of tons of substances that from some points of view can be characterized as poisons or even chemical weapons? From our present vantage point, knowing what we do about the potentials of green chemistry, isn't it somewhat strange that chemists (more than anyone else) chose not to activate these potentials earlier—a lot earlier?

With respect to nanoscience and technology, the dangers awaiting may or may not rival endocrine disruption and environmental cancer and species extinctions and the other effects of synthetic organic chemicals. However, it is difficult to miss the fact that the nanologists are busy creating nontrivial threats, some of which could be unprecedented. As Bill McKibben puts it, looking not just at nanotechnology but also at human biotechnology and related technologies, "a central question facing humanity in this century is whether by the end of it we still will *be* human."[55] So far, the nanoscientists are about as free of external restraint as were the brown chemists of the previous century. For the new political sociology of science, then, stepping outside the assumption that it is natural for scientists to pursue science makes it evident that something peculiar is going on, not unlike children playing with matches while the grownups are away—except, in this case, it is not clear who the grownups are.

Again, however, where is power in this story? Like a child who is not powerful can obtain a parent's gun and shoot up a school, scientists are not powerful in the ordinary sense of the word, yet they collectively have very substantial effects. A conventional way of thinking about the hazards created by technoscientists is in terms of unintended consequences, especially secondary and tertiary consequences. Social observers long have assumed such consequences to be a ubiquitous fact of social life, and it seems clear that complex new technoscientific endeavors are bound to entail outcomes that cannot be entirely foreseen. The dominant perspective acknowledges the predicament but treats it as just something to be regretted. That approach has plausibility, but it also steps over an interesting and important question: What is the nature of the power relationships around unintended consequences? Who can be construed as making the decisions?

Normally, the implicit answer is no one; vector outcomes emerge. However, as Langdon Winner says, unintended consequences are not

not intended.[56] Someone is at least implicitly deciding to go ahead even though going ahead entails unintended consequences—some of which are likely to be negative and perhaps quite potent. Whoever is participating in that choice can be said to be exercising authority over others, inasmuch as those ultimately suffering (or enjoying) the unforeseen results would not be doing so without those making the original choice having decided to proceed.

Many scientists and others with influence over unintended consequences would claim to be operating within a legitimate political-economic order and would claim that they have the consent required to proceed. Under contemporary law, that is correct. In every other way, however, the argument is silly, inasmuch as many of the future victims or beneficiaries are not yet born, live in countries other than the ones catalyzing the R&D, are utterly ignorant of the whole business, are outvoted by those who want relatively unfettered technoscience, have been led to suppose the activities entirely safe, or otherwise cannot meaningfully be said to have given consent. My purpose here is not moral blame or philosophical inquiry; it is merely to point out that enormous authority is being exercised in choosing to proceed with potent new technoscientific capacities, given the likelihood of unintended consequences. Political sociologists of science need to come to grips with this exercise of authority, not assume it away, as do most technoscientists, their allies, and even their opponents.[57]

A closely related exercise of power is that which goes into shaping agendas and shaping the governing mentalities via which both influentials and non-influentials think about issues.[58] This is one of the most important ways that power manifests, frequently leading to nondecisions rather than manifest controversies. Little noted to date in the literature on this "third face" of power, as Steven Lukes termed it, is the fact that technoscience excels at shaping human thought and the agendas based partly on it.[59] Thus, chlorine chemistry helped set the agenda for environmental problems: PCBs in thc Hudson River are poly*chlorinated* biphenols; DDT, dieldrin, and aldrin are *chlorinated* pesticides; CFCs that deplete the ozone layer are *chloro*fluorocarbons. Industry actually manufactured the chemicals, and millions purchased them, so chemists did not act alone; but they certainly created the potential that set the agenda.

In doing so, they inadvertently helped teach humanity (including future chemists) that unintended consequences are a normal part of technological innovation, part of what is termed the price of progress. As soon as a new capacity is developed, many people come to consider it unthinkable to "go back" to an earlier state—whether that is to a chemical state prior to organo-chlorines or to not manipulating matter at the nanoscale. More subtly, in the twentieth century, the potentials of brown chemistry and chemical engineering were so technically sweet and so agriculturally useful in "combating pests" that it was unthinkable to require that chemicals be proven safe prior to introducing megatons into the ecosystem. Creating unthinkability of these kinds arguably is the most subtle and most potent form of technoscientific power.

One of the main reasons that technoscientists can do this time after time, while maintaining their privileged position, is that they have considerable legitimacy and credibility in the eyes of most persons. Scientists in films sometimes are socially awkward and preoccupied with matters far from most people's everyday realities, but with the exception of renegades such as the dinosaur cloner in *Jurassic Park,* scientists normally are depicted as innocuous or as proceeding within acceptable norms. Except for aberrant situations such as the Tuskegee syphilis experiments, almost no one questions that scientists have a right to pursue the research they do—even if, on reflection, one might be hard-pressed to explain exactly from whence that "right" emanates. Similarly, although many people experience skepticism or impatience when expert opposes expert in testimony at a trial or other public venue, in the absence of such conflict most of us are disposed to assume that technoscientific experts pretty much know what they are doing within their fields of endeavor.

Restating that seemingly sensible and harmless assumption darkens its implications: university chemical and nanotechnology researchers going about their normal business are not questioned closely by outsiders and hence are not very accountable for their research or teaching (except, to a degree, to other members of their subfields). This helps insulate them from undue external influence—except perhaps from those awarding grants and contracts– but it also partially insulates them from *appropriate* external influence. Which verbs to use for describing the influence relationships? Have the scientists seized authority or been delegated it? Are they imposing outcomes or suggesting trajectories? Do they decide, or do they negotiate? My own sense is that many different influence-oriented verbs apply at various times to various facets of technoscientific choices and outcomes, and that the influence relationships are ineffably messy. What can be said with clarity is that some technoscientists have considerable latitude in their actions, that some of the actions interact so as to produce outcomes that sometimes have considerable impact on the world outside of science, and that technological civilization may lack both the discursive and institutional resources for holding the influentials accountable.

Discussion

Although those wielding scientific knowledge clearly exert enormous influence, the nanotechnology and green chemistry cases provide little or no evidence that scientists as individuals or as identifiable groups have the power to defeat other elites when there is manifest conflict. Military, industry, and government elites often give technoscientists quite a bit, of course; when there is overt bending to be done, however, it tends to be the researchers who do it. They want government or industry funding, because they lack the resources to evade or fight new government regulations, because universities cannot operate without business and governmental largesse, or because as employees the scientists and engineers are simply following the boss's orders.

Further reducing any power in the conventional sense that might be imputed to scientists is the fact that to a substantial degree they are creatures of the cultural assumptions of their societies. They think and act the ways they do partly because they are cognitive victims of their cultures: they simply have not gotten much help in revisioning what chemistry could be, nor in recognizing that rapid R&D and scale-up usually prove problematic, nor in thinking through the manifestly undemocratic implications of the privileged position of science.[60] If the mass media focus on conventional problems such as endangered species and chemical spills, rather than on restructuring molecules, scientists and engineers are among those whose mental landscapes become dominated by a focus on *symptoms,* who become unable to rethink the underlying causalities. If emerging or speculative gee-whiz capacities predominate in stories about technological futures, it is not only the attentive public whose thinking is thereby misshapen; technoscientists' thinking likewise is stunted.

For example, as of this writing, to my knowledge no mass media source or semi-popular science publication has ever pointed to the obvious similarity between nanoscience and green chemistry: both are about rearranging atoms and molecules to supposedly better serve (some) humans' purposes. Although chemists ought to be able to figure that out for themselves, in most respects they are just ordinary people operating according to standard procedures and cognitive schemas. Neither the cadre of chemistry-ignorant journalists nor the rest of technological civilization has given chemists much assistance in breaking out of those molds.

Another indicator of that lack was the failure in 2001 of environmental groups to contact a congressional committee considering tax credits to assist mom-and-pop dry cleaners in switching over to supercritical carbon dioxide in place of perchloroethylene, the dangerous solvent now widely employed. Dow Chemical, in contrast, stimulated an active letter-writing campaign by conventional (i.e., brown) dry cleaners, and the tax stimulus for chemical greening died in committee. The relatively toothless 2004 Green Chemistry R&D Act likewise expired in the Senate after passing the House, partly because it received modest endorsements from mainstream groups but no real push from environmentalists. More generally, even the major environmental NGOs with the partial exception of groups focusing on clean industrial production, are pretty much caught up in opposing various uses of brown chemistry rather than pushing for a green chemical industry.

It may seem a bit weird to say that chemists need to be informed by nonchemists, but it is a fact. As a vice president of Shaw Carpets, a business executive without an advanced degree, expressed the point at a recent congressional hearing: "I am the guy whose job it is to tell the chemists and chemical engineers that they *can* make carpets using environmentally friendly processes, because they come to the job from universities where they have not learned about it."[61] That kind of insistence has been altogether missing in the contexts in which most chemical R&D personnel have worked for the past century; the equivalent cuing is missing in the contexts now populated by nanologists.

That said, it nevertheless is worth reiterating that both technoscientists' failures to actively pursue green chemistry and their possibly misguided rush toward the nanoscale are occurring without the knowledge or consent of the vast majority of those potentially affected. Some people and organizations are exercising discretion on behalf of, or against, others, and hardly anyone seems troubled by the fact. Again, however, scientists have not gotten much help in noticing that this is true. They are content to proceed without consent partly because it often is in their interests, but also because they have not learned to do otherwise—the social order has not arranged to teach them. Green chemistry leader professor Terry Collins of Carnegie Mellon insists on chemistry ethics being central to the curriculum. I doubt whether that is nearly enough, because not only do business-oriented incentive systems give strong motivation for setting ethics aside, but longstanding, elite-dominated traditions more subtly impair just about everyone's capacities for thinking straight about emerging technoscience. One learns to accept that "private" businesses have a right to do whatever will sell profitably, to accept scientific research as the equivalent of free speech rather than just another social activity subject to periodic renegotiation, to believe that unlimited technological innovation is both inevitable and highly desirable, and to assume that unintended consequences are just a regrettable fact of life rather than a task to be mastered. These assumptions or myths align in ways that accede legitimacy and accord capacity to technoscientists and their allies, and they impair technoscientists' thinking and practice along with that of everyone else.[62]

In applying the above insights more specifically to the blocking of green chemistry and the acceleration of nanoscience, we can maintain symmetry by thinking in terms of a combination of technological, economic, and cognitive momentum. Brown chemists and their industrial allies and consumers of chemicals could be foregrounded, but it might be more accurate to de-individuate by saying that green chemistry has been marginalized by brown chemistry—literally, in the sense that textbooks brimming with brown-chemical formulas make it difficult for green chemical ideas to find a place in the curriculum or laboratory, but also in the more socially complex sense that industry engineers have been trained in brown chemistry, the industry has huge sunk costs in brown chemical plants and equipment, and the assumptions that go along with that momentum obscure the possibility or even the desirability of an alternative. Moreover, government regulatory procedures and laws are set up negatively—to limit the damage of brown chemicals—instead of positively and actively seeking the reconstruction of chemicals to be benign by design. Similar social processes are likely to interfere with sensible governance of nanoscience's emerging potentials.

The blending of structural position, quiescent public, and habits of thought that I have woven together fits with the general spirit of the more complex interpretations of power offered by third-wave feminists and by other recent observers of social life.[63] It fits as well with the multi-dimensional stories told by other contributors to this volume, especially that of Hess in chapter 5. The nuances are unfortunate, in a way, robbing us of a simple, black-and-white story that is easy to transport cognitively. A simplified but not simplistic condensed story of brown/green chemistry and nanotechnology might run along these lines: some technologists glimpse new capacities and begin to develop them; political-economic elites come to believe that these may serve their purposes, and they make additional funding available; other technoscientists move into the emerging fields, partly in order to obtain the new funds; the relevant scientific communities proceed to develop new capacities that then are scaled up by industry and government far too fast to allow the relatively slow learning from experience that humans and their organizations know how to do. There is no way that journalists, interest groups, government regulators, and the public can learn enough fast enough about the new capacities to institute precisely targeted protective measures in time. And, in fact, many of them actually have no disposition to "interfere."

Some STS scholars have attempted to depict the dicey relationship in terms of what they call "principal-agent theory."[64] The principals (e.g., government officials) who provide funds lack the requisite knowledge and other resources to prevent opportunism and shirking by the scientist-agents, so special controls need to be established, such as intermediary research institutions and research programs that can serve boundary functions in intermediating between researchers and those paying the bills.[65] Whether the putative "principals" are in any meaningful sense in control of the scientific enterprise is called into question by the nanoscience and green chemistry cases, among others.[66] More importantly, I doubt that government officials normally should be seen as the principals, inasmuch as they are in another sense putative agents of the general public, in a relationship that is deeply troubled. Government officials also have a proven willingness to invest heavily in dubious technoscientific projects coupled with an inability to catalyze meritorious R&D such as green chemistry.

Whether or not anything comes of the principal-agent approach, the green chemistry and nanotechnology cases offer an opportunity to reflect regarding which decisions appropriately can be left to scientists, which to industry and markets, and which really deserve broader scrutiny and deliberation by media, public-interest groups, independent scholars, government officials, and the public. How can scrutiny come early enough, how can it be sufficiently informed and thoughtful, and by what institutional mechanisms? Considering how profoundly scientific understanding, and lack of it, has affected everyday life in the case of synthetic organic chemistry, and probably will do so in the case of nanoscience, it seems to me that social thinkers may need to think harder about the "constitution" of technological civilization. For scientific knowledge and technical know-how are not mere tools; they actually help constitute and reconstitute social structures and behaviors.[67] Because decisions about technoscience have transformed everyday life at least as profoundly as what governments do, the new political sociology of science perhaps ought to avoid emphasizing nation-state decision making, as some contributors to this volume appear to be doing. Ought we not also—and perhaps more importantly—be on the lookout for authority relations embedded *in* the R&D system, innovations that lead to fundamental changes in the ways people spend their time, money, and attention? As Langdon Winner puts it, technology actually is a form of legislation, inasmuch as "Innovations are similar to legislative acts or political foundings that establish a framework for public order that will endure over many generations."[68]

In "deciding" that our built and natural environment would be shaped by plastics and other chemical products, governments were involved, certainly, but chemists and chemical engineers in conjunction with business executives arguably were the primary "policy makers."[69] Decision making, or non-decision making, of that sort now is shaping the future of nanofabrication and other emerging nanoscience. If experts are to participate more helpfully in the future in nongovernmental (as well as governmental) policy making, nontrivial revisions in the social relations of expertise almost certainly would be required.[70]

At present, scientists work according to agendas that are at least partly illegitimate—shaped without sufficiently broad negotiation, oriented substantially toward purposes many people would find indefensible if they understood them, and shaped without sufficient attention to the requisites for acting prudently in the face of high uncertainty. It is impossible to say how much of the power rests with scientists, how much with political-economic elites, and how much with systemic forces (if such actually can be separated out from those in the powerful roles), because the forces, vectors, and sectors influence one another in ways too complex to parse. What we can say with some assurance, as David Dickson expressed the point two decades ago, is that science is "a powerful tool that can help us understand the natural universe in potentially useful ways, but at the same time carries the seeds of human exploitation. How to tap the one without falling victim to the other is the key challenge. . . . [A] properly democratic science policy will be achieved only by a political program that directly challenges the current distribution of wealth and power."[71]

It seems to me that the brown/green chemistry and nanoscience cases offer opportunities to reinvigorate older, interest-based understandings of science and power by integration with the more recent constructivist tradition. A new, *re*constructivist understanding potentially awaits—one nuanced enough to appreciate micro-local realities within scientific laboratories and fields, and yet one that aims to reduce elite appropriation of knowledge and to direct technical capacities for more justifiable public purposes.[72]

Acknowledgments

My thanks to Brian Martin, Steve Breyman, Dan Sarewitz, Ed Hackett, Elisabeth Clemens and the editors for thought-provoking discussions and helpful comments.

Endnotes

1. An earlier analysis of nanotechnology decision making appeared as Daniel Sarewitz and Edward Woodhouse, "Small Is Powerful," in Alan Lightman, Daniel Sarewitz, and Christina Dresser, eds., *Living with the Genie: Essays on Technology and the Quest for Human Mastery* (Washington, DC: Island Press, 2003), 63–83.

2. A different version of the research on green chemistry is Edward J. Woodhouse and Steve Breyman, "Green Chemistry as Social Movement?" *Science, Technology, & Human Values,* 30 (2005): 199–222.

3. Edward Woodhouse, Steve Breyman, David Hess, and Brian Martin, "Science Studies and Activism: Possibilities and Problems for Reconstructivist Agendas," Social Studies of Science 32 (2002): 297–319.

4. On technology as legislation, see Langdon Winner, *Autonomous Technology: Technics-out-of-control as a Theme in Political Thought* (Cambridge, MA: MIT Press, 1977).

5. B. C. Crandall, ed., *Nanotechnology: Molecular Speculations on Global Abundance* (Cambridge, MA: MIT Press, 1996), 1.

6. The lecture is reprinted in Richard P. Feynman, *The Pleasure of Finding Things Out and the Meaning of It All* (New York: Perseus, 2002), 117–140.

7. K. Eric Drexler, *Engines of Creation* (Garden City, NY: Anchor Press/Doubleday, 1986).

8. Drexler, *Engines*, 171–172.

9. Bill Joy, "Why the Future Doesn't Need Us," *Wired* (April 2000): 37–51.

10. Michael Crichton, *Prey: A Novel* (New York: HarperCollins, 2002).

11. See Jacob West, "The Quantum Computer: An Introduction," April 28, 2000 (last updated May 30, 2000), www.cs.caltech.edu/~westside/quantum-intro.html (accessed November 17, 2004).

12. Columbia University Center for Electron Transport in Molecular Nanostructures, www.cise.columbia.edu/hsec (accessed June 9, 2005).

13. NanoInvestorNews, Company Profiles, Nanomix Inc., www.nanoinvestornews.com/modules.php?name=Company_Profiles&op=viewprofile&company=nanomix (accessed June 9, 2005).

14. For example, see Arizona State University's program in biomolecular nanotechnology, http://photoscience.la.asu.edu/bionano/index5.htm (accessed June 8, 2005).

15. Swiss Reinsurance Company, Royal Society and Royal Academy of Engineering.

16. Newt Gingrich, "We Must Fund the Scientific Revolution," *Washington Post*, October 18, 1999, A19.

17. M. C. Roco and W. S. Bainbridge, *Societal Implications of Nanoscience and Nanotechnology* (Arlington, VA: National Science Foundation, 2001); also available at www.wtec.org/loyola/nano/NSET.Societal.Implications/nanosi.pdf (accessed June 7, 2005).

18. Rensselaer Polytechnic Institute Nanotechnology Center, "Material Whirled," www.rpi.edu/change/ajayen.html (accessed May 13, 2005).

19. NanoBio Convergence, www.nanobioconvergence.org/ (accessed June 9, 2005).

20. News.NanoApex, "Northwestern University Receives $11.2 Million for Nanotechnology Research Center," September 19, 2001, http://news.nanoapex.com/modules.php?name=News&file=article&sid=854 (accessed June 7, 2005).

21. NanoBio Forum, www.nanosig.org/nanobio.htm (accessed November 20, 2003).

22. Carbon Nanotechnologies Incorporated, http://cnanotech.com/ (accessed October 7, 2004).

23. For further detail and a different conceptual approach, scc E. J. Woodhouse, "Change of State? The Greening of Chemistry," in *Synthetic Planet: Chemicals, Politics and the Hazards of Modern Life*, ed. Monica J. Casper, ed., pp. 177–193 (New York: Routledge, 2003).

24. See, for example, the *Journal of Carbohydrate Chemistry* and the *Journal of Lipid Research*.

25. Craig E. Colton and Peter N. Skinner, *The Road to Love Canal: Managing Industrial Waste before EPA* (Austin: University of Texas Press, 1995); Lois Marie Gibbs, *Love Canal: The Story Continues* (Gabriola Island, BC, Canada:

New Society, 1998); Jonathan Harr, *A Civil Action* (New York: Random House, 1995).

26. James Whorton, *Before Silent Spring: Pesticides and Public Health in Pre-DDT America* (Princeton, NJ: Princeton University Press, 1975).

27. C. A. Eckert, B. L. Knutson, and P. G. Debendetti, "Supercritical Fluids as Solvents for Chemical and Materials Processing," *Nature* 383 (1996): 313.

28. David Bradley, "Critical Chemistry," *Reactive Reports Chemistry WebMagazine*, no. 4, February 2000, www.reactivereports.com/4/4_1.html (accessed September 22, 2004).

29. "The Role of Polymer Research in Green Chemistry and Engineering: Workshop Report," University of Massachusetts, June 11–12, 1998, www.umass.edu/tei/neti/neti_pdf/Green%20Chemistry%20of%20Polymers.pdf (accessed November 10, 2004).

30. Barry M. Trost, "The Atom Economy: A Search for Synthetic Efficiency," *Science* 254 (December 6, 1991): 1471–1477.

31. On the role of counternarrative in public policy formation and in controversies more generally, see Emery Roe, *Narrative Policy Analysis: Theory and Practice* (Durham, NC: Duke University Press, 1994); and Nancy Fraser, *Justice Interruptus* (New York: Routledge, 1998).

32. Joe Thornton, *Pandora's Poison: Chlorine, Health, and a New Environmental Strategy* (Cambridge, MA: MIT Press), 2000.

33. Northwest Toxics Coalition, "Green Chemistry: A Workshop," video, Portland, Oregon, April 21–22, 2000.

34. University of Massachusetts, " Role of Polymer Research."

35. Thornton, *Pandora's Poison.*

36. Jeff Howard, "Toward Intelligent, Democratic Steering of Chemical Technologies: Evaluating Industrial Chlorine Chemistry as Environmental Trial and Error," PhD diss., Department of Science and Technology Studies, Rensselaer Polytechnic Institute, December 2004.

37. U.S. Environmental Protection Agency, "P3 Award: A National Student Design Competition for Sustainability Focusing on People, Prosperity, and the Planet," press release, December 15, 2003, http://es.epa.gov/ncer/p3/designs_sustain_rfp.html (accessed December 16, 2004).

38. Michael C. Cann and Marc E. Connelly, *Real-World Cases in Green Chemistry* (Washington, DC: American Chemical Society, 2000).

39. David Dickson, *The New Politics of Science* (Chicago: University of Chicago Press, 1984), 3.

40. Charles E. Lindblom, *Politics and Markets: The World's Political-Economic Systems* (New York: Basic, 1977).

41. Daniel Sarewitz, *Frontiers of Illusion: Science, Technology, and the Politics of Progress* (Philadelphia: Temple University Press, 1996), 10–11.

42. Henry Etzkowitz and Loet Leydesdorf, eds., *Universities and the Global Knowledge Economy: A Triple Helix of University-Industry-Government Relations* (New York: Thomson Learning, 1997).

43. U.S. House of Representatives, *The Societal Implications of Nanotechnology, Hearing before the Committee on Science,* 108th Cong., 1st sess., April 9, 2003, serial no. 108–113.

44. Dickson, *New Politics of Science,* 314.

45. Richard P. Feynman, "There's Plenty of Room at the Bottom: An Invitation to Enter a New Field of Physics," presented December 29, 1959, at the annual meeting of the American Physical Society, Pasadena, published in *Engineering and Science* (February 1960). Also available at www.zyvex.com/nanotech/feynman.html (accessed December 5, 2004).

46. Peterson is quoted in "Signed, Sealed, Delivered: Nano Is President's Prefix of the Day," *Small Times,* December 3, 2003, www.smalltimes.com/document_display.cfm?document_id=7035 (accessed March 29, 2004). Also see Christine L. Peterson, "Nanotechnology: From Feynman to the Grand Challenge of Molecular Manufacturing," *IEEE Technology and Society Magazine* 23 (Winter 2004): 9–15.

47. Quoted in Robert F. Service, "Is Nanotechnology Dangerous?" *Science* 290 (24 November 2000): 1526–1527, quote from 1527.

48. The Drexler-Smalley exchange is published as "Point-Counterpoint: Nanotechnology: Drexler and Smalley Make the Case for and against 'Molecular Assemblers,'" *Chemical and Engineering News* 81, no. 48 (December 1, 2003): 37–42.

49. 21st Century Nanotechnology Research and Development Act, Public Law 108–153, December 3, 2003. For commentary on the law's passage, see Small Times, "Signed, Sealed, Delivered," December 3, 2003, www.smalltimes.com/document_display.cfm?document_id=7035 (accessed June 10, 2005).

50. Alexander Huw Arnall, *Future Technologies, Today's Choices: Nanotechnology, Artificial Intelligence and Robotics; A Technical, Political and Institutional Map of Emerging Technologies* (London: Greenpeace Environmental Trust, July 2003).

51. ETC Group, "Size Matters! The Case for a Global Moratorium," April 2003, www.etcgroup.org/document/GT_TroubledWater_April1.pdf (accessed May 12, 2003); "Toxic Warning Shows Nanoparticles Cause Brain Damage in Aquatic Species and Highlights Need for a Moratorium on the Release of New Nanomaterials," April 1, 2004, www.etcgroup.org/documents/Occ.Paper_Nanosafety.pdf (accessed April 2, 2004).

52. For further discussion, see the symposium on social aspects of nanotechnology, *IEEE Technology and Society Magazine* 23 (Winter 2004).

53. A volume admirably devoted to analysis of science as a world-shaping influence, yet that fails utterly to come to grips with problematic scientific

activities such as chemistry and nanoscience, is Gili S. Drori et al., eds., *Science in the Modern World Polity: Institutionalization and Globalization* (Palo Alto, CA: Stanford University Press, 2003).

54. Richard Sclove, *Democracy and Technology* (New York: Guilford Press, 1995).

55. Back cover blurb of Lightman et al., *Living with the Genie;* an extended treatment of the theme is Bill McKibben, *Enough: Staying Human in an Engineered Age* (New York: Times Books, 2003).

56. Winner, *Autonomous Technology.*

57. For a very different take on this issue, see Steve Fuller, *The Governance of Science: Ideology and the Future of the Open Society* (Buckingham, UK: Open University Press, 2000).

58. Roger W. Elder and Charles D. Cobb, *Participation in American Politics: The Dynamics of Agenda-Building,* 2nd ed. (Baltimore: Johns Hopkins University Press, 1983). On governing mentalities, see Nancy D. Campbell, *Using Women: Gender, Drug Policy, and Social Justice* (New York: Routledge, 2000).

59. Steven Lukes, *Power: A Radical View* (London: Macmillan, 1974).

60. On the problems of rapid scale-up, see Joseph G. Morone and Edward J. Woodhouse, *The Demise of Nuclear Energy? Lessons for Democratic Control of Technology* (New Haven, CT: Yale University Press, 1989).

61. U.S. House of Representatives, Committee on Science, *Hearing H.R. 3970, Green Chemistry Research and Development Act of 2004,* 108th Cong., 2nd sess., March 17, 2004, serial no. 108–47 (Washington, DC: U.S. GPO), 2004.

62. On impairment generally, see Charles E. Lindblom, *Inquiry and Change: The Troubled Attempt to Understand and Shape Society* (New Haven, CT: Yale University Press, 1990). For analysis of myths around technoscience in particular, see Sarewitz, *Frontiers of Illusion: Science, Technology, and the Politics of Progress* (Philadelphia: Temple University Press, 1996).

63. See, for example, Fraser, *Justice Interruptus,* and Nancy D. Campbell, *Using Women: Gender, Drug Policy, and Social Justice* (New York: Routledge, 2000).

64. This is one version of the much-maligned but also widely heralded rational choice approach. An overview of the principal-agent approach in the context of social choice more generally is James S. Coleman, *Foundations of Social Theory* (Cambridge, MA: Harvard University Press, 1990).

65. David H. Guston, "Principal-Agent Theory and the Structure of Science Policy," *Science and Public Policy* 23 (August 1996): 229–240.

66. For a skeptical view of principal-agent theory based on analysis of several European research programs, see Elizabeth Shove, "Principals, Agents and Research Programmes," *Science and Public Policy* 30 (October 2003): 371–381.

67. Sclove, *Democracy and Technology.*

68. Langdon Winner, *The Whale and the Reactor: The Search for Limits in an Age of High Technology* (Chicago: University of Chicago Press, 1986), 29.

69. Frank N. Laird, "Technocracy Revisited: Knowledge, Power, and the Crisis of Energy Decision Making," *Industrial Crisis Quarterly* 4 (1990): 49–61; Frank N. Laird, "Participatory Analysis, Democracy, and Technological Decision Making," *Science, Technology, & Human Values* 18 (1993): 341–361.

70. Dean Nieusma, "Social Relations of Expertise: The Case of the Sri Lankan Renewable Energy Sector," PhD diss., Rensselaer Polytechnic Institute, Department of Science and Technology Studies, 2004.

71. Dickson, *New Politics of Science,* 336.

72. On reconstructivism, see Woodhouse et al., "Science Studies and Activism," and E. J. Woodhouse, "(Re)Constructing Technological Society by Taking Social Construction Even More Seriously," *Social Epistemology* 19 (April–September 2005): 1–17.

2

Science and Social Movements

7

When Convention Becomes Contentious

Organizing Science Activism in Genetic Toxicology

SCOTT FRICKEL

In the political history of U.S. environmentalism, 1969 stands as a year of pivotal institutional expansion and reform. The federal government initiated a series of highly visible policy actions addressing the biological and ecological impacts of synthetic environmental chemicals. Chief among them, the National Environmental Policy Act (NEPA) mandated the creation of the U.S. Environmental Protection Agency (EPA) and the Council on Environmental Quality and laid the institutional foundations for the major environmental bills signed into law during the early 1970s.[1] Other important federal-level developments in 1969 included major reports on the health and economic impacts of chronic chemical exposure (e.g., Food and Drug Administration Advisory Committee on Protocols for Safety Evaluation 1970; U.S. Department of Health, Education, and Welfare 1969) and the creation of the National Institute of Environmental Health Science (NIEHS) to direct basic research on "the effects of environmental factors, singly and in the aggregate, upon the health of man" (Research Triangle Institute 1965).

Outside the public spotlight of federal legislative action, a small group of geneticists worried that much of the government's regulatory and research initiatives were missing the deleterious *genetic consequences* that synthetic chemicals might have on human populations over the long term. Wanting to formalize their concern, these scientists created a new professional organization called the Environmental Mutagen Society (EMS) "to encourage interest in and study of mutagens in the human environment, particularly as these may be of concern to public health."[2] While less conspicuous than the state-sponsored measures of 1969, the response among geneticists and other research biologists was generally positive, if not resoundingly so. Within the year, EMS membership approached 500, the first issue of the *EMS Newsletter* had rolled off the presses, and 268 people—mostly scientists but also including a few public officials, lay citizens, and journalists—attended the first EMS conference held in Washington, D.C. By 1972, EMS societies had been established in Japan (JEMS) and Europe (EEMS), and national sections of the EEMS had formed in Italy, West Germany, and Czechoslovakia (Wassom 1989). By the time Congress passed the Toxic Substances Control Act (TSCA) in 1976 that "required testing and premanufacturing notification to EPA of all new chemical substances," the EMS had emerged as the organizational core of a broadly interdisciplinary science called genetic toxicology (Andrews 1999:243).[3] Then as today, genetic toxicology focused primarily on identifying genetically hazardous "environmental mutagens," describing those chemicals' specific effects in laboratory models, and assessing their potential risk to human populations (Preston and Hoffman 2001; see also Shostak 2003).

While few readers will be familiar with genetic toxicology and fewer still will have heard of the EMS, this case will be interesting to scholars who study boundary making and activism in science because the EMS provides a counter-intuitive illustration of each. Most studies of scientific boundary making focus on scientists' efforts to maintain clear and explicit distinctions between science and "politics," with the latter typically understood in terms of specific government policy or some more general action reflecting distinctly social values (e.g., Gieryn 1999; Jasanoff 1987; Kleinman and Kinchy 2003). Similarly, most studies of science activism demonstrate the hazards of crossing that line: popularizers of scientific ideas trade the respect of their peers for media fame, while "oppositional professionals" run grave professional and economic risks in siding with activist causes (Brown, Kroll-Smith, and Gunter 2000:19; B. Allen 2003; Goodell 1977; Moore and Hala 2002). Research in both areas demonstrates that conventional science and contentious politics do not easily mix.

In this chapter, I marshal evidence to demonstrate that within the orbit of the EMS's bid to establish genetic toxicology, successful boundary making and science activism were not mutually exclusive. Instead, institutionalization involved both.

Organizations, Boundaries, and Contentious Science

Recent research on activism and boundary making in science-based organizations lends a broader comparative perspective to the EMS case. Besides adopting an organizational level of analysis, studies by Moore and Hala (2002; see also Moore 1996) on the radical science organization Science for the People (SftP) and Kinchy and Kleinman (2003) on the Ecological Society of America (ESA), share with the present study several attributes that make for interesting comparisons.[4] SftP and the EMS were both founded in 1969 amid considerable social and political opposition to the war in Vietnam and to growing concern for ecological well-being, but these organizations responded to their respective challenges in strikingly different ways. The ESA was more than fifty years old when geneticists created the EMS, but in other respects are very much alike. Both are professional scientific societies, both represent scientific interdisciplines engaged in environmental research and policy, and both have struggled historically to define themselves in relation to environmentalism.

Moore and Hala (2002) want to understand institutional change. Their study of scientists' formation of a collective identity for "radical" science asks, "when and how do organizations challenge convention?" (318). They describe how activist physicists, rebuffed in their bid to mobilize opposition to U.S. foreign policy in Southeast Asia through the American Physical Society, created SftP and developed its radical identity through contentious collective action. In this case, science activism was radical in three respects. First, radical science activism was based on a critical reflexive analysis of institutional science and its relationship to state power. Activists argued that "the political values of scientists should be taken into account in judging the value of scientific claims" (325). Second, radical science activism was defined performatively, through overtly contentious behavior. Taking their cue from civil rights direct actions, SftP members disrupted scientific meetings by heckling speakers, performing guerrilla theater–style satire, and publicly denouncing their scientist peers. Third, radical science activism involved organizational innovation. The SftP provided a decentralized, autonomous, and grass-roots organizational model that stood apart from the conventional professional science associations that were and remain centralized, hierarchical, and governed from above. Through contentious discourse, performative dramaturgy, and innovative organization, SftP helped establish new boundaries distinguishing mainstream science from left politics that other politicized science organizations have since copied.

Kinchy and Kleinman's (2003) interest lies in the opposite direction of explaining institutional stasis. Their analysis of the ESA develops an explanation for the institutional conservatism they argue is characteristic of professional scientific societies. Particularly since World War II, they note, the ESA has worked "to establish a boundary between ecology and environmental politics, arguing that ecology is a value-free, objective science, while environmentalism is a political stance" (9). Drawing on evidence from several different episodes of contention, these authors examine the struggles involved in balancing claims of professional credibility in complement to claims of social utility. Recognizing the contextual and socially constructed nature of boundaries in science (Gieryn 1999), Kinchy and Kleinman argue that most boundary struggles among ESA members and officers have ended similarly, with the organization taking "pains to distinguish between ecology and environmentalism" (2003:9). For those who might find in professions a potential force for political reform, Kinchy's (2002:25) conclusion in a companion study is discouraging: "As long as credibility is perpetually linked to a refusal to advocate values, professional scientific organizations conform to a standard of neutrality that fails to promote social change."

The case study presented below both confirms and extends arguments about the institutional and ideological constraints to science activism presented in these studies. Where Kinchy and Kleinman want to understand why professional science organizations tend not to incubate science activism, and where Moore and Hala want to understand how activist scientists have responded to the institutional constraints within their profession, I am concerned with the conditions under which conventional practices become contentious and what impacts those politicized practices have on the organization of knowledge (for a different approach to a similar set of conceptual concerns, see Henke, this volume).

The analysis I present below can be summarized as follows: in building an interdisciplinary genetic toxicology, science activism and boundary making were intertwined processes conducted largely within and through the EMS. Convincing patrons, science administrators, and biologists from a range of disciplinary backgrounds that the non-obvious problem of "environmental mutagenesis" warranted immediate and systematic attention required both an organized campaign and the authority of credible science. Achieving both involved a two-part strategy in which publicly visible demarcations between objective science and environmental values were reinforced in such a way that epistemological and disciplinary boundaries *within science* loosened to facilitate the flow-through of people, money, practices, and ideas. Keeping a visible distance from popular environmentalism allowed the EMS to nurture less obvious but still effective variants of science activism within its expanding domain of influence. Genetic toxicology, the major outcome of these coordinated efforts, represents the institutional legitimation of politicized science.

In what sense can scientists' efforts to create genetic toxicology be reasonably understood as contentious politics? Within the context of discipline-based knowledge production, advocating for interdisciplinary knowledge was and remains a political act. Circa 1970, science activists working through the EMS advocated an interdisciplinary approach to mutagenesis research and its attendant public health implications at a time when biology was rapidly gaining ascendancy as a "hard" experimental science. In this sense, their call for the development of a public-service genetics that combined applied and basic approaches to knowledge production was doubly contentious. They urged geneticists to get involved in environmental health research and simultaneously to relinquish their exclusive claim to the problem of mutagenicity.

More centrally for this chapter, scientists' efforts to legitimate genetic toxicology involved a redistribution of disciplinary power. For Michel Foucault (1980), disciplining knowledge is primarily coercive, and the goal, ultimately, is social control. But as Christopher Sellers (1997:233) points out, discipline can also destabilize existing power structures. His research shows how, prior to 1960, industrial hygienists institutionalized a "scientific gaze" inside industrial workplaces making it possible for researchers to examine workers' bodies for visible signs of occupational disease and hold industry and government more accountable for workers' health (228). Similarly, genetic toxicologists worked to extend the scientific gaze even further inside the body, to the level of the gene. They also extended the scope of institutional concern beyond workers in factories to include consumers, communities, and the natural environment. They did so by experimenting on the genetic effects of "everyday" chemicals, developing new methods for estimating genetic risk, training toxicologists in genetics methods, creating new organizations, and engaging in public education and outreach. This conventional work had contentious implications. It gave new cultural and technical meaning to mutagenic agents and perforated institutional barriers that separated experimental work in genetics from public health and environmental knowledge. Their undercover approach to environmental politics suggests that science activism is not ephemeral to scientific practice. Rather, it reinforces science studies scholars' contention that in knowledge production systems, politics and practices are mutually constituted.

Historical Context

Chemical mutagens first entered geneticists' laboratories as research tools in the 1940s (Beale 1993). Potent chemicals were an inexpensive, efficient, and readily available means of generating specific types of mutations in living organisms, and their use became standard practice in genetic experiments (Auerbach 1963). Over the next three decades, researchers gradually expanded the list of known genetically hazardous compounds and deepened scientists' understanding of chemical-induced gene mutation (Auerbach 1962; Drake and Koch 1976). As late as 1966, however, the terms *genetic toxicology* or *environmental mutagenesis* were not part of professional life sciences discourse. This was not because geneticists lacked the requisite knowledge to make the necessary connections between chemical mutagens and public health (Lederberg 1997). Rather, it was because the organization of research on chemical mutagens remained highly decentralized, with few existing cross-laboratory research networks that might have served as mechanisms to generate sustained concern over mounting evidence that chemical mutagens could endanger the public health (de Serres and Shelby 1981). As a result, mutation research reflected the disciplinary and local institutional interests of individual scientists while sporadic attempts to raise awareness in research and policy circles by a concerned few repeatedly failed to attract widespread attention (Epstein 1974; Legator 1970; Wassom 1989).

Within a decade, however, chemical mutagens were transformed from tools of research in experimental genetics into an environmental problem of potentially global proportions. Although the EMS was central to that transformation, three broader crosscutting pressures conditioned genetic toxicology's consolidation as a new environmental health science. First, congressional budget cuts in federal R&D beginning in 1967 brought an abrupt end to scientists' perception of the golden age of federally sponsored research and increased competition among scientists for research funding (Gieger 1993:198; Dickson 1988; Pauly 2000). Second, an emerging environmental movement pressured public officials to enact environmental legislation (e.g., TSCA, introduced in Congress in 1971) and expand the infrastructure for environmental research and regulation, including establishing the NIEHS in 1969, the EPA in 1969, and the National Center for Toxicological Research in 1971. Third, although molecular biology and the "biomedicalization" of health and illness would not triumph for another decade or so (Clarke et al. 2003), as a new era of big biology commenced, genetics' disciplinary star was rising. As the professional status of genetics increased, so did the status of genetic toxicology's most ardent supporters, many of whom were midcareer geneticists with accomplished research records in university or government laboratories and firmly established careers in the mainstream of American life science (see Table 1).

These general pressures combined in a fortuitous way for genetic toxicology promoters. As the public's increasing demand for environmental knowledge made environmental health research an attractive slice of a shrinking budgetary pie, genetic toxicology advocates' newfound status provided some competitive advantage in mobilizing support for their new project. Acting collectively through organizations such as the EMS, geneticists could exploit to greater effect new opportunities to channel the interest of scientists and public officials and to capture needed financial, organizational, and institutional resources.

An Organizational Solution to an Ontological Dilemma

The EMS was created in response to concern among a growing number of life scientists that some unknown number of chemicals present in the human environment that were considered "safe" because they tested negative in standard toxicology screens were nevertheless responsible

Table 1. EMS Officers, August 1970

Name	Age	Institution	Position	Discipline/s
A. Hollaender (president)	72	ORNL, Biology Division	Sr. Research Advisor	Biochemistry
M. Messelson (vice president)	46	Harvard, Biology	Professor	Molecular Biology
M. S. Legator (treasurer)	44	FDA	Chief, Cell Biology Branch	Biochemistry, Bacteriology
S. S. Epstein (secretary)	43	Children's Cancer Research Foundation, Inc. (Boston)	Chief, Labs. of Carcinogenesis & Toxicology	Pathology, Microbiology
E. Freese (president, ex officio)	44	National Institute of Neurological Diseases and Stroke	Chief, Molecular Biology Lab	Biology
J. F. Crow (vice president ex officio)	54	U. Wisconsin, Genetics and Medical Genetics	Professor & Chair	Genetics

Source: American Men and Women of Science, 1971. vols. 1–6 (Physical and Biological Sciences), 12th ed. New York: R. R. Bowker.

for inducing minute changes in the genetic material that could have lasting and irreparable negative consequences. As one scientist enumerated them,

> These new man-made chemicals are everywhere. Some 2–3000 are used as food additives; 30 are used as preservatives; 28 as antioxidants; 44 as sequesterants; 85 as surfactants; 31 as stabilizers; 24 as bleaches; 60 as buffers, acids or alkalies; 35 as coloring agents; 9 as special sweeteners; 116 as nutrient supplements; 1077 as flavoring agents, and 158 for miscellaneous uses. Thousands of other compounds are used as drugs, narcotics, antibiotics, cosmetics, contraceptives, pesticides or as industrial chemicals.[5]

"[E]ven though the compounds may not be demonstrably mutagenic to man at the concentrations used," warned University of Wisconsin population geneticist James F. Crow (1968:113), "the total number of deleterious mutations induced in the whole population over a prolonged period of time could nevertheless be substantial." The potential public health crisis turned on a paradox: chemical compounds that caused the least amount of genetic damage to the individual held the most potential harm in terms of their long-term effects on the human population (Legator 1970; Neel 1970). Conversely, genetic damage most threatening to living individuals (i.e., somatic cell mutations typically implicated in carcinogenesis) was least problematic from an evolutionary perspective since by this metric only people, not populations, were threatened by somatic disease. As Harvard biologist Matthew Meselson (1971:ix) put it, "by its nature, this [minute] genetic damage can be cumulative over generations while even the most insidious non-genetic poison cannot accumulate in the body beyond the lifetime of an individual." In the context of President Nixon's declaration of war on cancer, and in the wake of the world's first Earth Day gatherings, however, these arguments by EMS members for an interdisciplinary genetic toxicology as a social solution to this pending "genetic emergency" did not garner immediate political or scientific support (Crow 1968:114). Unlike cancer deaths, fish kills, or urban smog, environmental mutagenesis was neither visible nor visibly eminent. As an environmental health problem, its ontological status remained in doubt.

Making environmental mutagenesis politically viable was an achievement that depended largely on the level of credibility that EMS officers were able to foster among organization members, patrons, and consumers of genetic toxicology information in industry, government, and universities and among the general public. In building an interdisciplinary research community, a stable funding base, and a market for genetic toxicology information and practices—all interrelated goals—it became important for the EMS as an organization to somehow rise above the often-competing interests of these diverse actors. As a professional scientific society, the EMS's own best interests lie in the appearance of disinterestedness.

On the other hand, success in the organization's substantive goal of building and populating a new interdiscipline relied heavily on overt political rhetoric (for a detailed analysis, see Frickel 2004a). In numerous published essays and public lectures delivered in the early 1970s, EMS members challenged the federal government's reluctance to fund chemical mutagenesis research (Lederberg 1969) as well as chemical regulatory policies that all but ignored the genetic impacts of environmental chemicals (e.g., Malling 1970). For their part, chemical and pharmaceutical companies contributed to the problem by pursuing "restricted approaches to toxicological problems—narrow questions, narrowly defined, narrowly posed, and often narrowly answered." The compartmentalization of research on chemical effects into separate federal and state agencies for dealing with the environmental, consumer, and occupational dimensions of environmental health problems also hampered the development of programs to train people who are not engaged in basic mutation research to competently conduct mutagenicity tests, since "each government agency [thinks] someone else should do it" (Hollaender 1973:232). "[C]onflicting regulatory, promotional, and research roles" also constrained environmental research at the national laboratories, ideal places for intensive research programs, had Cold War politics not confined their research missions so strictly to the study of radiation biology, and at universities where departmental structures impeded interdisciplinary interaction and the efficient and creative use of their uniquely concentrated resources (Lederberg 1969). Finally, EMS scientists charged that overblown distinctions between basic and applied research constrained environmental knowledge production at all of these institutions by reinforcing the belief that fundamental research must be conducted "piecemeal" even though directed research in mutagenicity testing had spun off and would likely continue to spin off basic genetic knowledge (Meselson 1971).[6]

Scientists in different research contexts were thus seen as being constrained in different ways by different configurations of bureaucratic and ideological boundaries, the general solution to which was the creation of new interdisciplinary organizations. "[T]his must be a cooperative effort," EMS president Alexander Hollaender (1973:232) told an audience at a mutagenicity workshop. "The people in industry, in the medical professions, in government laboratories, and investigators in research laboratories and universities must work together. Otherwise we won't get anywhere." Noting there were "enough examples of problems where simplistic unidisciplinary approaches to complex multidisciplinary problems have been unhelpful or even detrimental," EMS secretary Sam Epstein argued that "ideally, Universities should create new interdisciplinary and interdepartmental Units or Centers, with sole and primary responsibility in environmental, consumer, and occupational research."[7] In the meantime, the EMS would contribute "a vital function if it does nothing more than provide a channel for communication among a wide range of separate disciplines" (Lederberg 1969).

Encompassing the dual tasks of creating an interdisciplinary research infrastructure for genetic toxicology and raising the political saliency of chemical genetic hazards, the EMS engaged both its scientific and political missions without undermining either. The following sections illustrate how the EMS distanced its members and its institutional project of genetic toxicology from the largely external ideological politics associated with environmentalism and then used that credibility to insulate and control a different brand of environmental politics within its own organizational domain.

A PUBLIC FACE FOR A NEW SCIENCE

Scientists troubled by genetic hazards littering the human environment did not have to join the EMS to address the problem in their own research; yet many did. By 1976 EMS membership had surpassed one thousand. The modest legitimacy reflected in the steep rise in membership during the organization's first half-decade was not, however, derived solely from the nature of the threat to public health posed by environmental mutagens. Organizational credibility also mattered, and that was a thing to be earned, not given. Scientists' campaign to establish genetic toxicology depended in part on nurturing a generalized perception that the EMS embodied a spirit of scientific neutrality. To attract members and financial backing, the EMS strove to present itself as a society committed first and foremost to the production, rationalization, and dissemination of objective knowledge. In its official statements and in its routinized activities, the EMS kept environmental politics out of genetic toxicology data and information.

These efforts are illustrated by a "statement of activities" contained in an Internal Revenue Service application for tax-exempt status filed on behalf of the EMS in 1969. "Like most organizations of scholars," the report reads, "the EMS will, through scientific congresses, symposia, a journal and a newsletter, provide the traditional forums through which scientists of similar professional interest have for generations communicated with one another and with the public. Experimental data and new theories are shared and subjected to the inspection and critical review of informed colleagues." Tax exemption placed definite constraints on the kinds of political activities the EMS could legally pursue. A lawyer cautioned EMS secretary Samuel Epstein to "be wary of any participation in a public campaign during the adoption or rejection of specific legislation." These legal constraints enhanced the organization's credibility as one whose main, and perhaps only, formal interest was in the "inspection and critical review" of scientific knowledge.[8]

On that basis, formal relations with environmental groups, for example, were roundly discouraged. In reference to a letter from the Natural Resources Defense Council requesting information on environmental mutagenesis, the EMS Executive Council decided that the EMS would be "willing to function only as a resource facility, and not in the development of any action program." Five years later, EMS president Hollaender complained of frequent requests for information on "chemical toxicology," from the group Resources for the Future. There is little evidence that, in the interim, the EMS entered into relationships—formal or otherwise—with environmental organizations.[9]

Organizations that may be assumed to have harbored political and economic interests biased in the opposite direction received similar rebuffs. A proposal that the Association of Analytical Chemists be invited to review validity and reproducibility studies of mutagenicity tests was struck down on the grounds that it "has no special expertise in this matter." The same attitude guided relationships with firms having a direct economic interest in the production of genetic toxicology data. A report from an EMS Committee on Methods advanced the position that "the EMS should avoid putting itself into a position of certifying or providing an endorsement to any laboratory or test method. It should serve only as an assembly of scientists willing to provide individual expertise, upon request, to anyone requesting it." The committee advocated this position as a means of avoiding potential legal difficulties or conflict of interest charges. Such outcomes would threaten the EMS's appearance of organizational neutrality and undercut efforts by the EMS leadership to institutionalize the perception of ideological purity. The EMS did engage the public directly but did so through a carefully orchestrated strategy pursued largely through the Environmental Mutagen Information Center, or EMIC.[10]

SCIENCE IN THE PUBLIC SERVICE

EMIC began formal operations in September 1970. Housed in the Biology Division of Oak Ridge National Laboratory, EMIC served as an information clearinghouse for mutagenicity data. It employed a small technical staff charged with collecting published literature on chemical mutagenesis, condensing the data presented in those articles into uniform tabular abstracts, and building a computer database from that information that could be accessed via one of a number of standardized index codes as needed for specialized searches (Malling 1971; Malling and Wassom 1969).

EMIC's other primary task was disseminating the mutagenicity information that had been collected. The main mechanism for this was an annual literature survey that EMIC produced and distributed, mostly to members of the various EMS societies around the world. EMIC staff also published occasional "awareness lists"—short bibliographies of important subclasses of chemical compounds—in the *EMS Newsletter*. Far more frequently, EMIC staff attended to the specific requests for data by, as one report noted, "anyone who requested it." As this and other documents from the EMS archives make clear, keeping scientists but also "the general public informed about highly technical data" was EMIC's central concern and explicit mandate. In a letter written in 1970, for example, EMIC director Heinrich Malling mentioned that he had been answering questions on mutagenicity at "a rate of one per day." In 1971, EMIC staff reported receiving 222 individual requests for information. That report noted that "[t]he greatest proportion of these requests were from persons engaged in research, but some came from a variety of sources," including "city municipalities, high school students, free lance writers" and the occasional legislator.[11]

While it is reasonable to assume that under conditions of resource scarcity, requests from high school students or citizens' groups might not receive the same level of attention as those coming from scientists active in the mutation research field, the historical record makes clear that, in principle, EMIC—and, by direct extension, the EMS—was committed to serving the public interest as an impartial messenger of genetic toxicology information. That impartiality extended into the economic sphere as well. Although the EMS Executive Council several times considered billing industry and foreign researchers for chemical mutagenicity data as a means of offsetting tight budgets, such a policy never materialized.[12]

INSTITUTIONALIZING IMPARTIALITY

During its formative years, the EMS Executive Council pursued policies of conduct, scientific review, and public service that depended on and reinforced a strict division between the EMS as a scientific research organization and politics of various sorts—from environmental protest to the endorsement of particular testing protocols. As we've just seen, the organization's public boundary work was reflected in the absence of relationships with partisan groups and embodied most explicitly in the social service functions of EMIC. Together, these strategies described an explicit attempt to establish a distinct and publicly visible boundary between the producers and consumers of genetic toxicology knowledge.

Keeping the public face of genetic toxicology clean was in part a strategy born of organizational necessity. In 1970, the EMS was virtually alone in its commitment to raise awareness of the suspected dangers of environmental mutagens among scientists and policymakers. But organizing an attack against the problem meant coordinating research, data collection, and information exchange across geographically distant laboratories. And that required money.

At the time, genetic toxicology could boast few stable funding sources. As Malling told a European colleague in the fall of 1970, "The money situation in the U.S. is very tight. There is essentially no money for screening for the mutagenicity of harmful pollutants." Much of the funding at the time came in the form of federal budget line items during a period of general decline in the funding rate for basic research. Money for EMIC and support for other EMS projects were not at first easily obtained or readily recommitted. Numerous federal agencies, various chemical and drug companies, private foundations, the National Laboratories, and four or five of the National Institutes of Health contributed small sums to sponsor EMS conferences and workshops and to support the work conducted at EMIC, usually on a year-to-year basis. For fiscal year 1971, for example, EMIC was funded with a total budget of $40,000, with $10,000 contributed by NIEHS, FDA, NSF, and the AEC. That sum was less than was required, however, and a subsequent budgetary shortfall forced EMIC to temporarily curtail many of its data collection efforts and sent EMS officers scrambling to locate additional "emergency" funds to keep EMIC running. Given the heterogeneity of EMIC's patrons, and the resulting instability of the economic foundation underlying research and development in genetic toxicology, the boundary work conducted by the EMS can be seen as a means for ensuring continued access to scarce resources.[13]

The strategy worked. After two years of very precarious budgeting arrangements, in 1972 the funds committed to support EMIC from both the FDA and the National Institute of Environmental Health Sciences increased considerably. With this support these federal agencies gave the EMS and EMIC a stamp of legitimacy and provided a public endorsement of the importance of genetic toxicology research. By 1977, annual federal funding for EMIC through a single institution—NIEHS—had increased nearly fivefold, to $190,000 (National Institute of Environmental Health Sciences 1977:331).[14]

Organizational impartiality served a number of specific practical purposes. It helped secure EMS's tax-exempt status, reducing the young organization's economic burden even as it reinforced the science/politics boundary through restrictions on political lobbying and partisan endorsements. The rhetorical construction of "good" science in the interest of environmental health also attracted support from political elites such as EPA director William Ruckelshaus and Democratic senators Edmund Muskie and Abraham Ribicoff.[15] Inversely, the same spirit of neutrality gave drug and chemical companies little room to charge the EMS with environmentalist bias and therefore to avoid taking some responsibility in funding and participating in genetic toxicology's development.

Institutionalizing Contentious Science

Maintaining a strict distance from groups with clear-cut political or economic interests represented an organizational strategy for balancing the competing claims that research promoted by the EMS was at once socially relevant *and* unblemished by social bias (Kleinman and Solovey 1995). Policing that boundary, in turn, enhanced the legitimacy of the EMS's overall project—genetic toxicology—as well as the organization's own influence and autonomy. Within that orbit of influence, the politics of environmental knowledge took different forms. This section describes three specific examples that I call "activism by committee," "pedogogical activism," and "research activism."

On their face, these modes of collective behavior seem wholly routine, and in a sense they were. Scientists are supposed to critically review others' work, teach students, and make knowledge. When substantively infused with environmental values, however, committee work, curriculum development, and knowledge production became politicized and distinct from standard genetics practices. Committees reviewed, compiled, and distributed data on suspected chemical genetic hazards; workshops and symposia included training specifically for nongeneticists; research biologists were encouraged to engage more in policy-oriented work; and experimental models were redesigned so that the mutagenicity of more chemicals could be interpreted more efficiently. In this case, a more or less conventional organization promoted politicized research through conventional practices and in the process made interdisciplinary research credible in a domain where discipline-based knowledge reigned. I address some of the implications of this "undercover" activism for the political sociology of science in the final section.

ACTIVISM BY COMMITTEE

Committees and subcommittees, although often less formal and less enduring over time, undertook much of the behind-the-scenes work of building genetic toxicology. Some were those common to most professional societies—for example, a planning committee, a membership committee, a committee to handle publication of the newsletter, and another to deal with "public contact." Other committees reflected some goals specific to EMS. Minutes from the first formal meeting of the EMS leadership indicate the formation of a committee for establishing a chemical registry (what would become EMIC) and another for looking into the publication of a monograph on mutagenicity testing methods.[16] These committees embodied strategies for either meeting the basic requirements of any scientific society or the achievement of specific organizational goals. In that, they were entirely conventional.

Other EMS committees, however, took on tasks that bore implications far broader than its own organizational survival or identity and had fairly explicit and direct implications for the direction of research in genetic toxicology and chemical regulatory policy. The EMS Executive Council created and charged several committees with critically reviewing suspected or known environmental mutagens or mutagen classes. Examples from just the first few years include committees to study the mutagenicity of caffeine, cyclamate, mercury, hycanthone, and nitrosamines. Others were organized to establish recommendations for standard protocols for mutagenicity testing. In 1970 a "methods committee" was appointed "to critically assess recommended methodologies and also to recommend and evaluate future research and method development." The "cytogenetics committee" was another, more specific methods committee established the following year to essentially conduct the same set of tasks with respect to in vitro human cell tests that could be used to screen for chromosome aberrations in humans. The latter was in specific response to questions about the correlation between human genetic risk and positive mutagenicity in bioassays relying on lower-order organisms. Still other committees were more explicitly political. One committee, created in 1972, was charged with "extending [the Delaney Amendment] to mutagens and teratogens," and another, created in 1975, looked into issues of chemical protection for workers.[17]

These committees were formed within the EMS to accomplish very concrete tasks and upon their completion were generally disbanded. In that respect, they resembled the kinds of "action committees" common to social movement organizations more so than the specialty sections or interest groups commonly found in larger professional societies. Advisory committees in particular served important boundary-ordering functions. By reinforcing divisions between experts and non-experts even as they simultaneously redirected mutation research toward issues and questions relevant to environmental health and policymakers' attention toward the genetic bases of chemical insult, advisory committees sponsored by the EMS helped shape debate over proposed legislation to regulate the production and distribution of chemicals.

Such was the "Committee 17" report. As its name suggests, this was a seventeen-member scientific review body convened by the EMS Executive Council to review research in mutagen detection and in population monitoring, assess the risk implications of the data derived from these testing and monitoring systems, and recommend directions for future research and for chemical regulatory policy. The committee's final report was published in *Science* in 1975 under the title "Environmental Mutagenic Hazards" (Drake et al. 1975). It was essentially the EMS's position paper on research needs and regulatory responsibility for managing chemical genetic hazards. The report argued that "it is crucial to identify potential mutagens *before* they can induce genetic damage in the population at large" (504). Toward that end, the report urged federal regulatory agencies to undertake mutagenicity evaluation "in the same ways as are other toxicological problems" and to develop a "comprehensive system of regulation" with the burden of testing placed on chemical manufacturers and with risk-benefit evaluation determined by committees "composed of expert geneticists and toxicologists . . . economists . . . industrial safety evaluation personnel . . . [and] representatives of the public at large" (510). Human protection from environmental mutagenic insult, however, required both the development and validation of new methods for detecting environmental mutagens and the creation of a trained labor force to design, conduct, and interpret those tests.[18]

PEDAGOGICAL ACTIVISM

There were few ready-made niches for genetic toxicology in university science circa 1969. Although the study of mutation and mutagenesis was an established part of most genetics curricula (undergraduate as well as graduate), only those students working with scientists directly involved in environmental mutagenesis research would have been introduced to the latest developments in mutagen testing (Straney and Mertens 1969).[19] Moreover, Ph.D. candidates in genetics almost certainly would not have had training in toxicology or pharmacology, unless specializing in those areas. Conversely, specialized studies in genetics were not a common feature of the toxicology Ph.D. Graduate students in toxicology, itself a newly emerging field still very much tied institutionally to medical school pharmacology departments (Hays 1986), also would not have been introduced to genetics except perhaps to fulfill general life sciences requirements.

As a result, the scientists promoting genetic toxicology faced an immediate and acute shortage of people trained in the genetic principles underlying environmental mutagenesis and in genetic toxicology testing methods. "Although methods are presently available for evaluating mutagenic agents, and the demand for such testing is increasing, there is a critical shortage of trained personnel," EMS treasurer Marvin Legator urgently noted. "Industry, government agencies, and universities are seeking trained biologists who are familiar with some or all of the proposed methods for mutagenicity testing. There is no center where individuals can receive formal instruction in this area."[20]

Training in genetic toxicology developed eclectically, through a variety of mechanisms, most of which directly involved people associated with the EMS. One critically important training mechanism was the postdoctoral fellowship. An annual report issued in 1975 from the Environmental Mutagenesis Branch at NIEHS noted "the great shortage of scientists to do research in the area of environmental mutagenesis both in the United States and abroad has provided incentive for staff scientists to develop a training program at both the predoctoral and postdoctoral levels" (National Institute of Environmental Health Sciences 1975a:155).[21]

In the interim, the EMS played a key role in providing crash courses in mutagenicity testing methods, bioassay design, and genetic principles underlying the tests. A three-day "Workshop on Mutagenicity" convened at Brown University in the summer of 1971 was the first of several mini training sessions organized to introduce geneticists, toxicologists, and medical researchers to available testing and evaluation methods. A similar workshop, also sponsored by the EMS, was convened in Zurich, Switzerland. In 1972 the EMS Executive Council passed a resolution to establish "a comprehensive workshop on procedures for detection of chemically induced mutations." In his proposal, Legator called for monies to support a month-long workshop that would include a "series of lectures and intensive laboratory instruction to develop biologists with a working knowledge of principles in this field." Legator organized his envisioned courses into morning lectures, afternoon laboratories, and evening working sessions on topics such as cytogenetic slide reading, interpretation of statistical data, and the integration of mutagenicity tests into standard toxicological screening programs. The proposed nine-member organizing committee included seven EMS members. Although funds for this workshop did not materialize, smaller week-long workshops did continue to be offered fairly regularly in Italy, Canada, Great Britain, India, and the United States. By 1976, EMS-sponsored workshops included foreign participants and reflected the EMS's efforts to promote genetic toxicology training in Argentina, Brazil, Egypt, Japan, and Mexico.[22]

RESEARCH ACTIVISM

Historically, geneticists interested in questions about gene structure and function had been attracted to research organisms whose physiology and behavior complemented those research interests (e.g., bacteria, yeast, fungi, simple plants, and insects) (G. Allen 1975; Drake 1970). The right set of tools for studying the mechanisms of inheritance, however, were not necessarily the best tools for producing realistic estimates of the genetic risk from human exposure to chemical mutagens. Submammalian mutagenicity bioassays were quick, sensitive, and inexpensive to run, but the results could not reliably be extrapolated to humans. This was because "some pollutants in the environment are neither mutagenic nor carcinogenic by themselves but can be converted by mammalian metabolism to highly reactive and genetically active metabolites. Microorganisms have only a fraction of the toxification-detoxification mechanisms that a mammal has" (Malling 1977:263). Scientists' interest in generating experimental results that could reasonably be generalized to humans favored more expensive and time-consuming mutagenicity bioassays that took mammalian metabolic processes into account. In the rush to develop test systems that optimized requirements for speed and sensitivity on the one hand, and generalizability to humans on the other, a hotly competitive test development economy emerged, as EMS members designed and adjudicated among these competing systems.

Two bioassay designs defined the area of most intense competition. One sought to create in vitro bacterial systems that incorporated humanlike metabolism. These were the microorganism bioassays with metabolic activation, of which the Ames test is most renowned (Ames et al. 1973). The other sought to create in vivo mammalian systems that incorporated bacterial indicators. These were the "host-mediated" bioassays in which bacterial cultures were surgically implanted in live animals, who were then treated with chemicals and sacrificed, and the bacterial cultures were examined for mutagenicity (Legator and Malling 1971). Besides these two general strategies that mark the center of a swirling storm of activity, scientists designed and promoted numerous other systems, techniques, and methods for testing compounds for various mutagenic endpoints. The 111 articles that were published between 1970 and 1986 in the ten-volume series *Chemical Mutagens: Principles and Methods for Their Detection* represent probably less than half of the more than 200 test systems developed during that period (Hollaender and de Serres 1971–1986). These test systems—some of which are still in use, but many which have drifted into obsolescence—stand as the material embodiment of research activism.

In this case, the infusion of environmental values into normal bench science fueled an important shift in the logic of knowledge production that had guided mutation research for thirty years. Where scientists had once focused on a handful of highly potent chemical mutagens—because either their modes of action or their mutational effects offered insight into genetic-level phenomena—scientists now calling themselves genetic toxicologists focused increasingly on the broad spectrum of potentially mutagenic chemicals and their attendant environmental risks. Where experimental design had once emphasized a few theoretically interesting chemical mutagens, production goals in genetic toxicology, shaped by EMS commitment to prevent an increased frequency of human genetic disease, came to emphasize the rapid identification of as many chemical mutagens as possible.

What and Where Is Science Activism?

The EMS was a central player in a campaign to institutionalize a new order of environmental inquiry. In this chapter I have argued that genetic toxicology's rise during the 1969–1976 period may be traced in part to the EMS's effectiveness at maintaining a publicly visible boundary between environmental science and environmental politics while simultaneously subverting that same boundary within its own organizational domain as EMS officers and council members laid the groundwork for a new public service genetics.

This study provides additional evidence to support findings from previous research on the politics of boundary making and activism within science-based organizations. In concert with research by Moore and Hala (2002), by Kinchy (2002), and by Kinchy and Kleinman (2003), the EMS's role in making genetic toxicology reaffirms that (1) activism and boundary making in science are intertwined processes; (2) organizations play central roles in coordinating scientist collective action; and (3) to maintain credibility, scientists' organizations craft explicit boundaries between science and politics. The latter seems to hold whether the organization's identity is based on radical politics or a conservative doctrine of political neutrality.

The EMS case also differs from these earlier studies in two important respects. Where boundary work by ESA officials has been conducted to protect the interdiscipline from "environmentalism," the EMS also subverted that boundary internally as a means of legitimating both cross-disciplinary interaction and politicized science.[23] Moreover, unlike the activist-oriented SftP, which also reconfirmed the science/politics boundary in identifying its objective as "radical science," the EMS accomplished its political and institutional goals through largely conventional means. To some extent, these processes were interdependent. Stabilizing the porous and difficult-to-control boundaries that defined genetic toxicology required the authority of credible science, just as mounting a multidisciplinary attack on environmental mutagenesis similarly required a credible political campaign to make a non-obvious problem a viable issue for policy debate.

At first blush, the science activism that helped establish genetic toxicology appears utterly mundane when compared to the more familiar examples emanating from socialist movements of the 1920s and 1930s (Kuznick 1987; McGucken 1984), the student movements of the 1960s and 1970s (Moore forthcoming), the anti-nuclear movements of the 1980s (Martin this volume), and the community-based research movements of today (Moore, this volume). Looks can be deceiving, however, and the present study suggests that conventional behavior in the pursuit of contentious goals can produce significant sociopolitical change. I conclude by advancing three observations that are meant to encourage broader and more systematic thinking about what counts as science activism and where we should look for it.

First, although set in a period of political and cultural turmoil, the science activism that created genetic toxicology was not led by students politicized by the war in Vietnam or by Earth Day protests, but by seasoned professionals—geneticists mostly, along with a few biochemists, toxicologists, and pharmacologists, whose reputations and careers were firmly established. We know from social movements research that activism is risky, and so it makes sense that those scientists who engage in activism would tend to be those best positioned structurally to withstand opposition from within the profession. In science, comparative advantage tends to go to those in positions to best withstand the storms of controversy—older researchers with tenure, established reputations, economic stability, and time for politics.

Second, with a few notable exceptions, the scientists who created the EMS did not identify with radical politics. They did not form unions or march with workers, nor did they all oppose the war in Southeast Asia. The majority would not have recognized themselves as activists inclined toward what some saw as the "irrational" views promoted by domestic environmentalists. Instead, these scientists were engaged in a more oblique form of environmental activism, one that attacked the unintended consequences of chemical pollution from the lab bench and the lectern.[24] Rather than take their politics to the streets, scientist activists tailored strategies, tactics, and modes of collective action in relation to the structural conditions specific to their work. Theirs was a movement that sought to redirect research and science policy, not undermine or remake it. To that end, most of their organizational and promotional efforts were focused within their scientific communities and networks, and as often as not, inside their own laboratories and in classrooms, where it crucially mattered.

Finally, if individuals were instrumental in making genetic toxicology, organization and collective action made those individual actions effective. Although we know very little about how science activism is organized, accumulating case study evidence suggests that the emergence of something akin to protest networks among scientists is far more widespread than either science activists or science studies scholars generally assume (Frickel 2004c). If laboratories and classrooms are key sites for science activism, there can be little doubt that research to date has missed most of the action. As the histories of genetic toxicology (Frickel 2004b) and green chemistry (Woodhouse, this volume; Woodhouse and Breyman 2005) both suggest, quiet science activism organized largely beneath the public's radar can and does occasionally result in deceptively profound changes in the institutions that produce and certify knowledge.

Acknowledgments

This chapter is based on an earlier version of "Organizing a Scientists' Movement" in *Chemical Consequences: Environmental Mutagens, Scientist Activism, and the Rise of Genetic Toxicology,* by Scott Frickel. Copyright 2004, by Scott Frickel. Reprinted by permission of Rutgers University Press. I thank the RUP for permission to use this material and the National Science Foundation for funding the research under grant SBR-9710776. I also thank the scientists who took time to speak with me, in particular John Wassom, who provided me with access to key documents from his personal files. My gratitude also goes to Steven Wolf, who invited me to present my arguments at a seminar at Cornell University. Those arguments have been greatly aided by comments on earlier versions from Elisabeth Clemens, Nicole Hala, Abby Kinchy, Michael Lounsbury, Kelly Moore, and Ned Woodhouse.

Endnotes

1. NEPA was passed by Congress on December 24, 1969, and signed into law by President Nixon on January 1, 1970. Congressional environmental protection action subsequent to NEPA included the Clean Air Act (1970), the Resource Recovery Act (1970), the Water Pollution Control Act (1972), and the Federal Insecticide, Fungicide, and Rodenticide Act Amendments (1972). See Andrews 1999 and Gottlieb 1993:124–128.

2. Mutagens are exogenous chemical, radioactive, or viral agents that induce genetic change. Minutes, Meeting of the Ad Hoc Committee of the Environmental Mutagen Society, January 8, 1969; News Release, March 1, 1969, both in EMS Archives (EMSA).

3. TSCA gave the EPA administrator authority to fund and develop research programs "directed toward the development of rapid, reliable, and economical screening techniques for carcinogenic, mutagenic, teratogentic, and ecological effects of chemical substances and mixtures" as well as for research establishing "the fundamental scientific basis" of chemical screening and population monitoring techniques (Toxic Substances Control Act of 1976).

4. These studies, which examine the explicit boundaries drawn by organizational actors between science and politics, are distinct from "boundary organizations" that, as David Guston defines them, are instrumental in internalizing those same boundaries in order to smooth communication and facilitate rational formulation of science or environmental policy (Guston 2000, 2001). As I develop it here, the EMS falls somewhere between these organizational functions. I emphasize the former literature mainly because my main focus is on science and activism, rather than science and policy.

5. Frederick J. de Serres, "Detecting Harmful Genetic Effects of Environmental Chemical Pollutants," mimeographed lecture (n.d.), MS-1261, Box 3, Folder 15, Radiation Research Society Archives (RSSA).

6. Samuel S. Epstein, "The Role of the University in Relation to Consumer, Occupational and Environmental Problems," transcript of talk given at Case Western University, January 15, 1971, MS-1261, Box 3, Folder 15, RSSA; quotes are from pages 5 and 8. Alexander Hollaender, "Thoughts on Pollution," March 11, 1970, MS-1261, Box 3, Folder 16, RRSA.

7. Epstein, "Role of University," p. 11.

8. EMS report to IRS (draft, no date), MS-1261, Box 3, Folder 9, "EMS Legal Correspondence (1969)," RRSA. As EMS lawyers explained to Samuel Epstein, tax law at the time precluded organizations with tax exempt status from "utilizing a 'substantial part' of its activities in 'attempting to influence legislation,'" but that "an organization will not fail to qualify [for tax exemption] 'merely because it advocates, as an insubstantial part of its activities, the adoption or rejection of legislation.'" Blinkoff to Epstein, October 2, 1969, MS 1261, Box 3, Folder 9, "EMS legal correspondence (1969)," RRSA. On the chilling effect that the 1969 changes in tax-exemption status for nonprofit organizations had on left-wing political movements during this period, see Jenkins 1987.

9. Minutes, EMS Council meeting, July 27, 1971, Malling Papers, EMSA; Hollaender to Sobels, April 29, 1976, MS-1261, Box 3, Folder 13, RRSA. Individual EMS officers and council members could and did join environmental organizations.

10. Quoted passages are from Minutes, EMS Council meeting, October 17, 1972, Malling Papers; and Zeiger to Drake, 5 November 1976, both in EMSA.

11. EMIC Annual Report to EMS Council, March 22, 1971, John Wassom personal files; Minutes, EMS Council meeting, March 26, 1972, Malling Papers, EMSA; "Answers to SEQUIP questionnaire," March 27, 1970, John Wassom, personal files; Malling to Peters, June 4, 1970, Malling Papers, EMSA; EMIC Annual Report to EMS Council, March 22, 1971, John Wassom personal files.

12. Minutes, EMIC Register meeting, March 25, 1970; Minutes, EMS Council meeting, March 26, 1972, Malling Papers, EMSA; Minutes, EMIC Program Committee, December 18, 1970, John Wassom personal files.

13. Malling to Sobels, October 1, 1970, EMSA; Minutes, EMS Council meeting, September 18–19, 1970, Malling Papers, EMSA. On emergency funding, see Hollaender to Ruckelshaus, January 4, 1971; Kissman to Davis, January 11, 1971; Memo, Malling to EMIC staff, February 11, 1971, all in John Wassom personal files.

14. On EMS funding in 1972, see Minutes, EMS Council meeting, March 26, 1972, Malling Papers, EMSA. Genetic toxicology gained a strong institutional ally in the NIEHS in 1972 with the creation of a Mutagenesis Branch headed by former Oak Ridge geneticist and future EMS president Frederick J. de Serres. By 1974, de Serres would rename his laboratory the Environmental Mutagenesis Branch, which evolved under his leadership as an increasingly important venue for the promotion, funding, and execution of genetic toxicology research (see Frickel 2004b:59–62).

15. Ruckelshaus and Muskie both advocated for genetic toxicology to public health scientists, whereas Ribicoff recruited EMS members for policy advice. See mimeographed copy of William D. Ruckelshaus, "An Address to the American Society of Toxicology," March 9, 1971, RRSA, MS1261, Box 3, Folder 15; U.S. Senate 1971; Muskie 1969.

16. Minutes, EMS, February 8, 1969, Folder: "EMIC Archives," EMSA.

17. On mutagen review committees, see Minutes, EMS, February 8, 1969, Folder: "EMIC Archives," EMSA. Cyclamate was an artificial sweetener banned by the FDA in 1969. Hycanthone is a drug used to combat the tropical disease schistosomias; nitrosamines are found in food preservatives. On methods committees, see Minutes, EMS Council meeting, September 18–19, 1970, Malling Papers, EMSA; Nichols to Sparrow, October 2, 1971, MS-1261, Box 4, Folder 12; Nichols to Mooreland, October 2, 1971, MS-982, Folder: "Environmental Mutagen Society," RRSA. On Delaney Amendment committee, see Minutes, EMS Council meeting, October 17, 1972, Malling Papers, EMSA. The 1958 amendment to the Food and Drug Act required bans on food additives testing positive for carcinogenicity. Chu to de Serres, October 24, 1975, MS-1261, Box 1, Folder 18, RRSA.

18. Not surprisingly, there were committees created for this as well. In 1972 a subcommittee was formed by EMS to plan "for training workshops on a regular, ongoing basis" that were to be "directed at the bench scientist level" and which would "concern themselves with problems of interpretation besides techniques." Minutes, EMS Council meeting, March 28, 1972, Malling Papers, EMSA.

19. Representative data on genetics departments and graduate-level genetics curricula are difficult to track down. Graduate-level textbooks published during the 1960s and early 1970s do, however, support the contention that while mutagenesis was a basic component of genetics course work, the environmental or public health implications of chronic exposure to chemical mutagens were not.

20. Marvin Legator, "Workshop at Brown University for Testing and Evaluating Chemicals for Mutagenicity," (n.d.), p. 3, MS-1261, Box 3, Folder 2, RRSA.

21. In 1977 the NIEHS instituted extramural training programs in environmental toxicology, environmental pathology, environmental epidemiology and biostatistics, and environmental mutagenesis. The latter program, the smallest of the group, involved "minor parts" of several institutional awards in environmental toxicology at universities and medical schools receiving NIEHS training grants, and two postdoctoral fellowships (National Institute of Environmental Health Sciences 1977:13).

22. Program, "Workshop on Mutagenicity," July 26–28, 1971, Brown University; Program, "International Workshop on Mutagenicity Testing of Drugs and Other Chemicals," October 2–5, 1972, University of Zurich, both in MS-1261, Box 3, Folder 2, RRSA; Marvin Legator, "Workshop at Brown University for Testing and Evaluating Chemicals for Mutagenicity," (no date), p. 3, MS-1261, Box 3, Folder 2, RRSA; Hollaender to Garin, May 11, 1976, MS-1261, Box 3, Folder 13, RRSA; on foreign recruitment of genetic toxicology trainees, see Hollaender to Legator, March 23, 1976; Hollaender to Brown, February 23, 1976; Hollaender to Tazima, April 28, 1976, all in MS-1261, Box 3, Folder 13, RRSA.

23. This difference may be spurious for one of two reasons. Kinchy and Kleinman's (2003) work does not examine unpublished internal ESA documents and so it is uncertain whether the kinds of internalized boundary struggles evident in EMS documents also occurred within the ESA. Alternatively, we may be seeing an age effect. Following the institutional isomorphism thesis, older established professional societies may tend to adopt positions on politics that are publicly and internally consistent, while newly created organizations that are constructing institutional and disciplinary identities at the same time they attempt to gain credibility as a voice for science may exhibit less consistency.

24. My thanks to Lis Clemens for suggesting the term "oblique activism."

References

Allen, Barbara. 2003. *Uneasy Alchemy: Citizens and Experts in Louisiana's Chemical Corridor Disputes*. Cambridge, MA: MIT Press.

Allen, Garland E. 1975. "The Introduction of *Drosophila* into the Study of Heredity and Evolution, 1900–1910." *Isis* 66:322–333.

Ames, Bruce N., et al. 1973. "Carcinogens Are Mutagens: A Simple Test System Combining Liver Homogenates for Activiation and Bacteria for Detection." *Proceedings of the National Academy of Science* 70:2281–2285.

Andrews, Richard N. L. 1999. *Managing the Environment, Managing Ourselves: A History of American Environmental Policy*. New Haven, CT: Yale University Press.

Auerbach, Charlotte. 1962. *Mutations: An Introduction to Research on Mutagenesis*. Edinburgh, Scotland: Oliver and Boyd.

———. 1963. "Past Achievements and Future Tasks of Research in Chemical Mutagenesis." In *Genetics Today (Proceedings of the XI International Congress of Genetics, The Hague, The Netherlands)*, vol. 2, ed. S. J. Geerts, pp. 275–284. New York: Pergamon Press.

Beale, G. 1993. "The Discovery of Mustard Gas Mutagenesis by Auerbach and Robson in 1941." *Genetics* 134:393–399.

Brown, Phil, Steve Kroll-Smith, and Valerie J. Gunter. 2000. "Knowledge, Citizens, and Organizations: An Overview of Environments, Disease, and Social Conflict." In *Illness and the Environment: A Reader in Contested Medicine*, ed. S. Kroll-Smith, P. Brown, and V. J. Gunter, pp. 9–25. New York: New York University Press.

Clarke, Adele E., et al. 2003. "Biomedicalization: Technoscientific Transformations of Health, Illness, and U.S. Biomedicine." *American Sociological Review* 68:161–194.

Crow, James F. 1968. "Chemical Risk to Future Generations." *Scientist and Citizen* 10:113–117.

de Serres, Frederick J., and Michael D. Shelby. 1981. "Comparative Chemical Mutagenesis." In *Environmental Science Research*, vol. 24. New York: Plenum Press.

Dickson, David. 1988. *The New Politics of Science*. Chicago: University of Chicago Press.

Drake, John W. 1970. *The Molecular Basis of Mutation*. San Francisco: Holden-Day.

Drake, John W., et al. 1975. "Environmental Mutagenic Hazards." *Science* 187: 503–514.

Drake, John W., and R. E. Koch. 1976. *Mutagenesis*. Stroudsburg: Dowden, Hutchinson & Ross.

Epstein, Samuel S. 1974. "Introductory Remarks to Session on 'Mutagens in the Biosphere.'" *Mutation Research* 26:219–223.

Food and Drug Administration Advisory Committee on Protocols for Safety Evaluation. 1970. "Panel on Reproduction Report on Reproduction Studies in the Safety Evaluation of Food Additives and Pesticide Residues." *Toxicology and Applied Pharmacology* 16:264–296.

Foucault, Michel. 1980. *Power/Knowledge: Selected Interviews and Other Writings 1972–1977*. New York: Random House.

Frickel, Scott. 2004a. "Building an Interdiscipline: Collective Action Framing and the Rise of Genetic Toxicology." *Social Problems* 51:269–287.

———. 2004b. *Chemical Consequences: Environmental Mutagens, Scientist Activism, and the Rise of Genetic Toxicology*. New Brunswick, NJ: Rutgers University Press.

———. 2004c. "Just Science? Organizing Scientist Activism in the US Environmental Justice Movement." *Science as Culture* 13:449–469.

Geiger, Roger L. 1993. *Research and Relevant Knowledge: American Research Universities Since World War II*. New York: Oxford University Press.

Gieryn, Thomas F. 1999. *Cultural Boundaries of Science: Credibility on the Line*. Chicago: University of Chicago Press.

Goodell, Rae S. 1977. *The Visible Scientists*. Boston: Little, Brown.

Gottlieb, Robert. 1993. *Forcing the Spring: The Transformation of the American Environmental Movement*. Washington, DC: Island Press.

Guston, David H. 2000. *Between Science and Politics: Assuring the Integrity and Productivity of Research*. New York: Cambridge University Press.

———, ed. 2001. "Special Issue on Environmental Boundary Organizations." *Science, Technology, & Human Values* 26(4).

Hays, Harry W. 1986. *Society of Toxicology History, 1961–1986*. Washington, DC: Society of Toxicology.

Hollaender, A., and F. J. de Serres. 1971–1986. *Chemical Mutagens: Principles and Methods for Their Detection*, vols. 1–10. New York: Plenum Press.

Hollaender, Alexander. 1973. "General Summary and Recommendations for Workshop on the Evaluation of Chemical Mutagenicity Data in Relation to Population Risk." *Environmental Health Perspectives* 6:229–232.

Jasanoff, Sheila S. 1987. "Contested Boundaries in Policy-Relevant Science." *Social Studies of Science* 17:195–230.

Jenkins, J. Craig. 1987. "Nonprofit Organizations and Policy Advocacy." In *The Nonprofit Sector: A Research Handbook*, ed. W. W. Powell, pp. 296–318. New Haven, CT: Yale University Press.

Kinchy, Abby J. 2002. "On the Borders of Advocacy: The Organizational Boundary-Work of the Ecological Society of America." Paper presented at the Fourth Triple Helix Conference, Copenhagen, Denmark, November 6–9.

Kinchy, Abby J. and Daniel L. Kleinman. 2003. "Organizing Credibility: Discursive and Organizational Orthodoxy on the Borders of Ecology and Politics." *Social Studies of Science* 33:1–28.

Kleinman, Daniel Lee, and Abby J. Kinchy. 2003. "Boundaries in Science Policymaking: Bovine Growth Hormone in the European Union." *Sociological Quarterly* 44:577–595.

Kleinman, Daniel Lee, and Mark Solovey. 1995. "Hot Science/Cold War: The National Science Foundation after World War II." *Radical History Journal* 63:110–139.

Kuznick, Peter J. 1987. *Beyond the Laboratory: Scientists as Political Activists in 1930s America.* Chicago: University of Chicago Press.

Lederberg, Joshua. 1969. "Environmental Chemicals' Hazards Still Little Known." *Washington Post, p.* A 15, November 1.

———. 1997. "Some Early Stirrings (1950 ff.) of Concern about Environmental Mutagens." *Environmental and Molecular Mutagenesis* 30:3–10.

Legator, Marvin S. 1970. "Chemical Mutagenesis Comes of Age: Environmental Implications." *Journal of Heredity* 61:239–242.

Legator, Marvin S., and Heinrich V. Malling. 1971. "The Host Mediated Assay, a Practical Procedure for Evaluating Potential Mutagenic Agents in Mammals." In *Chemical Mutagens: Principles and Methods for Their Detection,* vol. 1, ed. A. Hollaender. New York: Plenum Press.

Malling, Heinrich V. 1970. "Chemical Mutagens as a Possible Genetic Hazard in Human Populations." *Journal of the American Hygiene Association* 31:657–666.

———. 1971. "Environmental Mutagen Information Center (EMIC) II: Development for the Future." *EMS Newsletter* 4:11–15.

———. 1977. "Goals and Programs of the Laboratory of Environmental Mutagenesis." *Environmental Health Perspectives* 20:263–265.

Malling, H. V., and J. S. Wassom. 1969. "Environmental Mutagen Information Center (EMIC) I: Initial Organization." *EMS Newsletter* 1:16–18.

McGucken, William. 1984. *Scientists, Society, and the State: The Social Relations of Science Movement in Great Britain, 1931–1947.* Columbus: Ohio State University Press.

Meselson, Matthew. 1971. Preface to *Chemical Mutagens: Principles and Methods for Their Detection,* vol. 1, ed. A. Hollaender, pp. ix–xii. New York: Plenum.

Moore, Kelly. 1996. "Organizing Integrity: American Science and the Creation of Public Interest Science Organizations, 1955–1975." *American Journal of Sociology* 101:1592–1627.

———. Forthcoming. *Disruptive Science: Professionals, Activism, and the Politics of War in the United States, 1945–1975.* Princeton, NJ: Princeton University Press.

Moore, Kelly, and Nicole Hala. 2002. "Organizing Identity: The Creation of Science for the People." In *Social Structure and Organizations Revisited,* vol. 19, ed. M. Lounsbury and M. J. Ventresca, pp. 309–335. Amsterdam: JAI Press.

Muskie, Edmund S. 1969. "Chemicals, the Toxicologist, and the Future of Man." *Forum for the Advancement of Toxicology* 2:1, 3.

National Institute of Environmental Health Sciences. 1975. "Annual Report." National Institute of Environmental Health Sciences, Research Triangle Park, NC.

———. 1977. "Annual Report." National Institute of Environmental Health Sciences, Research Triangle Park, NC.

Neel, James V. 1970. "Evaluation of the Effects of Chemical Mutagens on Man: The Long Road Ahead." *Proceedings of the National Academy of Sciences* 67:908–915.

Pauly, Philip J. 2000. *Biologists and the Promise of American Life*. Princeton, NJ: Princeton University Press.

Preston, Julian R., and George R. Hoffman. 2001. "Genetic Toxicology." In *Casarett and Doull's Toxicology: The Basic Science of Poisons,* ed. C. D. Klaassen, pp. 321–350. New York: McGraw-Hill.

Research Triangle Institute. 1965. "Recommendations for the Development and Operation of the National Environmental Health Sciences Center." Department of Health, Education, and Welfare, U.S. Public Health Service, Bureau of State Services (Environmental Health), Research Triangle Park, NC.

Sellers, Christopher C. 1997. *Hazards of the Job: From Industrial Disease to Environmental Health Science*. Chapel Hill: University of North Carolina Press.

Shostak, Sara. 2003. "Disciplinary Emergence in the Environmental Health Sciences, 1950–2000." Ph.D. diss., Department of Social and Behavioral Sciences, University of California–San Francisco.

Straney, Sister Margaret J., and Thomas R. Mertens. 1969. "A Survey of Introductory College Genetics Courses." *Journal of Heredity* 60:223–228.

Toxic Substances Control Act of 1976, Public Law 94–469, 94th Cong., 2nd sess. (11 October 1976).

U.S. Department of Health, Education, and Welfare. 1969. *Report of the Secretary's Commission on Pesticides and Their Relationship to Environmental Health*. Washington, DC: U.S. GPO.

U.S. Senate. 1971. "Chemicals and the Future of Man: Hearings before the Subcommittee on Executive Reorganization and Government Research." 92nd Cong., 1st sess. Washington, DC: U.S. GPO.

Wassom, John S. 1989. "Origins of Genetic Toxicology and the Environmental Mutagen Society." *Environmental and Molecular Mutagenesis* 14, Supplement 16:1–6.

Woodhouse, Edward J., and Steve Breyman. 2005. "Green Chemistry as Social Movement?" *Science, Technology, & Human Values*, 30:199–222.

8

Changing Ecologies

Science and Environmental Politics in Agriculture

CHRISTOPHER R. HENKE

What does it take to make a change? Applied science, by definition, is meant to change things in the world, but is science an appropriate and practical mode of social change? This is a pressing question for agricultural science, an area of research that, over the past century, has contributed to radical changes in how food is produced, but at the same time has also played a major role in the creation of environmental problems related to agricultural production. Mechanized farm equipment, synthetic pesticides and fertilizers, and intensive production practices have allowed ever fewer farmers to produce ever more food. These same techniques and technologies, however, have led to a number of environmental problems, especially related to pesticide and nitrate contamination of water sources (National Research Council 1989: chapter 1). As a result, agricultural scientists now find themselves in a strange position: they are faced with the challenge of reordering a system that they have constructed and promoted as especially efficient and rational, while mitigating its environmental impact.

This chapter explores the politics of change in agriculture and the elements that enable and constrain agricultural scientists as agents of environmental change. I seek to understand how scientists who work closely with industrial interests balance social order and change: is it possible for scientists to work within a structure and, at the same time, change it? Social theorists hold mixed views on this question. Perhaps the most well-known of these theorists is Ulrich Beck, whose "risk society" thesis holds a very dim view of the ability of scientists to address environmental problems (Beck 1992; Beck, Giddens, and Lash 1994). Beck claims that science and technology are institutions at the heart of industrial modernity, responsible for generating environmental risks in the first place. Further, and more sinisterly, science and technology are often deployed as a kind of "'counter-science' gradually becoming institutionalized in industry" and beholden to the interests of capital (Beck 1992:32; Van Loon 2002; see also Rycroft 1991; Luke 1999; and Fischer 2000). In contrast to Beck's risk society thesis, ecological modernization theory places science and technology at the center of attempts to create a "green" modernity, presuming that scientists and other experts will, in concert with vast changes to the regulatory state, be a vanguard of environmental change (Spaargaren and Mol 1992; Mol and Spaargaren 1993; Hajer 1995; Mol 1996).[1]

And yet, despite these concerns about the ability of science and technology to address environmental problems, there has been very little empirical research on the role of experts in environmental conflicts.[2] How do scientists really deal with environmental problems? Though there have been attempts to theoretically bridge risk society and ecological modernization theories (Cohen 1997), I argue that it is necessary to empirically explore the actual factors that either promote or constrain scientists as agents of environmental change. Overall, science has been treated as an institutional black box in the theoretical debates over modernity and the environment: instead of critically examining the place of science in specific environmental conflicts, many of these authors have assumed an overly simplistic view of scientific practice (Wynne 1996).

My aim in this chapter is to open the black box and investigate the role of scientists in environmental conflicts, drawing on my ethnographic fieldwork with a group of agricultural scientists and their attempts to improve the environmental sustainability of farming practices in California. I argue that agricultural scientists working to mitigate environmental problems related to farming can act as agents of social change, but that their work in this regard is tightly circumscribed by a heterogeneous "ecology" of sociomaterial elements (Star and Griesemer 1989; Star 1995; Henke 2000a). This ecology structures the relationships between actors and institutions, defining the power relations among them, and provides a context for potential change. Agricultural scientists use this context strategically, manipulating elements in the larger institutional ecology of agriculture to frame the perception of environmental problems. Though this strategy is a fundamentally conservative technique of social change, it also allows scientists to "go local," and develop close relationships with some of the largest contributors to farm-based pollution.

To make this case, I draw on data from my ethnographic fieldwork with University of California (UC) Cooperative Extension "farm advisors," agricultural scientists employed by the university system but stationed in specific counties throughout the state.[3] These advisors are charged with advising their local farm clientele on ways to improve farming practices in their county, including ways to make farming more environmentally friendly. In fact, the farm advisors with whom I worked (described in more detail below) devoted a considerable amount of their research efforts to studying and promoting more environmentally friendly alternatives to the farm industry. Therefore, the advisors provide a window onto the relations between scientists and business interests vis-à-vis environmental conflicts. In the following section I describe in more detail the ecological framework I use to understand farm advisors' work and provide additional context on the history and politics of farming and farm advising.

The Sociomaterial Ecology of Farming and Farm Advising

The modern U.S. history of environmental problems in agriculture is also a history of the relationships between agricultural scientists and growers, especially their cooperative attempts to increase farm production. During the approximately 150-year history of professional agricultural research, scientists have focused on increased farm output as their main ambition. This goal has led to a system of farming based on rationalized farm practices and the use of technologies such as mechanized equipment and synthetic pesticides and fertilizers. At the same time, the growers who remain in U.S. agriculture are largely those who have supported and adopted this system.[4] In this context, then, agricultural scientists and the farm industry are more than "cozy"—they are two sides of the same coin, partners in creating the massive built environment that is modern U.S. agriculture (Busch and Lacy 1983; Marcus 1985; Kloppenburg 1988; Kloppenburg and Buttel 1987).

This close collaboration, however, has not precluded conflict from the grower-scientist relationship; the more specific history of Cooperative Extension shows that tensions have always been inherent in attempts to change farm practices. Cooperative Extension was created as a national program in 1914 through the U.S. Department of Agriculture (USDA). Attached to the land grant university systems in most states, farm advisors are funded through the "cooperative" allocation of funds from the USDA, the land grant universities, and the specific counties where advisors are stationed. Cooperative Extension is intended to improve the productivity of American agriculture, bringing the "university to the people" and thereby applying the discoveries of modern agricultural science to farms throughout the nation. Despite these intentions, Cooperative Extension has often received a mixed response from rural folk, especially in the decades just after its creation (Scott 1970; Danbom 1979, 1995; Rasmussen 1989; Fitzgerald 1990; Kline 2000). As a result, farm advising remains a somewhat awkward profession, where grower interest and expert advice do not always neatly match. Unlike the USDA's inspection agencies, which regulate food processing and the movement of food products into and out of the United States, farm advisors have no regulatory power—they just give advice. Charged with helping their local farm communities but given no real authority to direct the practice of farming, advisors need to find methods of producing consent around new techniques and technologies that do not rely on formal (legal-rational) authority and power relations.

For my own fieldwork, I worked with a group of farm advisors stationed in California's Salinas Valley (see Figure 1).[5] Known to many as the birthplace of John Steinbeck and the setting for many of his stories, the Salinas Valley is also one of the most productive vegetable-growing regions in the world. Though the valley contains only about 200,000 acres of farm land, it accounts for a majority of the U.S. production for several vegetable crops. Overall, Monterey County, which comprises most of the Salinas Valley, has an agricultural industry that grosses about 2 billion dollars a year, and about a fourth of that (around 500 million dollars) is from iceberg lettuce alone. Table 1 shows the importance of Monterey County as a leading producer of vegetable crops for California and the nation.

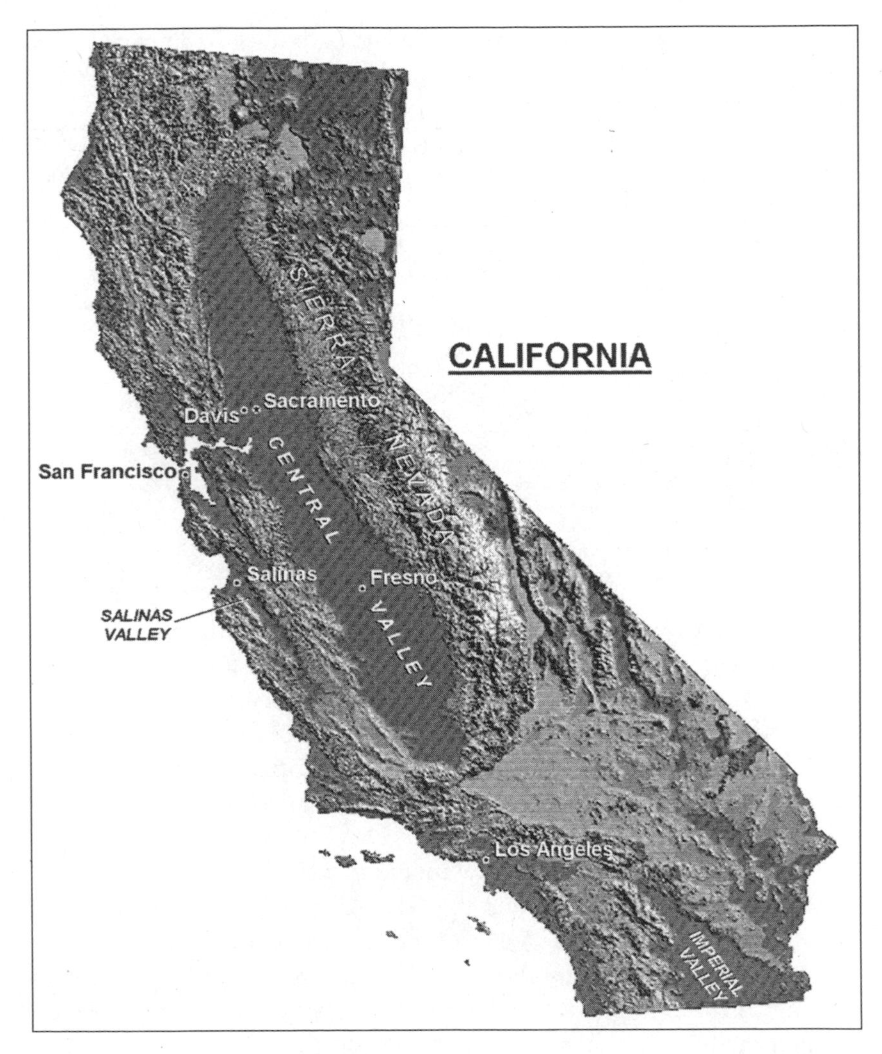

Figure 1. California. Map template courtesy U.S. Geological Survey, www.flag.wr.usgs.gov/USGSFlag/Data/maps/CaliforniaDEM.html.

I term Salinas Valley vegetable production an agricultural "industry" purposefully, to emphasize the intensity of the production practices used and the high level of capitalization of the growers. Although many of the major companies that dominate farming in the Salinas Valley are family-owned, they are not "family farms" in the sense that we usually use that term. Most of the farm industry firms have complex divisions of labor, just like any other medium- to large-sized business. Growers in

Table 1. Vegetable Crops in Which California Is the Predominant U.S. Producer, Including Production Share, Cash Value, and Leading California County, 1998

Commodity	CA Share of U.S. Prod. (%)	Total Value ($1,000)	Leading County in California	County's Share of CA Prod. (%)
Artichokes	99	62,899	Monterey	71.4
Broccoli	92	502,875	Monterey	53.1
Cauliflower	87	196,696	Monterey	56.4
Celery	93	207,515	Ventura*	49.6
Lettuce, all	70	1,114,295	Monterey	52.6
Spinach, fresh	80	86,640	Monterey	69.5

Source: California Agricultural Statistics Service, "California Agricultural Statistics," www.nass.usda.gov/ca/bul/agstat/indexcas.htm (accessed June 20, 2003); "Summary of County Agricultural Commissioners' Reports," www.nass.usda.gov/ca/bul/gross/indexacv.htm (accessed June 20, 2003).

*Monterey County ranks second in celery, with 33% of California's total production.

the most lucrative crops often form their own "commodity boards," which serve to self-tax the industry and collect money for marketing and research. For example, there is a Lettuce Board with offices in the city of Salinas that collects a small amount of money from each carton of lettuce its members produce. The Lettuce Board then distributes these funds to university and private researchers through a competitive grants program.

This money spent for research includes substantial support of UC agricultural scientists, including the farm advisors. In the case of the farm advisors in the Salinas Valley, the majority of their research funds came from various commodity boards and other industry sources of funding. Although the UC pays for their salaries and the county pays for their facilities, the farm industry funds almost all of the advisors' research expenses. In all, although the farm advisors are scientists employed by the state, they often have much more contact and closer ties with people in the farm industry than with other UC researchers in other parts of the university system. Therefore, farm advisors make a very good case for discussing the kinds of tensions involved when scientists work on environmental problems, trying to understand and work within a community of industrial interests, promoting change.

The question then remains: how do they do it? In previous work, I developed an "ecological" conception of agriculture and agricultural science to understand how farm advisors use place-bound forms of experiment called "field trials" to demonstrate the value of a new practice, attempting to produce consent around it (Henke 2000a).[6] Advisors use field trials to account for the specific character of local farming ecologies, which include a very heterogenous set of elements, such as land, plants, labor, technologies, and practices. A successful field trial incorporates these elements of the farm ecology and uses them to influence change, most often in just one part of the overall system. In order to convince growers of the benefit of a new practice or technology, advisors need to "see" their farm community as a sociomaterial ecology and strategically position their research within this network to influence change.

Environmental problems in agriculture further complicate this process and require an expanded sense of what constitutes a farm ecology, including institutionalized and emerging relationships between scientists, the farm industry, and the regulatory state. These relationships are themselves a kind of ongoing, negotiated order, and the story of farm advising in the late-twentieth and early-twenty-first centuries is in large part the story of the shifting politics surrounding farming and environmental problems. Although political and economic factors at the meso- and macrosociological level have always been important factors for advisors to consider when promoting new farming methods (Henke 2000a:492–493), the increasing role of the state in regulating farm practices means that advisors must further expand how they see the farm ecology, to include the connections between local practice and state power in their work.[7] By their nature, environmental problems are systemic and often defy an easy solution related to small, incremental changes. Further, there can often be disagreement about the ontology of the "problem," where growers may see an environmental problem as a "political" problem, abstracted from the local ecology of farming. In this context, advisors still promote change through the demonstration of new techniques, but they also frame this work against the "threat" of action by the regulatory state. The following section describes this process of problem definition in more detail.

Defining "Environmental Problems" in Agriculture

So far, I have used the category of an "environmental problem" somewhat unproblematically, and a clarification of this term will help establish how both growers and agricultural scientists understand the meaning of these problems. Like any social problem, environmental problems in agriculture are defined through a diverse set of interests, including the scientists and growers who are the main actors in this story, but also governmental regulatory agencies, environmental activist groups, and community organizations. These groups alternately promote and decry an issue in "public arenas," shaping its definition as a problem (or not). Thus, an environmental problem follows a kind of life course that attracts attention from both these groups as well as, perhaps, a wider public (Hilgartner and Bosk 1998).

This struggle to define environmental problems in public arenas is always connected back to those specific local practices that are at the heart of the debate: Does some way of farming lead to pollution or other forms of environmental degradation? If so, is there an appropriate regulatory response? The connection between problem definition, the particularities of a farm ecology, and regulatory action makes the definition of environmental problems a subject of intense interest among growers. Because agriculture is a complex sociomaterial system, just about any kind of problem associated with it *could* count as an environmental problem. In the case of the farm industry, however, their perception of environmental issues as problems often takes a different form than their perception of problems associated with insects and plant diseases. If a bug is eating a grower's crop, then this is an immediate threat that may ruin his or her yield. Environmental problems, though, are not always this apparent; there is often conflict over whether a problem exists at all, much less what the best solution is. Of course, environmental issues in agriculture are directly related to issues of production, because the problems are the effect and production is the cause. For the major environmental problems related to farming in Monterey County, however, there appears to be no direct impact on production. Instead, it is the threat of increased governmental regulation, loss of certain chemical controls and negative publicity that are the real menace in the eyes of the farm industry. In this respect, it is not really the environment that is a problem for growers, but rather environment*alism*. Therefore, farm industry responses to environmental problems are often just as much concerned with spinning a positive image of agriculture and deflecting potential regulation as they are about actually addressing the problems.

This does not necessarily mean that growers are always hard-hearted when it comes to environmental issues, but it merely reflects their own struggles to maintain control over production practices and the larger sociomaterial ecology of their industry. Recognizing this position is essential for understanding the tricky relationship between agricultural scientists and the farm industry with respect to environmental issues. Many of the growers I interviewed explicitly termed environmental problems as political problems and cited the state as the source of unwanted—and, in their view, often unneeded—intervention in the practices of the farm industry. For example, this grower[8] told me a story about testifying before a California state commission, asking approval to use a pesticide in his industry:

> CRH: Were you involved with any other problems that weren't necessarily related to production—say political issues or things like that?
>
> GROWER: Well, a little bit. When I first started in the pesticide field, we used to go to Sacramento and testify on why we needed [approval to use a pesticide]. . . . After doing this two or three times, I happened to have an opportunity to speak to one of the people sitting up in front of us—I guess you could call it the board or whoever was evaluating [our request]. And I happened to bump into him after that, and I said, "What happened up there? You sat on the panel up there, and I'm just curious as to what happened. Because the state never did ask us anything, so what are we doing here?" [laughs]. And this guy was a qualified person. He said something like, "Hey, they had their minds made up before you got here." Now that's [just] politics, [but] I didn't do it [again] after that.
>
> CRH: It just seemed like a waste of time?
>
> GROWER: Well, to me it was a waste of time because . . . [the state] has their own people and they make their own decisions, but, according to the law, they have to have hearings. And I think they have hearings sometimes just [to satisfy] the law.

Similarly, in another interview a different grower implicated the UC system itself as being overly involved with the politics of environmental change. In this excerpt, the grower complains that urban politicians with a dilettantish interest in nonconventional farming have better access to the UC than the farm industry:

> GROWER: [The university will] listen to some dingbat from San Francisco who wants to have sustainable organic agriculture

> on the roofs of the apartment buildings in San Francisco. Because [the dingbat] will vote against you if he doesn't get that in. And so you will devote political time and devote extension time to do that, because that's where the vote is. I understand that's the site of the problem. You know, I just think that somebody in the hallowed halls of Berkeley or Davis could say, "Screw you—we're not doing that—go away." But they never do.

In each of these cases, the growers portrayed state actors as "political" agents who based their decisions on political convenience, rather than an understanding of the ecology of the farm industry. I often heard these complaints expressed in terms of a distinction between those actors "inside" the farm industry, with knowledge of the local conditions of farm production, versus "outside" actors, without this knowledge. Typically, growers portrayed government regulators as well-intentioned but naive outsiders at best, and manipulative zealots at worst (in the case of the grower's quote above). These outsiders, the insiders argued, lacked a full understanding of the local conditions of agricultural production, acting as regulators who wanted to change the ecology of agriculture without understanding how politics related to practice. Thus, the growers argued, the state's regulatory solutions were based not on information or familiarity, but instead were grounded in uninformed, politically motivated decisions handed down from above.

This way of framing environmental problems represents a problem in its own right for the UC researchers who are working on solutions to some of the these problems—especially the farm advisors. As employees of UC Cooperative Extension, the advisors are charged with improving certain areas of agricultural production, and environmental problems are a significant part of their efforts. At the same time, however, they are not an "official" part of the regulatory state, even though they are employed by the state. Growers' attitudes further complicate this situation: though advisors may feel that they are acting in the "interest" of the farm industry by working on environmental problems, there is a danger that growers might conflate advisors' work on environmental problems with a kind of political agenda identified with environmental activism and the regulatory state. In practice, this means that advisors' research must be perceived either as "neutral" science, providing key information to influence new practices, or as a kind of moderate environmental politics based on "leadership," an attempt to guide the industry through the perils of change for its own good.

As a consequence of these dynamics, one might expect that advisors would act in accordance with the cynical view of science and environmental politics—essentially acting as an appendage to industry and conducting research that supports its short-term interests (as in Beck's view). After all, despite the structural and financial links advisors have to the local farm industry, their status as "insiders" is not granted automatically. On one hand, their local position in the Salinas Valley helps to ensure them insider standing, but on the other hand, a strong association with environmental concerns and activism is one of the easiest ways to mark oneself as an outsider. Overall, in a situation where the boundaries have been drawn this starkly, the simplest choice would likely be to position oneself on the "safe" side of this boundary, especially given the close relationship that advisors have with the farm industry (and their relative lack of contacts with other, campus-based UC researchers).

Instead, however, during my research with the farm advisors I found that their work—and the way that they talked about it—was more complex than this. On one hand, advisors spend a great deal of their time working on environmental problems, testing and promoting new, "environmentally friendly" techniques of production to their grower clientele. On the other hand, they strove to moderate their interest in environmental problems by claiming that their work on these issues was itself in the long-term interest of the farm industry. Like the growers I quoted above, they often justified this claim by contrasting their approach with the methods of the regulatory state. To better understand how advisors balanced their work with the farm industry and their attention to environmental issues, I asked them whether they considered themselves "environmentalists" in interviews. I expected that they might feel somewhat uneasy applying the term to themselves, given the kind of radical connotations often attached to it.[9] In fact, all the advisors were at pains to express their environmentalist sensibilities as in balance with other considerations. Here are two interview excerpts that illustrate this point:

CRH: Do you think of yourself as an environmentalist . . . ?

GRAPE ADVISOR: I think those principles of being concerned about what you're doing—if you want to call that an environmentalist—I guess when you say environmentalist a lot of people would have a different view of what that may potentially be.

CRH: People out there chaining themselves to trees?

GRAPE ADVISOR: I'm not out there chaining myself to someone's spray rigs so they won't spray a pesticide [CRH laughs]. That's what a lot of people have envisioned as environmentalists, that they're a bunch of crazies. But I think if your definition is someone who is concerned about what is happening . . . I guess yeah. We have a lot of words like that that mean different things to a lot of different people [laughs].

• • •

CRH: So, do you think of yourself as an environmentalist?

PLANT PATHOLOGY ADVISOR: I suppose so, although those are loaded terms in the political world. I'm not an environmentalist in the true sense of the science. I'm a plant pathologist, and to me an environmentalist in the science world is more systems-oriented in their research. But if you are referring to the political term of someone who is promoting and defending the environment, certainly, I share those concerns and would put myself in that circle, in that broader sense. [By] the same token, I'm not really an activist—I choose not to be too involved politically and actively in terms of those issues.

In both cases, the advisors recognized themselves as environmentalists but were careful to avoid a definition of the term that was overtly "political"—each framed his environmental conscience as an agenda that did not cross into protest or other forms of public statement. The key strategy for advisors is to couch their concerns about agriculture's environmental impact in terms of a kind of balanced "concern," sensitive to each side of an issue but giving full allegiance to neither. This excerpt, taken from the same interview with the Grape Advisor who, above, qualified his identity as an environmentalist, further explores this sense of concern:

GRAPE ADVISOR: Yeah, [environmental issues] are a big concern. I think we do get drawn in—sometimes, they can be somewhat political issues. But if you're drawn in, you're drawn in as kind of an unbiased source. Not too long ago [the county] had some concerns about [farming on] hillsides. And [I got] contacted from growers and . . . from the [county] Planning Department. You get involved in—I guess you could call them political actions—but it's more because of what you know. Not because you're out there trying to promote one side or the other. I think most of us should stay pretty much neutral in those political type of issues.

Though both of these advisors staked out positions that precluded what they perceived as protest or other radical measures, their sense of concern, nonetheless, manifested itself in a form of response. Although they did not necessarily characterize this reaction to environmental problems as "activism," it is clear that the advisors thought out the bounds of their concern, an appropriate set of possible responses, and how to "sell" these possibilities to the farm industry. In sum, advisors typically framed this process as "leadership"—subtly, yet actively, leading the farm industry toward more environmentally sustainable practices. In another interview excerpt, an advisor specializing in entomology described his view of leadership on environmental problems:

> CRH: Can you say generally your sense of how environmentalism, the environmental movement, and environmental regulations from the state have had an impact on your job?
>
> ENTOMOLOGY ADVISOR: That's a really good issue because it's a situation where you can either be a leader or you can just follow along. And I think if you're not willing to accept that environmental issues are important and that they are going to change the way agriculture is done, then you're just going to be following along and you're going to get left behind. I think, in this time, you have to realize that there are changes occurring and you have to help the growers deal with those. And you have to find a way to present those issues to the growers that makes them realize that this is something they have to deal with whether they like it or not. . . . As long as you explain to [growers] that you're doing this because you feel it's a future direction and something that's of importance instead of just saying, "What you're doing is wrong," I think it can be done very carefully. . . . And I think that the farm advisors should be responding to [environmental issues] and trying to be in a proactive mode rather than strictly a reactive mode.

In this excerpt, the advisor describes his leadership as both environmental and political; he implicitly invoked the role of the state and environmental regulation as a "future direction" of California agriculture. At the same time, he acknowledged that his own place in the ecology of farming requires that this work be "done very carefully," given that he is an advisor, not a regulator. In this sense, he tried to make the case for environmental change by treating it as an impending reality for which he could provide leadership, shepherding the industry through a period of turbulent political change.

In each of the statements above, the advisors strove to balance their interests in addressing environmental problems with the interests of the farm industry. This is not to say that each group's interests can be neatly sorted out, or even that there is a consensus among, say, growers, on the definition of environmental problems and the appropriate response. But the overall definition of environmental problems and appropriate solutions called for an understanding of the larger farm ecology, an integration of local practice with moderate politics. Further, the advisors' attempts to cultivate a balanced "concern" are themselves the consequence of a complex set of interests, perhaps the most important of which is preserving their local connections with and influence among the farm industry. In sum, this complexity troubles any overly simplistic view of how agricultural scientists conceptualize and respond to environmental problems related to farming. In the next section, I treat a specific example in more depth.

"A HARD SELL": THE CASE OF NITRATE CONTAMINATION IN GROUNDWATER

In 1998 the U.S. Environmental Protection Agency (EPA) designated agriculture as the largest nonpoint source[10] of water pollution in the United States (U.S. EPA 1998), and pollution from pesticides and fertilizers was thereby seen as a "problem" by many environmental groups and state regulatory agencies. At the time of my research, these problems were also present in the Salinas Valley, where most of the land used for growing vegetable crops is farmed very intensively, producing two or three crops per year. This level of production is highly dependent on synthetic fertilizers and pesticides, and the environmental impacts associated with these products brought unfavorable attention to the farm industry. The example I describe in this section revolves around farm advisors' attempts to decrease the use of fertilizers in the valley. Groundwater taken from wells throughout the region was found to have unsafe levels of nitrate contamination, the primary ingredient in synthetic fertilizers. High levels of nitrate, when consumed in drinking water, can interfere with respiration, especially in the very young and very old. Wells throughout the valley were designated unfit for drinking water, and during my main period of research, in 1997–1998, the nitrate contamination issue was receiving considerable attention in the local press. Residents in some of the poorest areas of the valley were generally affected most by the contamination, and the county was trucking in drinking water for their use. Local community groups were petitioning the county to drill new, deeper wells, and the issue was eventually publicized in the *San Francisco Chronicle*, to the dismay of many in the farm industry (McCabe 1998a, 1998b). Because of the severity of the problem and the publicity, some kind of regulation seemed imminent, and the vegetable industry appeared to be one of the most likely sources of the contamination.[11]

California's vegetable industry, in a sense, grew up with the fertilizer industry, especially beginning in the 1930s and 1940s, when synthetic fertilizers became more widely available. At the time, Cooperative Extension farm advisors were among the greatest champions of this new technology, using field-based demonstration trials to show that, for a modest cost per acre (relative to the potential sale price of the crop), growers could boost their yields and potential profits substantially. For vegetable growers in the Salinas Valley, whose crops are very expensive to produce but can also bring large profits when commodity prices are favorable, the use of synthetic fertilizers was promptly adopted throughout the industry. Further, as growers gained more experience using fertilizers, they learned that overapplication of the product does not harm vegetable crops but instead acts as a kind of "crop insurance," maximizing yields for a slightly increased cost of inputs. Therefore, the treatment of vegetable crops with fertilizer, often several times per crop season, quickly became a standard practice throughout the vegetable industry. This practice also became institutionalized in the fertilizer industry, which offers growers an application service, spraying fertilizer on the grower's crop at fixed intervals throughout the growing season. The environmental downside to this practice, however, is that any nitrogen or other nutrients that are not absorbed by the plants' roots may eventually leach into the water tables below the fields.

The solution to this problem seems deceptively simple: just get growers to use less fertilizer on their crops, and then less nitrate will leach into the valley's water supply. This, in fact, was the approach taken by a team of researchers working to address the nitrate problem in the late 1990s. The research team consisted of one of the Salinas Valley farm advisors, a UC extension specialist, and a few other researchers working with county-level administrative agencies; they were supported in part by grants from a statewide fertilizer industry association. The team developed and pushed for growers to begin using a "quick test" soil sampling system for monitoring their soil and making better judgments about the need for fertilizer at a given time. As the name implies, the quick test system is a relatively cheap and simple way to check the level of the nitrate already present in the soil, thereby allowing growers to fertilize only when required by the physiological needs of the plant, rather than a fixed program of fertilizer application that would be applied regardless of need. In this way, the quick test was designed as a new form of "management" that would serve as an alternative to the application schedules sold by the fertilizer industry.

Based on field trials of the quick test system on several farms throughout the valley and other lettuce-growing regions of California, the researchers found that the system could be used to effectively reduce fertilizer use while maintaining the same yields. Despite the promise of the field tests, however, and the potential to save a bit on fertilizer bills, the research team had a difficult time convincing growers to implement the quick test system in their own fields, and few growers ultimately adopted it. The quick test had run up against the standard practice of lettuce growers, which is to overfertilize as a kind of crop insurance, and the research team could not guarantee that risk had been completely eliminated. Thus, as this grower explained, many people in the industry remained skeptical of the new approach, even when field trials using the quick test method had shown initial success:

CRH: It seems like on [certain] problems, that Extension can be pretty helpful, but are there other kinds of things where it's not always as clear whether [Cooperative Extension] is gonna be helpful or not?

GROWER: Yeah, I think in some cases—Monterey County has a nitrate problem, and the state is rattling their saber about that. But it's a problem that does have to be cured and solved. Extension probably is not as capable of solving that type of problem because it's not a day or night situation. It's very long-term, because what you did two years ago probably still affects what you're doing today. And, it's very subtle, what your results are gonna be. And so they probably do a poorer job of that than they do on other stuff. [UC Specialist's] data shows that if you test [the soil] . . . you may be able to reduce how much nitrogen you put on by as much as 50 percent in

> some cases. . . . He's done this experiment two or three times, and it hasn't been accepted by the industry. Because nobody's gonna skip a $40 fertilizer application and possibly lose everything they got out there. They would rather make sure they have enough or too much. And, it's hard to sell that on paper. . . . Because everybody knows, well you know: that was this time but what about next time? You don't get paid for blowing a $2500 crop because you were trying to save $40. It's not their fault that Extension isn't effective—the research has been good and the results have been proven. But it's subtle, and it's very difficult for Extension to do that well.

This grower described the quick test's results as "subtle," but what makes them subtle? For this grower and many others, the quick test represented an "edge" of change, marking off the institutionalized, industry practice (overfertilize as insurance) and the threat of more drastic change instigated by regulatory agencies. The subtlety stemmed from an uneasy sense of being placed on this edge without a clear and easy decision. One might argue that this grower and others were simply responding to an economic calculation about the risks and benefits of change, and it is true that the grower made a very direct reference to the costs saved from using the quick test compared with the larger potential sales price of the crop. This decision, however, is highly context dependent. The definition of the quick test as either a cursory act of maintenance or a more radical transformation is shaped by the larger ecology of the farm industry as a whole—especially the grower's sense of risk in terms of regulation. In a sense, this grower was already thinking about the realities of the nitrate problem and even mentioned that the state was "rattling their saber," alluding to the threat of increased regulation. Further, the fertilizer industry itself funded much of the research and development costs of the quick test system, again, because of concerns about state regulation of fertilizer use. In all, the "subtleness" pointed to this grower's concerns about a changing farm ecology, where risks of change and the risks of regulation made an uncertain balance between the institutionalized practices of fertilizing and the politics of environmental change.

The members of the research team understood that the quick test was on this edge of change, and, in fact, this was actually their argument for the system's implementation. They argued that the quick test was a balanced approach that accounted for grower practice but also made changes that would appeal to regulators, perhaps even forestalling regulatory action. This next excerpt, from an interview with the state specialist working on the quick test team, illustrates the balance the team was trying to achieve and also the context-dependence of the choices:

> UC SPECIALIST: It's less compelling for a farmer to make changes in fertilizer use [with the quick test method] than if I was to tell them that they could change and increase production. I'm telling them that by monitoring their use of [nitrogen] they can save about 50 dollars per acre on a crop that may cost around 2000 dollars per acre and yield much more. This is a lot different than telling them that there is a new irrigation method that could increase their yield by 15 percent. This would really make them pay attention. On the other hand if the California EPA hauls [the industry] into court on a class-action suit, or starts taxing water use, or some other punitive method of dealing with nitrates, then I will suddenly be their best friend.

In this excerpt, the specialist pointed to the potential for more radical state regulatory action to change growers' attitudes toward the reduction of fertilizer use, and the specialist also used this point when speaking to growers about the nitrate issue. In one meeting that I attended, the specialist explicitly raised the possibility of imminent state regulation as an implicit reason that growers should begin managing their own fertilizer use, before the state forced them to do so. In this way, the specialist echoed the advisor quoted in the previous section, who talked about the need to "carefully" lead growers toward change. Though the specialist here portrayed himself as *potentially* the industry's "best friend," should the state decide to regulate fertilizer use, he also implied that he was already working in the best interests of the industry.

Ironically, although this specialist and the farm advisor working on the project both agreed that regulation would be the quickest way to change grower practices, neither of them was very confident in the ability of the state to implement a program that would effectively balance the needs of both the environment and the farm industry. Just as growers expressed a wariness regarding the intentions of regulators and their potential misunderstanding of the larger ecology of agriculture, so the researchers also declared similar worries:

> SOIL & WATER: The state will probably step in pretty soon, because something like half of the groundwater in the southern

part of the valley is unsafe to drink. The problem is, you can't just stop putting fertilizer on crops. It doesn't matter how clean your water is if you don't have any food to eat.

• • •

CRH: What about the reaction [to the initial success of the quick test system] from the regulatory agencies—they must be sitting up and saying, "We were right."

UC SPECIALIST: They are excited by it, but I've tried to work with them to get them to see the realities of production ag. These agencies are mostly staffed with people who have backgrounds in biology or chemistry, but have no experience or sense of the industry. They need to have an idea of how growers live their lives and do their work. The growers are not Luddites or [intentional] polluters. They have minimal profits and already have a lot of regulations and paperwork to deal with.

These two research team members implied that an outright ban on fertilizers would be impractical, regardless of growers' hesitance in adopting quick tests. Once again, the researchers described their roles as leaders for change, but for balanced change. In addition, they also talked about providing leadership to both the industry *and* the regulatory state, trying to make them see the "realities" of the full farm ecology. Just as the grower I cited was on what he called a "subtle" edge of change, the research team hoped to use this edge itself as a selling point and to renegotiate the meaning of the quick test method as a reasonable and balanced mode of change. This process can be frustrating: at this stage in the negotiations over environmental change, the quick test did not satisfy the ultimate ideals of either the industry or the regulators.

When I returned to conduct follow-up interviews with the Salinas Valley farm advisors in the spring of 2003, little had changed. Though nitrate contamination of groundwater was still a legitimate problem in the valley, other problems related to water and agriculture had moved to the fore and diminished the importance and urgency of the nitrate problem.[12] While at the beginning of my research, both the regulatory agencies and the farm industry were worried about lawsuits and negative publicity stemming from the nitrate contamination problem, these newer issues were now the focus of growers' and regulators' worries. As a result, growers were still showing little interest in the quick test system, and a grower survey conducted by a Monterey County water resources agency in 2001 showed that the quick test was rarely used as a management tool for regulating the use of fertilizers—although 78 percent of respondents claimed that they used some method of testing to assess the amount of residual nitrogen in their soils, only 3 percent of respondents used the quick test. This suggested that most growers were using slower laboratory-based testing, most likely only once per season, instead of the continual "management" approach suggested by the quick test's promoters. Further, more than half the growers said that they used private consultants for advice on fertilizer use, including 50 percent of growers who reported a reliance on consultants from the fertilizer dealers themselves (Monterey County Water Resources Agency 2002).

In addition, the farm advisor who specialized in soil and water problems in the Salinas Valley had resigned, and a new farm advisor was hired to work on these issues. This new advisor had few hopes for the use of the quick test as a method of reducing fertilizer use, claiming that growers were "not comfortable with it" and felt it was too "low-tech" when compared with the sophisticated systems they use on their valuable crops. In this sense, the quick test system failed as an edge of change in at least two respects. First, it seemed too simplistic and potentially unreliable to growers whose bottom line depended much more on the appearance and size of their crops than the consequences of fertilizers leaching into the valley's water supply. Second, and very much related to the first point, the quick test no longer represented the "cutting edge" politically—the advisor and others concerned about groundwater contamination found it difficult to sell the quick test as a balance between overfertilization as crop insurance and more radical regulation from the state. In the end, the "subtle" choice faced by growers was eliminated through a shift in the definitions of local environmental problems, and maintenance of the status quo became an easy choice.

Conclusion

Despite the apparent failure of the quick test system to reduce fertilizer use among vegetable growers, the example points to a complex ecology of institutionalized practices, environmental politics, and state regulation that informs growers' choices about environmental change. The scientist who hopes to intervene in this tangle of interests needs to engage with actors and produce consent around some means of environmental change. But what form of environmental politics does this kind of engagement entail? Do scientists develop a parallel course to current practices, finding some method of "greenwashing" an industry's environmental impacts (Athanasiou 1996; Austin 2002)? Or, do they instead take a sharp turn, pushing for a more radical transformation? Clearly, each of these extremes must play out in some cases, and the possibility of multiple responses from scientists in diverse contexts troubles a casual use of broad terms such as "science" and "scientists."

My case has shown that farm advisors, as scientists charged with a mission of application, do not really take either of these strategies. Instead, they often strive to walk a more conservative line that marks off these two possibilities, creating a kind of edge position that marks the boundary between the stability of locally embedded practices and the threat of change—especially change looming from "the outside." The location of these positions, such as use of the quick test, are often chosen so that order and change can be balanced, making change an option, but not necessarily a requirement. As such, the kind of engagement that Cooperative Extension farm advisors have with their grower clientele points to two interrelated conclusions about their ability to effect environmental change. First, the creation of edge positions depends crucially on the "location" of the advisors: their integration into the local community of growers they advise gives them a special entree that outsiders do not have and cannot easily acquire. Because they are situated in this space, familiar with the local farm community and its systems of production, advisors have the potential to create networks and devise new techniques that may not be apparent to either the industry or to regulators. They have "gone local" and can "see" the ecology of the farm industry in ways that others may not. This location, however, is also their Achilles heel: being local means having both special access and special pressures, including the temptation to choose the easier path of compliance with growers' desires to avoid change. No doubt the advisors' wariness when describing their identities as "environmentalists" and their distrust of regulators, as shown in the examples above, has at least some basis in their very localized domain of influence. At the same time, advisors form close connections with some of the biggest contributors to water pollution, providing them a unique chance to create the conditions for environmental change. If advisors can make even a relatively modest change in the practices of the industrial growers who provide a majority of the produce for the United States, then their work could have a larger overall impact.

Second, advisors' ability to work an "edge" between order and change depends crucially on the existence of some kind of "carrot" or "stick" to make change seem compelling; this incentive usually comes through state regulation. In this sense, ecological modernization theory's coupling of science-based environmental activism with increased regulation appears to be a practical blueprint for change, but my case shows the potential difficulty of this approach. In essence, the farm advisors are "policymakers" with no regulatory authority, except their ability to point to "outside" threats of impending regulation and to "demonstrate" potential responses, such as the quick test. Awarding the advisors regulatory powers would perhaps increase their ability to effect short-term change, but it would immediately take away their status as local "insiders" in the eyes of the growers. At the same time, their location in a particular community gives them unique insights into the development of solutions that account for the needs of both the environment and industry. Though many of our most pressing environmental problems are "global risks" (Beck 1996), their roots are often based in the specific practices and culture of a more local community of actors. Thus, the creation of effective environmental policy likely requires a linking of the local and global (Gould, Schnaiberg, and Weinberg 1996). Although it is not unreasonable to imagine that regulatory agents themselves could serve as this bridge, they would find it difficult to develop the same kinds of networks that advisors have been able to make with their local grower clientele. Again, advisors' unique position comes with its own hazards, but a division of labor appears to be the correct strategy for promoting this mode of change. In a case such as the quick test, its development, testing, and promotion needed an external push from regulatory agencies to get growers to adopt it in larger numbers. If the political will could have been found to make the choice less "subtle" for growers, pressing them to adopt the quick test on a wider scale, it would likely have been a successful solution. In this sense, the quick test example points to the need for a better integration of local knowledge and political influence for ecological modernization to succeed.

In environmental conflicts like the ones I have described here, it can be difficult to make this kind of integration work. Even among coalitions of actors who are seemingly "on the same side"—such as the advisors and growers—there can be hidden conflicts, masking the kinds of interests behind public statements and proposals. Given the complex politics surrounding an environmental problem like nitrate contamination of groundwater, policymakers may have doubts about whom they should listen to, but there is a promising candidate: they should listen to me. The need for integration of the local and global in environmental policymaking points to the value of ethnography as a means of garnering local knowledge and linking it to larger networks of political and economic power (Burawoy 1991). Though critics have rightly questioned the authority of the fieldworker to make strong claims about his or her case (Clifford and Marcus 1986; Clifford 1988), I have confidence in the comprehensiveness and symmetry of my data (Bloor 1991). I have spoken to many of the stakeholders and understand their opinions and interests, the constraints on their actions, and the possibilities of change. My years spent studying farm advisors allowed me to "go local," like the farm advisors, but also to see the links between practice and politics, to understand the ecology of agriculture.

If we acknowledge that solving environmental problems will require engagement between science, industry, and the state, then why should not ethnographers be a part of that engagement? It remains for ethnographers to create the conditions for this kind of applied influence.

Acknowledgments

First and foremost, my gratitude goes to those California farm advisors and growers who, though they remain anonymous here, made this research possible. Previous versions of this chapter were presented to Cornell University's Department of Science and Technology Studies, Colgate University's Environmental Studies Program, at the 2001 American Sociological Association annual meeting, and at the 2003 Society for Social Studies of Science annual meeting. I thank all participants at these events for their feedback. In addition, I received invaluable feedback and comments from Florian Charvolin, Dimitra Doukas, Joshua Dunsby, Leland Glenna, Steve Hilgartner, Carolyn Hsu, Francois Melard, Karen Oslund, and Elizabeth Toon. Finally, my thanks to editors Scott Frickel and Kelly Moore for many useful comments on prior drafts of this manuscript.

Endnotes

1. Though I have cited the most relevant English sources on ecological modernization theory here, the theory was first developed by German sociologist Joseph Huber.

2. Some recent exceptions include Tesh (2000) and Frickel (this volume).

3. In many states, Cooperative Extension employees are called "county agents." I will use the Californian variant "farm advisor" throughout this chapter. In addition, I also adopt the term "grower," which is generally used to denote a farmer in California.

4. Less than 2 percent of Americans worked in farming at the beginning of the twenty-first century, down from about 40 percent at the start of the twentieth century (USDA, NASS 2004).

5. My research with the Salinas Valley farm advisors began in 1997, when I moved to Salinas and began working with the advisors on a daily basis. Over the course of approximately one year, I conducted a program of participant observation, interviews, and historical research on the process of farming advising; two rounds of follow-up research were conducted again in the spring of 1999 and 2003. I conducted a total of seventy interviews with advisors, retired advisors, growers, retired growers, and other persons associated with farming in the valley, using a "snowball" sampling method to obtain informants. The interviews were conducted in person, and nearly all of them were tape-recorded with the consent of the interviewee. The interviews were semi-structured, following a loose list of questions I prepared beforehand, but with room for new subjects introduced through our interaction. Each interview was also transcribed to a computer file after its completion. With both my fieldnotes and interview transcripts in digital text, I was able to use Ethnograph qualitative data software to code and analyze the data.

6. This ecological approach draws on the actor-network approach advocated by Latour and others (Latour 1987, 1988, 1993; Callon 1986; Law 1987, 2002), but is a modified form first developed by Star and Griesemer (1989). Using ecology as a metaphor for sociomaterial networks, this approach blends actor-network theory's focus on networks and the work of enrollment with an interest in institutions and boundaries found in the "social worlds" cultural theory of scientific communities (Gieryn 1983, 1999; Clarke and Fujimura 1992; Pickering 1992; Star 1995). The main advantages of this framework vis-à-vis actor-network theory stem from its attention to objects as sites of negotiation among diverse actors. In other words, an ecological approach treats the process of consent making less in terms of the efforts of the "great man," such as Louis Pasteur (Latour 1988), and attends instead to the creation of a "negotiated order" among multiple actors and communities (Fine 1984; Henke 2000b).

7. The chapters by Scott Frickel and Steven Wolf, both in this volume, also treat this connection between environmental politics and change. Frickel's chapter, in particular, describes a case that is perhaps the closest to mine here, where geneticists concerned about the effects of exposure to mutagenic chemicals in the early 1970s attempted to blur the boundaries between science and environmental activism. In addition, the ecological approach I develop here

also has similarities to the "political ecology" framework used by environmental geographers. Space considerations limit how much I can say about these links, but political ecology also seeks to integrate the study of place with analysis of institutions and politics.

8. All names of interview subjects are confidential. For growers, I simply refer to them with the generic name Grower. For advisors and other researchers, I use a name that refers to their profession or specialization.

9. My use of this line of questioning was at least partly influenced by Rose's (1994) study of hip- hop artists. Rose finds that many women rappers are hesitant to call themselves "feminists," even though their music often contains lyrics that voice common feminist concerns about gender politics and relationships (Rose 1994: chapter 5). By asking a leading question like this, I run the risk of biasing the advisors' responses, but I believe that this risk is minimal in this case, for two reasons. First, I had been working with the advisors for more than six months when I conducted these interviews, and advisors were by then comfortable with me and seemed to feel no need to impress me. Second, and perhaps most importantly, advisors spend a lot of time working on environmental problems; they are not making up an environmental identity out of nothing.

10. A "nonpoint" source of pollution (such as polluted water running off an irrigated vegetable field) differs from point sources (such as a pipe spilling wastewater directly into a river) in that the distribution of pollutants is more diffuse.

11. Though I am using the past tense to describe these events, as I write this chapter, in the fall of 2005, most of the conditions I describe here are still the same. Nitrate contamination of groundwater is still prevalent in many wells throughout the Salinas Valley, and many wells are still designated as unsafe for drinking.

12. The most important of these problems was related to intrusion of seawater into groundwater supplies along the Monterey Bay coast. Although this problem had existed for decades, increased pumping of groundwater for crop irrigation had quickly brought seawater further inland during the late 1990s and early 2000s, nearly reaching the very valuable farming lands just outside the city of Salinas. In addition, there were many concerns about pesticide and fertilizer runoff into Monterey Bay, a federally protected Marine Sanctuary.

References

Athanasiou, Tom. 1996. "The Age of Greenwashing." *Capitalism, Nature, Socialism* 7:1–36.

Austin, Andrew. 2002. "Advancing Accumulation and Managing Its Discontents: The US Antienvironmental Countermovement." *Sociological Spectrum* 22:71–105.

Beck, Ulrich. 1992. *Risk Society: Towards a New Modernity*. London: Sage.

———. 1996. "World Risk Society as Cosmopolitan Society? Ecological Questions in a Framework of Manufactured Uncertainties." *Theory, Culture, and Society* 13:1–32.

Beck, Ulrich, Anthony Giddens, and Scott Lash. 1994. *Reflexive Modernization: Politics, Tradition, and Aesthetics in the Modern Social Order*. Stanford, CA: Stanford University Press.

Bloor, David. 1991. *Knowledge and Social Imagery*. 2nd ed. Chicago: University of Chicago Press.

Burawoy, Michael. 1991. "The Extended Case Method." In *Ethnography Unbound: Power and Resistance in the Modern Metropolis*, ed. Michael Burawoy, et al., pp. 271–290. Berkeley, CA: University of California Press.

Busch, Lawrence, and William B. Lacy. 1983. *Science, Agriculture, and the Politics of Research*. Boulder, CO: Westview Press.

California Agricultural Statistics Service. 2001a. "California Agricultural Statistics." www.nass.usda.gov/ca/bul/agstat/indexcas.htm (accessed June 20, 2003).

———. 2001b. "Summary of County Agricultural Commissioners' Reports, 2001." www.nass.usda.gov/ca/bul/agcom/indexcav.htm (accessed June 20, 2003).

Callon, Michel. 1986. "Some Elements of a Sociology of Translation: Domestication of the Scallops and the Fishermen of St. Brieuc Bay." In *Power, Action, and Belief: A New Sociology of Knowledge?* ed. John Law, pp. 196–233. London: Routledge, Kegan, and Paul.

Clarke, Adele E., and Joan H. Fujimura, editors. 1992. *The Right Tools for the Job: At Work in Twentieth Century Life Sciences*. Princeton, NJ: Princeton University Press.

Clifford, James. 1988. "On Ethnographic Authority." In *The Predicament of Culture: Twentieth-Century Ethnography, Literature, and Art*, ed. James Clifford, pp. 21–54. Cambridge, MA: Harvard University Press.

Clifford, James, and George E. Marcus, editors. 1986. *Writing Culture: The Poetics and Politics of Ethnography*. Berkeley: University of California Press.

Cohen, Maurie J. 1997. "Risk Society and Ecological Modernisation: Alternative Visions for Post-Industrial Nations." *Futures* 29:105–119.

Danbom, David B. 1979. *The Resisted Revolution: Urban America and the Industrialization of Agriculture, 1900–1930*. Ames: Iowa State University Press.

———. 1995. *Born in the Country: A History of Rural America*. Baltimore: Johns Hopkins University Press.

Fine, Gary Alan. 1984. "Negotiated Orders and Organizational Cultures." *Annual Review of Sociology* 10:239–262.

Fischer, Frank. 2000. *Citizen, Experts, and the Environment: The Politics of Local Knowledge*. Durham, NC: Duke University Press.

Fitzgerald, Deborah. 1990. *The Business of Breeding: Hybrid Corn in Illinois, 1890–1940*. Ithaca, NY: Cornell University Press.

Gieryn, Thomas F. 1983. "Boundary-Work and the Demarcation of Science from Non-Science: Strains and Interests in Professional Ideologies of Science." *American Sociological Review* 48:781–795.

———. 1999. *Cultural Boundaries of Science: Credibility on the Line*. Chicago: University of Chicago Press.

Gould, Kenneth A., Allan Schnaiberg, and Adam S. Weinberg. 1996. *Local Environmental Struggles: Citizen Activism in the Treadmill of Production*. New York: Cambridge University Press.

Hajer, Maarten A. 1995. *The Politics of Environmental Discourse: Ecological Modernization and the Policy Process*. New York: Oxford University Press.

Henke, Christopher R. 2000a. "Making a Place for Science: The Field Trial." *Social Studies of Science* 30:483–512.

———. 2000b. "The Mechanics of Workplace Order: Toward a Sociology of Repair." *Berkeley Journal of Sociology* 44:55–81.

Hilgartner, Stephen, and Charles L. Bosk. 1988. "The Rise and Fall of Social Problems: A Public Arenas Model." *American Journal of Sociology* 94:53–78.

Kline, Ronald R. 2000. *Consumers in the Country: Technology and Social Change in Rural America*. Baltimore: Johns Hopkins University Press.

Kloppenburg, Jack, Jr. 1988. *First the Seed: The Political Economy of Plant Biotechnology, 1492–2000*. Cambridge: Cambridge University Press.

Kloppenburg, Jack, Jr., and Frederick H. Buttel. 1987. "Two Blades of Grass: The Contradiction of Agricultural Research as State Intervention." *Research in Political Sociology* 3:111–135.

Latour, Bruno. 1988. *The Pasteurization of France*. Cambridge, MA: Harvard University Press.

———. 1987. *Science in Action: How to Follow Scientists and Engineers through Society*. Cambridge, MA: Harvard University Press.

———. 1993. *We Have Never Been Modern*. Cambridge, MA: Harvard University Press.

Law, John. 1987. "Technology and Heterogenous Engineering: The Case of Portuguese Expansion." In *The Social Construction of Technological Systems: New Directions in the Sociology and History of Technology*, ed. Wiebe E. Bijker, Thomas P. Hughes, and Trevor J. Pinch, pp. 111–134. Cambridge, MA: MIT Press.

———. 2002. *Aircraft Stories: Decentering the Object in Technoscience*. Durham, NC: Duke University Press.

Luke, Timothy. 1999. "Eco-Managerialism: Environmental Studies as a Power/Knowledge Formation." In *Living with Nature: Environmental Politics as Cultural Discourse*, ed. Frank Fischer and Maarten A. Hajer, pp. 103–120. New York: Oxford University Press.

Marcus, Alan I. 1985. *Agricultural Science and the Quest for Legitimacy: Farmers, Agricultural Colleges, and Experiment Stations, 1870–1890*. Ames: Iowa State University Press.

McCabe, Michael. 1998a. "Nitrate-Laced Water Sickens Town; Monterey County Warns Residents Not to Use Taps." *San Francisco Chronicle,* May 12, 1998, p. A1.

———. 1998b. "Monterey County OKs Well for Its Poorest Town." *San Francisco Chronicle,* May 26, 1998, p. A11.

Mol, Arthur P. J. 1996. "Ecological Modernisation and Institutional Reflexivity: Environmental Reform in the Late Modern Age." *Environmental Politics* 5:302–323.

Mol, Arthur P. J., and Gert Spaargaren. 1993. "Environment, Modernity, and the Risk Society: The Apocalyptic Horizon of Environmental Reform." *International Sociology* 8:431–459.

Monterey County Water Resources Agency. 2002. "2001 Nitrate Management Survey Results Report." Salinas, CA: MCWRA.

National Research Council, Board on Agriculture. 1989. *Alternative Agriculture.* Washington DC: National Academy Press.

Pickering, Andrew, editor. 1992. *Science as Practice and Culture.* Chicago: University of Chicago Press.

Rasmussen, Wayne. 1989. *Taking the University to the People: Seventy-Five Years of Cooperative Extension.* Ames: Iowa University Press.

Rose, Tricia. 1994. *Black Noise: Rap Music and Black Culture in Contemporary America.* Hanover, NH: Wesleyan University Press.

Rycroft, Robert W. 1991. "Environmentalism and Science: Politics and the Pursuit of Knowledge." *Knowledge: Creation, Diffusion, Utilization* 13:150–169.

Scott, Roy V. 1970. *The Reluctant Farmer: The Rise of Agricultural Extension to 1914.* Urbana: University of Illinois Press.

Spaargaren, Gert, and Arthur P. J. Mol. 1992. "Sociology, Environment, and Modernity: Ecological Modernization as a Theory of Social Change." *Society and Natural Resources* 5:323–344.

Star, Susan Leigh, editor. 1995. *Ecologies of Knowledge: Work and Politics in Science and Technology.* Albany: State University of New York Press.

Star, Susan Leigh, and J. R. Griesemer. 1989. "Institutional Ecology, 'Translations,' and Boundary Objects: Amateurs and Professionals in Berkeley's Museum of Vertebrate Zoology, 1907–1939." *Social Studies of Science* 19:387–420.

Tesh, Sylvia Noble. 2000. *Uncertain Hazards: Environmental Activists and Scientific Proof.* Ithaca, NY: Cornell University Press.

U.S. Department of Agriculture, National Agricultural Statistics Service. 2004. "Historical Data." www.usda.gov/nass/pubs/histdata.htm (accessed November 12, 2004).

U.S. Environmental Protection Agency. 1998. "National Water Quality Inventory: 1998 Report to Congress." Washington, DC: EPA.

Van Loon, Joost. 2002. *Risk and Technological Culture: Towards a Sociology of Virulence*. New York: Routledge.

Wynne, Brian. 1996. "May the Sheep Safely Graze? A Reflexive View of the Expert-Lay Divide." In *Risk, Environment, and Modernity: Towards a New Ecology*, ed. Scott Lash, Bronislaw Szerszynski, and Brian Wynne, pp. 44–83. Thousand Oaks, CA: Sage.

9

Embodied Health Movements

Responses to a "Scientized" World

RACHEL MORELLO-FROSCH, STEPHEN ZAVESTOSKI,
PHIL BROWN, REBECCA GASIOR ALTMAN,
SABRINA MCCORMICK, and BRIAN MAYER

Historically, health social movements (HSMs) have been an important political force in the United States, specifically around issues of health access and quality of care. Previous research has focused on individual cases of health social movements; elsewhere we consider them as a collective group that has been critical for promoting social change (Brown, Zavestoski, McCormick, Mayer, et al. 2004). Scholars have written about individual social movements dealing with health, including such areas as occupational safety and health, the women's health movement, AIDS activism, and environmental justice organizing. Other scholars, who focus more generally on changes in the health care system, point to the significance of these movements in medical history (Porter 1997) and health policy (Light 2000). Despite this significant body of research, scholars have not examined the forces that gave rise to the wide array of health social movements, nor carried out comparative analysis of these movements' different strategic, tactical, and political approaches. Generally, scholars have not explored the collective development and impact that the myriad of health social movements have had on public health, medical research, and health care delivery. We believe there is an analytical benefit to consider the origins and impacts of HSMs collectively. Highlighting how this unique group of social movements has developed its own discourse and recruitment strategies to shape social policy has important implications for social movement theory (Brown, Zavestoski, McCormick, Mayer, et al. 2004). We define HSMs as collective challenges to medical policy, public health policy and politics, belief systems, research, and practice that include an array of formal and informal organizations, supporters, networks of cooperation, and media. Health social movements, as a class of social movements, are centrally organized around health, and address access to, or provision of, health care services; disease, illness experience, disability, and contested illness; and health inequality and inequity based on race, ethnicity, gender, class, or sexuality.

For this chapter we focus specifically on a new type of health social movement, known as "embodied health movements" (EHMs), whose unique strategies and characteristics have not previously been central to the study of health movements. We first examine contextual and temporal factors that explain the rise of this new class of movements in the last decade. Next, we examine the characteristics that make EHMs unique compared to other movements involving health. Finally, we explore their impact on scientific research, medical practice, and democratizing the production of scientific knowledge and policymaking generally.

Science and technocratic decision making have become an increasingly dominant force in shaping social policy and regulation in the United States. Indeed, the insatiable quest for "better science" in policymaking has become a significant and powerful tool used to support dominant political and socioeconomic systems. Through this "scientization" of decision making, industry exerts considerable control over debates regarding the costs, benefits, and potential risks of new technologies and industrial production by deploying scientific experts who work to ensure that battles over policymaking remain scientific, "objective," and effectively separated from the social milieu in which they unfold (Beck 1992). The end result of this process is threefold: first, questions are posed to science that are virtually impossible to answer scientifically due to data uncertainties or the infeasibility of carrying out a study. Second, the process inappropriately frames political and moral questions (i.e., "transcientific" issues) in scientific terms, thus limiting public participation in decision making and ensuring that it becomes the purview of "experts" (Weinberg 1972). Third, the scientization of decision making delegitimizes the importance of those questions that may not be conducive to scientific analysis. All of these processes can exclude the public from important policy debates and diminish public capacity to participate in the production of scientific knowledge itself. Jenny Reardon provides an excellent example of this through her analysis of the conflicts between scientists and indigenous communities engaged with the Human Genome Diversity Project in chapter 13 of this volume. Finally, the ongoing quest for "sound science" can ultimately slow down and weaken policymaking by ensuring "regulatory paralysis through (over) analysis."

In their efforts to counter this trend, what we term "embodied health movements" (EHMs) have utilized medical science and public health as important vehicles through which to marshal resources, conduct research, and produce their own scientific knowledge. Until recently, most health social movements have focused on expanding access to health care and improving the quality of health care. As the latest emerging crop of HSMs, embodied health movements are highly focused on the personal understanding and experience of illness, while often addressing some of the access concerns from earlier movements. The AIDS/HIV movement highlights this unique characteristic that is particular to EHMs (Epstein 1996).

The emergence of EHMs has been catalyzed by growing public awareness about the limits of medical science to solve persistent health problems that are socially and economically mediated, the rise of bioethical issues and dilemmas of scientific knowledge production, and ultimately the collective drive to enhance democratic participation in social policy and regulation. By using science to democratize knowledge production, embodied health movements engage in effective policy advocacy, challenge aspects of the political economy, and transform traditional assumptions and lines of inquiry regarding disease causation and strategies for prevention. Drawing on case studies of the environmental breast cancer movement and environmental justice activists working on asthma, we discuss the three major foci or strategies of EHMs and how they have responded to this "scientization" process by defining their embodied experience of illness as a counter-authority to technocratic decision making and by engaging directly with scientific knowledge production itself.

The Politics of Health Social Movements and Their Impact on Social Policy

While this chapter focuses on EHMs, it is important to understand the broader context from which these particular struggles have emerged. Health social movements, especially embodied health movements, affect our society in three main ways. First, they produce changes in the health care and public health systems, both in terms of health care delivery, social policy, and regulation. Second, they produce changes in medical science, through the promotion of innovative hypotheses, new methodological approaches to research, and changes in funding priorities. Activists have fought against hospital closings, against curtailment of medical services to the poor, and against restrictions by insurers and managed care organizations (Waitzkin 2001). Self-care and alternative care activists have broadened health professionals' awareness of the capacity of laypeople to deal actively with their health problems (Goldstein 1999). Disability rights activists have garnered major advances in public policy, including enhanced accessibility to public facilities and protection against job discrimination (Shapiro 1993). Toxic waste activists have brought national attention to the adverse health effects of chemical, radiation, and other hazards, helping to shape the development of the Superfund Program and related right-to-know laws (Szasz 1994). Environmental justice activists, who are centrally concerned with environmental health, have publicized the links between physical health and the socioeconomic and political environment and have shown that health improvement and disease prevention require attention to, and reform of, a variety of social sectors, such as housing, transportation, land use planning, and economic development (Bullard 1994; Shepard et al. 2002). Occupational health and safety movements have brought medical and governmental attention to a wide range of ergonomic, radiation, chemical, and stress hazards in many workplaces, leading to extensive regulation and the creation of the Occupational Safety and Health Administration and the National Institute of Occupational Safety and Health (Rosner and Markowitz 1987). Physicians have organized doctor-led organizations to press for health care for the underserved, to seek a national health plan, and to oppose the nuclear arms race (McCally 2002).

Third, health social movements produce changes in civil society by pushing to democratize those institutions that shape medical research and policymaking. Specifically, health social movements have advanced progressive approaches to promoting public health and disease prevention, helped expand and improve health services, and fostered democratic participation in social governance. For example, women's health activists have greatly altered medical conceptions and treatment of women, expanded reproductive rights, expanded funding and services in many areas, altered treatment protocols for diseases disparately impacting women (e.g. breast cancer), and pushed to enhance community oversight in how medical research is conducted (Ruzek 1978; Ruzek, Olesen, and Clarke 1997; Morgen 2002). Similarly, AIDS activists have obtained expanded funding for treatment research, fought for medical recognition of alternative treatment approaches, and obtained major shifts in how clinical trials are conducted (Epstein 1996). Mental patients' rights activists radically altered mental health care delivery by ensuring the protection of patients' civil rights that used to be inferior to those of prisoners. These rights have been expanded to include the right of patients to demand better treatment and the right to refuse treatments (Brown 1984). Many health social movements have achieved these political victories by forging strategic intersectoral connections between health and other social sectors (such as the environment), which enable them to overturn ineffective policies and push for more stringent regulation of industrial production that moves from a pollution control to a pollution prevention framework (Morello-Frosch 2002).

Embodied Health Movements' Challenge to Scientization and Authority Structures

Embodied health movements have emerged as a unique subset of HSMs through their strategy of reshaping and democratizing the production and dissemination of scientific knowledge. This raises the question of what is occurring in civil society that stimulates the emergence and rapid proliferation of embodied health movements.

As outlined above, there has been a gradual scientization of policymaking, in which a growing array of socioeconomic and political phenomena are addressed through scientific analyses and technical approaches. The continuing advance of societal rationalization raises the role of objective scientific expertise above that of public knowledge for most social issues. Ironically, however, the quest for better science to inform public decisions is often a veiled attempt to hide the politicization of the policy process (Weiss 2004). This has been recently demonstrated by the persistent opposition of the U.S. government to the widening scientific consensus on the question of global climate change and its attendant ecological and human health impacts (Gelbspan 1997).

Popular conceptions of the key role of science in society can be seen in how often the public views itself as an actual or potential object of scientific study (Epstein 1996). This also has had the effect of increasing the myriad of ways in which the public acquires and shares scientific information for personal use and ultimately leverages data to promote policy change. People obtain extensive knowledge through increased interpersonal sharing of health concerns in self-help and support groups. Information is also obtained through major dissemination of scientific knowledge by the media (primarily print) and by wide access through the Internet to medical databases, research studies, and regular news coverage of the challenges of research in the world of medical science. Nevertheless, the public has become more aware of the limitations of modern medical science to effectively address the persistent and most challenging health issues of the day; indeed, medical technology advances also go hand-in-hand with an increased risk of medical errors and microbial resistance to drug treatments, as well as iatrogenic complications that can lead to adverse health outcomes (Garrett 1994). Not surprisingly this has led to the increased popularity of combining traditional medical treatments with alternative medicine (e.g., acupuncture, meditation, and body work). Hence, the public's overall faith in medical science is tempered by their use of other avenues to disease treatment and health.

Public health research constantly reminds us that advances in modern medicine have not by themselves made the largest difference in improving the health status of diverse populations (McKeown 1976). Biomedical research emphasizing how individual risk factors predict diseases receives the major share of regulatory attention, despite a significant body of research demonstrating that the health status of populations is largely determined by structural features such as race, class, income distribution, geography, and other environmental factors (Krieger et al. 1993; Berkman and Kawachi 2000). This is exemplified by the increased prevalence of chronic diseases that are now impacting new population groups (e.g., diabetes and obesity in children) and the resurgence of infectious disease that the public and medical authorities had previously assumed had been eradicated or at least brought under control.

Finally, debates on ethical questions involving medical science have captured public attention. The advent of genetic testing for a variety of diseases has raised valid concern about the use of that information by employers to discriminate in the workplace and justify the exclusion of certain workers from insurance coverage (Schulte et al. 1995; Schulte et al. 1997). Ethical concerns have been further fueled by revelations of conflict of interest in the support of research by pharmaceutical firms and the increasing power of private corporations to direct university research (Krimsky 2003). Public confidence in the integrity of scientific research has been challenged by revelations of corporations and federal agencies that violate the peer-review process or suppress the dissemination of research results that conflict with corporate economic interests (Ong and Glantz 2001; Rosenstock and Lee 2002; Greer and Steinzor 2002).

In addition, a spate of recent, highly publicized adverse events in human biomedical research has generated concerns about the ethical implications of using human subjects as research volunteers and has implicated the institutional culture at several prestigious research institutions in failing to adequately protect study participants. The 1999 death of an eighteen-year-old patient in a gene therapy trial at the University of Pennsylvania and the 2001 death of a healthy twenty-four-year-old volunteer in an asthma study at Johns Hopkins University are the two cases that have received the greatest level of attention (Steinbrook 2002). In the Hopkins case, one of the criticisms leveled at the investigator and the university charged that because the volunteer was an employee in the laboratory conducting the experiment, she may have been subtly and inappropriately pressured by her employer or by her colleagues to participate in the experiment ("Laxity in the Labs" 2001). These ethical concerns highlight how easily the thin veil of objectivity in medical research can be breached, despite the cardinal claim that science is essentially an objective endeavor. For the public, these problems highlight the polemical intersection of science, society, and institutional culture, which ultimately paves the way for a critique of the dominant political economy and its adverse effect on the public's health.

Because health concerns are so pervasive throughout society, people are more likely to focus many grievances through a lens of health. For example, during an economic recession and periods of high unemployment, it is understandable that people will make demands for broader and better health insurance and for expansion of coverage to include the uninsured. Similarly, in an industrial society where environmental degradation is increasingly visible and in which the government has begun to roll back decades of environmental regulation and protection, it becomes easier for the public to connect health with socioeconomic, political, and institutional concerns and to begin pushing for increased regulation of industrial production and enhanced community participation in the formation of environmental policy.

Some of this struggle for democratization is due to social trends that provide a medium for the growth of health social movements; some of it is due to the achievements of health social movements themselves. The key is that increasing numbers of people presently believe they have the right and the authority to influence health policy, including access issues, quality of care, clinical interaction, and federal funding of research. Although some of that pressure occurs at the individual level, the most effective forms originate from the collective efforts of health social movements.

Understanding Embodied Health Movements

In their personal understanding and experience of illness, embodied health movements share three characteristics. First, they introduce the biological body to social movements in central ways, and they leverage the body and embodied experience of illness as a counter-authority to challenge medicine and science. This leads to the second feature. EHMs typically include challenges to existing medical and scientific knowledge and practice. Third, EHMs often involve activists collaborating with scientists and health professionals in pursuing treatment, prevention, research, and expanded funding.

Many health social movements in fact have one or even two of these characteristics, but embodied health movements are unique in possessing all three. This section explains these unique features and further argues how EHMs are a distinct breed of health social movements. While the simultaneous possession of these three characteristics makes EHMs unique, they are nevertheless much like other social movements in that they depend on the emergence of a collective identity as a mobilizing force. In the case of illness, people often relate to existing health institutions that treat their health problems as an individualized one. When these institutions—science and medicine—fail to offer disease accounts that are consistent with individuals' own experiences of illness, or when science and medicine offer accounts of disease that individuals are unwilling to accept, people may adopt an identity as an aggrieved illness sufferer. Such an identity emerges first and foremost out of the biological disease process happening inside a person's body. But, in a further step, people begin to view their disease experiences within a socioeconomic and political context, and they start to collectivize their experiences and move toward shared action.

Although the body is also implicated in other social movements, especially identity-based movements, these are typically movements that originate from a particular ascribed identity that leads a group of people to experience their bodies through the lens of social stigma and discrimination. Such is the case with the women's movement and the lesbian and gay rights movements. With EHMs, a disease identity results from processes happening within the body. This identity represents the intersection of social constructions of illness and the personal illness experience of a biological disease process. The body, combined with people's embodied experience, thus becomes a counter-authority in which social movement groups can base their critique of medical science.

Another consequence of this embodied experience lies in how it constrains the options available to a movement once mobilized. Illness sufferers can work either within or against their target, in this case the system of the production and application of scientific and medical knowledge. They are less free, depending on the severity of their condition, to simply exit the system. Though some illness sufferers seek alternative or complementary therapies, many others either need or seek immediate care and are forced to pursue solutions within the system they perceive as failing their health needs. Most importantly, people who have the disease have the unique experience of living with the disease process, its personal illness experience, its interpersonal effects, and its social ramifications. Their friends and family, who share some of the same experiences, may also join them in collective action. These personal experiences give people with the disease or condition a lived perspective that is unavailable to others. It also lends moral legitimacy to the mobilized group in the public sphere and scientific world.

Challenges to existing medical and scientific knowledge and practice are a second unique characteristic of EHMs, whether those movements work within the system or challenge it from the outside. After all, if the aims of EHMs are to bring the social constructions of the disease more in line with the actual disease process, and if disease processes are understood through scientific investigation, then EHMs become inextricably linked to the production of scientific knowledge and to changes in practice. Just as EHMs are not the only movements that involve the physical body, they also are not the only movements to confront science and scientific knowledge and practice. Environmental groups, for example, often confront scientific justifications for risk management strategies, endangered species determination, global warming, or resource use, by drawing on their own scientific evidence for alternative courses of action. However, many environmental disputes can also prioritize nature and the value placed on it by opposing interests. In these cases, some environmental groups can abandon scientific arguments, for example, appealing instead to the public's desire to protect open spaces for psychological or spiritual reasons, or to preserve resources for enjoyment by future generations. However, it is not necessarily EHMs' challenge of science that sets them apart from other movements, but *how* they go about doing it. EHM activists often judge science based on intimate, firsthand knowledge of their bodies and illness experience.

Furthermore, many EHM activists must simultaneously challenge *and* collaborate with science. While they may appeal to people's sense of justice or shared values, they nevertheless remain dependent to a large extent upon scientific understanding and continued innovation if they hope to receive more effective treatment, and elucidate effective strategies for prevention. For example, when little was known about AIDS, activists had to engage the scientific enterprise in order to prod medicine and government to act quickly enough to approve new drug protocols (Epstein 1996). Even EHMs that focus on already understood and treatable diseases are dependent upon science. Although they may not have to push for more research, they typically must point to scientific evidence of causation in order to demand public policies for prevention. For example, asthma activists who demand better transportation planning for inner cities and who seek better-quality and affordable housing do so knowing that they are supported by solid scientific evidence linking outdoor and indoor air quality to asthma attacks (Brown, Mayer, et al. 2003).

EHMs' dependence on science leads us to a third characteristic—activist collaboration with scientists and health professionals in pursuing treatment, prevention, research, and expanded funding. Lay activists in EHMs often strive to gain a place at the scientific table so that their personal illness experiences can help shape research design, as Epstein

(1996) points out in his study of AIDS activists. Even if activists do not get to participate directly in the research enterprise, they often realize that their movement's success will be defined in terms of scientific advances, or in terms of transformation of scientific processes. Part of the dispute over science involves a disease group's dependence on medical and scientific allies to help them press for increased funding for research and to raise money to enable them to run support groups and get insurance coverage. The more that scientists testify to those needs, the stronger the claims of patients and advocates. The above points indicate that science is an inextricable part of EHMs and thus are in a fundamentally different relationship with science than other movements. Equally important is the fact that as the capacity of EHM organizations increase, they often move from models of lay-expert collaboration to essentially designing and conducting scientific research themselves, thus blurring the line between lay activists and experts (Tesh 2000). The scientific work of Silent Spring Institute, a community-based organization that conducts research on environmental breast cancer causation, exemplifies this culmination and is discussed further below.

EHMs are a heterogeneous group of movements. They differ in the challenges they pose to the status quo, in their choice of strategy and tactics to challenge the dominant political economy, and in their direct involvement in science. By drawing on case study material of the environmental breast cancer movement and environmental justice activists working on asthma, we discuss how some EHMs center the body as a counter-authority, engage science, and challenge science more generally.

Case Studies: Exploring the Characteristics of EHMs

Case study material derives from our broader project that studies three diseases or conditions and the social movements involving them: asthma (with a specific focus on environmental justice groups involved in asthma), breast cancer (with a specific focus on the environmental breast cancer movement), and Gulf War illnesses. Details on the data, methods, and findings regarding this ongoing project can be found elsewhere (Brown, Zavestoski, McCormick, Mandelbaum, et al. 2001; Brown, Zavestoski, Mayer, et al. 2002; Brown, Zavestoski, McCormick, Mayer, et al. 2004; McCormick, Brown, and Zavestoski 2003). Here, we draw specifically from our work on asthma (Brown, Mayer, et al. 2003) and environmental breast cancer activism (McCormick, Brown, and Zavestoski 2003).

INTRODUCTION TO CASE STUDIES

Breast Cancer Activism and the Environmental Breast Cancer Movement

Since the early 1990s, a new subset of the breast cancer movement emerged that combined tactics from AIDS activists and activists from the environmental, women's, and general breast cancer movements (Klawiter 1999). We call this new subset the environmental breast cancer movement or ECBM (McCormick, Brown, and Zavestoski 2003). These activists were motivated by the lack of evidence of breast cancer etiology and by the potential link to environmental contaminants, the latter of which was discovered as women noticed higher than average concentrations of cases in several geographic areas around the country—Long Island, New York, the San Francisco Bay Area, and Massachusetts. Studies of breast cancer and other illnesses have devoted large amounts of research dollars to analyzing individual and behavioral risk factors, such as lifestyle and diet, but this line of inquiry has yielded equivocal answers. Despite major investments in research, morbidity appears to be increasing, and mortality rates have remained more or less unchanged (Ries et al. 2002). These disappointing results for breast cancer raise concern that the "war on cancer" has not yielded promising results for understanding causation and lowering incidence. This situation has compelled advocates in these three geographic areas to push for new approaches to studying environmental causes of breast cancer in order to elucidate possible pathways toward true prevention of the disease. Their efforts have grown into a successful, national-level movement with the following four goals: 1) to broaden public awareness of potential environmental causes of breast cancer; 2) to increase research into environmental causes of breast cancer; 3) to create policy that could prevent environmental causes of breast cancer; and 4) to increase activist participation in research.

Silent Spring Institute (SSI) is a valuable example of how this activism penetrates the scientific realm. SSI was founded by the Massachusetts Breast Cancer Coalition (MBCC) as a research organization to study environmental factors in breast cancer, with a commitment to involving women with breast cancer not just as subjects in the scientific process but as co-creators of scientific knowledge.

Asthma Activism and Environmental Justice

Asthma rates have increased so much in the United States that medical and public health professionals invariably speak of it as a new epidemic. Asthma strikes lower-income and minority populations more than other groups. Many minority people now report asthma as one of their chief health problems. In many low-income urban communities of color, rates are significantly higher than the national average (U.S. EPA 2003). As the number of cases has increased, medical and public health professionals and institutions have expanded their treatment and prevention efforts, environmental and community activists have made asthma a major part of their agenda, and media coverage has grown.

Several activist groups that define themselves as environmental justice organizations have made asthma a core part of their agenda. Many of these activists groups have entered into coalitions with academic research centers, health providers, public health professionals, and even local and state governmental public health agencies. Asthma has become for many people a "politicized illness experience" whereby community-based environmental justice organizations show people with asthma how to make direct links between their experience of asthma and the social determinants of their poor health. These groups view asthma within the larger context of community well-being, and they emphasize the unequal distribution of environmental risks and hazards according to race and class, and the reduction of environmental factors that they believe are responsible for increased asthma in their communities. In their approaches to asthma, such groups combine general education about asthma with political actions to reduce local pollution sources.

Environmental justice activists adopt an intersectoral approach, linking health to neighborhood development, economic opportunity, housing policy, planning and zoning activities, transportation accessibility, sanitation, social services, and education. In this way, they promote intervention strategies that focus on alleviating the socioeconomic and environmental factors that affect health. This perspective is supported by research suggesting that community environmental interventions can reduce the risk of severe respiratory diseases, including asthma (LeRoux, Toutain, and Le Luyer 2002).

Here we focus on two environmental justice organizations that are working on asthma issues, Alternatives for Community and Environment (ACE) based in the Roxbury-Dorchester area of Boston, and West

Harlem Environmental Action (WE ACT) based in New York City. ACE began in 1993 as an environmental justice organization, and one of its earliest actions was a successful mobilization to prevent an asphalt plant from being permitted in Dorchester. ACE had initially expected to focus on issues such as vacant lots and did not intend to focus on asthma, but a year of talking with the community showed ACE that residents saw asthma as their number one priority. ACE believes that to address asthma requires addressing housing, transportation, community investment patterns, access to health care, pollution sources, and sanitation, as well as health education. As one staff member notes, "Everything we do is about asthma."

Similarly, West Harlem Environmental Action was founded in 1988 in response to environmental threats to the community created by the mismanagement of the North River Sewage Treatment Plant and the construction of the sixth bus depot in northern Manhattan. WE ACT quickly evolved into an environmental justice organization with the goal of working to improve environmental protection and public health in the predominately African American and Latino communities of northern Manhattan. They identified a wide range of environmental threats, including air pollution, lead poisoning, pesticides, and unsustainable development. WE ACT has continued to grow and expand, extending its reach beyond West Harlem to other northern Manhattan communities.

Exploring the Characteristics of EHMs

Using the three primary characteristics of embodied health movements, namely: 1) the centrality of the body and lived experience of illness, 2) challenges to existing medical/scientific knowledge and practice, and 3) collaborating with scientists and health professionals in pursuing research, we explore how these elements are integrated into the organizing and advocacy efforts of the environmental breast cancer movement and environmental justice asthma activists.

CENTRALITY OF THE BODY AND THE LIVED EXPERIENCE OF ILLNESS

EHM activists understand the limitations of the biomedical model, which treats disease as a discrete entity occupying the body, and the body as a discrete entity separate from the person occupying it (Freund, McGuire, and Podhurst 2003). Despite the physical disease process, the primary focus of medical professionals, activists with asthma or breast cancer embody their illness and all its social implications. Indeed, individuals with any disease typically have a bodily experience of illness that differs from medical understanding of the disease. This is the classic distinction between health and illness long made by medical sociologists. In the case of asthma and breast cancer, race, class, gender, and the social context in which these factors interact shape the disease experience itself.

A central tenet of the breast cancer movement, derived in part from the women's health movement that preceded it, has been the inadequacy of the patriarchal medical profession to understand a breast cancer illness experience that is deeply rooted in being a woman. The body is central to the breast cancer movement in that it is the objectified entity upon which medical experts gaze, and it is the same body whose socially constructed meaning shapes the experience of a woman with breast cancer. Objectification of the body can best be seen in the past practice of anesthetizing a woman to perform a biopsy, and then, while she is still under anesthesia, performing a radical mastectomy on the malignant tumor without prior consent. In 1983 the mainstream breast cancer movement succeeded in passing informed consent legislation that gave women the option to choose their preferred treatment.

The environmental breast cancer movement relies on the body in constructing a politicized collective illness identity in two ways. First, it criticizes medical objectification of the female body, and its treatment of women's breasts as objects of research independent of women's bodies and their locations in toxic environments. As an activist noted, the relationship between the environment and breast cancer is "sentinel" in women's health and "remind[s] people that reproduction occurs in the context of a wider world." The criticism is also captured by one scientist who argued that researchers need to "think about the biologic processes happening in the breast in relationship to the world in which the woman is walking who happens to have those breasts so that there's not just this disembodied breast that's hanging out somewhere."

Second, the environmental breast cancer movement links the bodily experience of breast cancer to a social structure that exposes people to many environmental burdens, yet the dominant social and medical structures emphasize treatment over prevention. Though environmental breast cancer activists acknowledge the importance of treatment research, they contend that prevention ought to be the main priority of the mainstream movement. Focusing on prevention presses breast cancer researchers to begin asking difficult questions about whether there is something about women's lives in modern consumer societies that increases susceptibility to breast cancer. For instance, environmental breast cancer activists have long questioned the wisdom of prescribing estrogen, a suspected cause of breast cancer, in hormone replacement therapy. At Silent Spring Institute, researchers are investigating the chemicals women are chronically exposed to on a daily basis. Through these and other strategies, environmental breast cancer activists transform the individual illness experience into a politicized collective illness identity that focuses policy on promoting the health of populations or communities rather than simply the health of individuals (Rose 1985).

Environmental justice activists, such as those at ACE and WE ACT, deploy similar strategies and collectively frame asthma as a social justice issue, and they therefore transform the personal experience of illness into a collective identity aimed at discovering and eliminating the underlying socioeconomic causes of asthma. When people view asthma as related to both air pollution and to the living conditions of poor neighborhoods, they reconstruct asthma narratives differently than the narrative reconstruction that occurs with other chronic illnesses. Rather than emphasizing personal responsibility or attributing asthma to elements of one's work, family, relationships, or schooling, these narratives emphasize the social production of asthma through inequalities in housing standards, transportation access, neighborhood development, income, and enforcement of government regulations. In adopting such a politicized illness experience, the poor and communities of color have used asthma to emphasize that social inequalities are a key determinant of health.

CHALLENGES TO SCIENCE

In their efforts to reshape the production and dissemination of scientific knowledge, both the EBCM and environmental justice activists have leveraged the reality of scientific uncertainty to break regulatory and policy gridlock and promote social action. For example, while studies of environmental causation of breast cancer have produced conflicting results, the EBCM has used this situation to promote more stringent regulation of industrial production. Environmental breast cancer activists have effectively argued that in the never-ending quest for better data and unequivocal proof of cause and effect, some scientists and regulators have lost sight of a basic public health principle—the importance of disease prevention. As a result, activist organizing has pushed for the implementation of the precautionary principle in environmental regulation. Practically, this means that in the face of uncertain but suggestive evidence of adverse environmental or human health effects, regulatory action is needed to prevent future harm. The precautionary principle perspective seeks to mobilize environmental and public health policy-making that otherwise can be paralyzed when implementation is too dependent on scientific certainty. The precautionary principle also seeks to shift the burden of hazard assessment, monitoring, and data generation activities onto those who propose to undertake potentially harmful activities or chemical production (Raffensperger and Tickner 1999). In this vein, activists push for research that more systematically examines possible environmental etiologies of breast cancer as much as traditional inquiries into lifestyle and genetic risk factors.

EBCM groups have challenged assumptions that undergird mainstream breast cancer science and prevention strategies. For example, they have criticized the genetic determinism that permeates etiologic research. The discovery of the BRCA-1 gene mutation focused media attention on genetic causes even though it has since been recognized that genetic risk factors only account for 5 to 10 percent of all breast cancer cases (Davis and Bradlow 1995). Activists rightly point out that the genome does not evolve rapidly enough to account for the dramatic increase in breast cancer incidence over the last thirty years. In 1964, a woman's lifetime risk of contracting breast cancer was 1 in 20, and now it is 1 in 7. Activists have also challenged the traditional research emphasis on lifestyle and behavioral variables that subtly blames women for risk factors over which they have little control, such as late childbearing and diet. Especially in poor communities of color, supermarkets with fresh fruits and vegetables are often few and far between, and so too are the means of transportation that would allow community members to travel to areas where such food is available. Similarly, individuals have little control over whether food is treated and processed in unhealthy ways. EBCM advocates have criticized government and nonprofit institutions such as the American Cancer Society and the National Cancer Institute for promoting the erroneous message that mammography is an effective means of breast cancer prevention. Activists have exposed the obvious contradiction that once a tumor is detected through mammography, prevention has clearly failed. These activists have also pushed the scientific community to thoroughly examine possible environmental links to breast cancer, such as endocrine-disrupting chemicals that are ubiquitous in the environment (Davis 2002; Rudel et al. 2003).

Similarly, environmental justice activists' work on asthma has pushed the scientific and regulatory communities to better address the structural issues that drive racial inequities in community exposures to pollutants and that exacerbate or cause asthma. Epidemiological research has uncovered equivocal evidence linking air pollution to asthma. However, a large body of evidence shows that existing cases of asthma in children can be exacerbated by exposure to air pollution, such as ozone and particulates (Gilliland, Berhane, et al. 2001; Gent et al. 2003; Fauroux et al. 2000; Mortimer et al. 2002). Traffic studies also suggest an adverse impact on respiratory health to nearby residents (Brauer et al. 2002; Künzli et al. 2000). The risk of developing respiratory allergies from exposure to diesel emissions has a strong genetic component. Researchers estimate that up to 50 percent of the United States population could be in jeopardy of experiencing health problems related to diesel air pollution due to genetic susceptibility (Gilliland, Li, et al. 2004). Finally, a recent study surprised the scientific community by implicating the role of outdoor ozone exposure in potentially *causing* asthma among children who participate in outdoor sport activities. The study showed that children spending large amounts of time exercising outdoors in areas with higher ozone levels were more likely to develop asthma (McConnell et al. 2002). These effects were seen at ozone concentrations significantly below the national standard.

This scientific evidence, combined with rising community concerns about childhood asthma, has compelled activists to focus broadly on air pollution as a key factor in asthma, both by pushing for more research on asthma causation and by working toward prevention by advocating for transportation policy that improves air quality. For example, since ACE identifies diesel buses as a problem for community environmental health, they have also taken up transportation issues more broadly rather than focusing narrowly on air pollution regulation. Charging "transit racism," ACE argued that the estimated 366,000 daily bus riders in Boston were being discriminated against by the over $12 billion of federal and state money being spent on the "Big Dig" highway project, while the Massachusetts Bay Transit Authority (MBTA) refused to spend $105 million to purchase newer, cleaner buses and bus shelters. In tying dirty buses to higher asthma rates, ACE successfully framed an issue of transit spending priorities into one of health, justice, and racism. In 2000 the Transit Riders' Union, largely created by ACE, got the MBTA to allow free transfers between buses, since the many inner-city residents who relied on two buses for transportation had to pay more than those who received free transfers on subways. A key component of ACE's education and empowerment efforts is reflected in its Roxbury Environmental Empowerment Project. REEP teaches classes in local schools, hosts environmental justice conferences, and through its intern program trains high school students to teach environmental health in schools.

Similarly, WE ACT has identified diesel exhaust as a major factor behind the disparate burden of asthma experienced in their community. Using publicity campaigns such as informative advertisements placed in bus shelters, public service announcements on cable television, and a direct mailing, WE ACT has reached a vast number of community residents and public officials and let them know that diesel buses could trigger asthma attacks. Though their efforts increased public awareness of WE ACT and its work to reduce asthma, the media campaign did not lead to a shift in the New York Metropolitan Transit Authority's (MTA) policy toward diesel buses. Thus in November 2000, WE ACT filed a lawsuit against the MTA with the federal Department of Transportation claiming that the MTA advances a racist and discriminatory policy by disproportionately siting diesel bus depots and parking lots in minority neighborhoods.

Challenges made by environmental justice groups to scientific approaches to asthma have led to concrete results in health policy, especially in terms of health tracking, academic-community collaboration, and stronger air quality regulation. The Trust for America's Health (formerly the Pew Environmental Health Commission) has pointed to asthma as one of the major reasons the United States needs a national health-tracking system, and has garnered much scientific and governmental support for this approach, including recent passage of an environmental health-tracking bill in Congress and the creation of a funding stream for such efforts that are administered by the Centers for Disease Control. Innovative academic-community collaborations sponsored by federal agencies have developed in recent years as well, with asthma becoming a main focus because there are such strong community organizations available to do joint work with researchers. Last, the growing power of environmental justice activism, combined with a solid base of environmental health researchers and a public health focus on eliminating social disparities in health, holds the potential to promote more effective regulatory strategies and policies to improve air quality.

ENGAGING AND DOING SCIENCE

Both environmental breast cancer and environmental justice activists have pioneered innovative strategies to promote and institutionalize lay collaboration on scientific research to answer challenging environmental health questions. EBCM groups, for example, have been very effective in getting the government to do more breast cancer research and in pursuing their own scientific studies. Silent Spring Institute represents a major achievement of this movement in terms of doing science. SSI is a scientific organization that is consciously part of a broader social movement for environmental health. The Massachusetts Breast Cancer Coalition established SSI as a research institute with a mission to not only merge activism and science, but to also nourish lay-professional collaborations. Activist and lay input is built into the institutional structure of SSI. That is, SSI has structured input channels at all levels of the research process to ensure that lay participation is not overshadowed by scientists' knowledge. This environment fosters novel thinking because the activists generate solutions unconstrained by dominant scientific research paradigms or methodologies. SSI also includes the public presentation of hypotheses, data, and analysis to lay audiences as part of its mission.

In 1993, Long Island activists garnered backing from other well-known scientists and held a conference on the issue of breast cancer, at which the Centers for Disease Control, Environmental Protection Agency, and National Cancer Institute were major presences. This enabled the passage of a bill that planned and funded the Long Island Breast Cancer Study Project through the National Cancer Institute. Congress appropriated $32 million dollars in federal funding for study of potential environmental causes of breast cancer. In 1992, Breast Cancer Action and the Breast Cancer Fund joined with the National Alliance of Breast Cancer Organizations to push for lay involvement on research proposals funded by the California Breast Cancer Research Program. This proved to be a model for the well-regarded citizen participation in the Department of Defense's breast cancer research program and has included some projects on environmental factors.

Environmental justice activists have also played a central role in shaping federal grant programs on environmental health research through their participation on advisory boards to develop new research programs that focus on social justice concerns and their role in the review process for funding allocation to support scientific research on environmental health and justice issues. Nevertheless, environmental justice organizations are also wary of the pitfalls of engaging in resource-intensive epidemiological studies, over which communities can fail to exert any control and that can yield equivocal results. For example, ACE understands the need for scientific evidence and legitimacy and recognizes the long-term importance of establishing links between air pollution and asthma. Yet, the organization's use of an environmental justice frame means that it is not wedded to science as a primary mode of improving community health. ACE's decision to use science selectively, and to insist on the role of science in empowering community residents, is central to their asthma work.

Support from some researchers at the Harvard School of Public Health and the Boston University School of Public Health provides an opportunity for ACE to work with science in its own way. ACE's AirBeat project monitors local air quality and then analyzes the relationship between air quality and medical visits. ACE mobilized researchers and government agencies to install a monitor at their Roxbury office. Community members are also directly involved in the planning and implementation of these studies, as evidenced by the involvement of REEP students in identifying data types to be collected from community clinics. On one level ACE collaborates with scientists to produce quantifiable outcomes they hope will lead to greater understanding of air pollution and asthma. This has resulted in jointly authored articles on air particulate concentrations, published in major environmental health journals. However, AirBeat is useful in other ways as well. ACE derives legitimacy from the involvement of government agencies and scientists in the process, including the presence of Harvard scientists and the then-EPA Region 1 head John Devillars at the press conference when the air monitor was unveiled.

WE ACT has been much more eager than ACE to work with university-based scientists. They are partners with Columbia University School of Public Health in a federally funded academic-community research collaboration that involves public outreach and advocacy. They have published several journal articles and in 2002 edited a supplement of the prestigious *Environmental Health Perspectives* on "Community, Research, and Environmental Justice" (which contained an article by ACE staff members). WE ACT conducted a conference in 2002 on "Human Genetics, Environment, and Communities of Color: Ethical and Social Implications," co-sponsored by the National Institute of Environmental Health Sciences (NIEHS), the NIEHS Center for Environmental Health at the Columbia School of Public Health, and the Harlem Health Promotion Center. The conference brought together community advocates, policymakers, and scientists from across the United States to educate one another and deal with critical ethical, legal, and social implications of human genetics for communities and people of color. Both ACE and WE ACT believe they are pushing their scientific allies to be continually more community oriented in defining problems and designing research and interventions.

Conclusion

Studying health social movements offers insight into an innovative and powerful form of political action aimed at transforming the health care system, modifying people's experience of illness, and addressing broader social determinants of health and disease of diverse communities. We have shown here how embodied health movements, a more recent form of health social movements, lead people to develop a politicized illness experience by making the socioeconomic drivers of health explicit and central to their struggle. In this way, EHMs challenge state, institutional, and cultural authorities in order to increase public participation in social policy, to enhance regulation, and to democratize the production and dissemination of scientific knowledge in medical science and public health research.

In order to achieve their goals, EHMs deploy an array of strategies and are nimble in the way they shift arenas of struggle. The intersectoral nature of EHMs strengthens their capacity to impact the scientific and policy realms as they forge strategic alliances with movements targeting other sectors, such as the environmental movement. Finally, EHMs utilize a broad range of tactics: they engage in the legal realm, shape public health research, promote new approaches to medical science, employ creative media tactics to highlight the need for structural social change and true disease prevention, and engage within the policy arena to enhance public power to monitor and regulate industrial production.

Perhaps the most powerful point of engagement for EHMs has been their efforts to democratize science and to enhance public engagement in knowledge production. Both the environmental justice movement and the breast cancer movement have helped raise money and redirect federal research dollars to support new scientific avenues of research that more purposefully address how social context and economic inequities impact public health (Morello-Frosch et al. 2002). For example, environmental justice activists have played a central role in shaping the National Institute of Environmental Health Sciences' grants programs on environmental justice through participation on advisory boards examining ethical issues related to toxicogenomic research and through their role in reviewing how resources are allocated to support scientific research and public health communication on environmental health and justice issues. Similarly, environmental breast cancer activism has achieved lay-professional collaborations to advance scientific research on breast cancer, and these efforts have been institutionalized in the Department of Defense's breast cancer research program, the NIEHS program for Centers of Excellence on Breast Cancer Research, and in some state programs that ensure participation by breast cancer activists in shaping the allocation of research dollars (McCormick, Brown, and Zavestoski 2003). These institutional shifts in medical science and public health have helped nurture a new cadre of scientists who support activist efforts to democratize knowledge production through their roles as program officers in federal funding agencies and as researchers working collaboratively with EHM organizations to address persistent challenges in disease treatment and prevention.

Despite these positive developments, the benefits of EHM activism in scientific knowledge production are not without pitfalls or potential contradictions. Although engaging in scientific endeavors is important, this process can also sap energy and staff time that might otherwise be directed toward political and community organizing. Engaging in scientific activities may cause dissension among movement groups, especially if those working on collaborations with academic researchers begin to attain far more resources and institutional access than other groups. Thus, it is important to keep in mind that as activists begin to take science into their own hands, they must grapple directly with some of the same polemical issues and contradictions that they had previously criticized. For example, some health movement groups have major disagreements over whether to take corporate funding to support their work. This issue has been particularly controversial for the environmental breast cancer movement, where groups have debated whether to accept funding from major pharmaceutical firms. Some activists have argued that accepting such corporate funding can create a real or perceived conflict of interest and undermine the credibility of an organization to reliably analyze and disseminate scientific information, especially data regarding clinical trials for new drug protocols. Other groups must address ethical quandaries, such as Native American groups that work with scientists to analyze the presence of persistent contaminants in human breast milk. In carrying out this research, activists have sought to develop informed consent procedures that address the needs of the community and not just individual community members, and they must negotiate appropriate ways to report individual and collective study results to the community (Schell and Tarbell 1998). Despite these challenges, EHMs have successfully leveraged their embodied experience of illness and forged a new path for how social movements can effectively engage in scientific knowledge production. Thus, EHMs serve as a critical counter-authority aimed at democratizing and reshaping social policy and regulation in a way that transforms the socioeconomic and political conditions that underlie unequal distributions of health and disease in the United States.

Acknowledgments

This research is supported by grants to the third author from the Robert Wood Johnson Foundation's Investigator Awards in Health Policy Research Program (Grant #036273) and the National Science Foundation Program in Social Dimensions of Engineering, Science, and Technology (Grant # SES-9975518).

References

Beck, Ulrich. 1992. "From Industrial Society to the Risk Society: Questions of Survival, Social Structure and Ecological Enlightenment." *Theory, Culture and Society* 9:97–123.

Berkman, Lisa, and Ichiro Kawachi. 2000. *Social Epidemiology*. Cambridge: Oxford University Press.

Brauer, Michael, Gerard Hoek, Patricia Van Vliet, Kees Meliefste, Paul H. Fischer, Alet Wijga, Laurens P. Koopman, et al. 2002. "Air Pollution from Traffic and the Development of Respiratory Infections and Asthmatic and Allergic Symptoms in Children." *American Journal of Respiratory and Critical Care* 166:1092–1098.

Brown, Phil. 1984. "The Right to Refuse Treatment and the Movement for Mental Health Reform." *Journal of Health Policy, Politics, and Law* 9:291–313.

Brown, Phil, Brian Mayer, Stephen Zavestoski, Theo Luebke, Joshua Mandelbaum, and Sabrina McCormick. 2003. "The Politics of Asthma Suffering: Environmental Justice and the Social Movement Transformation of Illness Experience." *Social Science and Medicine* 57:453–464.

Brown, Phil, Stephen Zavestoski, Brian Mayer, Sabrina McCormick, and Pamela Webster. 2002. "Policy Issues in Environmental Health Disputes." *Annals of the American Academy of Political and Social Science* 584:175–202.

Brown, Phil, Stephen Zavestoski, Sabrina McCormick, Joshua Mandelbaum, Theo Luebke, and Meadow Linder. 2001. "A Gulf of Difference: Disputes over Gulf War–Related Illnesses." *Journal of Health and Social Behavior* 42: 235–257.

Brown, Phil, Stephen Zavestoski, Sabrina McCormick, Brian Mayer, Rachel Morello-Frosch, and Rebecca Gasior. 2004. "Embodied Health Movements: Uncharted Territory in Social Movement Research." *Sociology of Health & Illness* 26:1–31.

Bullard, Robert, ed. 1994. *Confronting Environmental Racism: Voices from the Grassroots.* Boston: South End Press.

Davis, Devra. 2002. *When Smoke Ran Like Water: Tales of Environmental Deception and the Battle against Pollution.* New York: Basic Books.

Davis, Devra and H. Leon Bradlow. 1995. "Can Environmental Estrogens Cause Breast Cancer?" *Scientific American* 273:167–172.

Epstein, Steven. 1996. *Impure Science: AIDS, Activism, and the Politics of Knowledge.* Berkeley: University of California Press.

Fauroux, B., M. Sampil, P. Quénel, and Y. Lemoullec. 2000. "Ozone: A Trigger for Hospital Pediatric Asthma Emergency Room Visits." *Pediatric Pulmonology* 30:41–46.

Freund, Peter, Meredith McGuire, and Linda Podhurst. 2003. *Health, Illness, and the Social Body: A Critical Sociology,* 4th ed. Englewood Cliffs, NJ: Prentice-Hall.

Garrett, Laurie. 1994. *The Coming Plague: Newly Emerging Diseases in a World Out of Balance.* New York: Farrar, Straus and Giroux.

Gelbspan, Ross. 1997. *The Heat Is On: The Climate Crisis, the Cover-Up, the Prescription.* Cambridge, MA: Perseus.

Gent, Janneane F., Elizabeth W. Triche, Theodore R. Holford, , Kathleen Belanger, Michael B. Bracken, William S. Beckett, and Brian P. Leaderer.

2003. "Association of Low-Level Ozone and Fine Particles with Respiratory Symptoms in Children with Asthma." *Journal of the American Medical Association* 290:1859–1867.

Gilliland, Frank D., Kiros Berhane, Edward B. Rappaport, Duncan C. Thomas, Edward Avol, W. James Gauderman, Stephanie J. London, et al. 2001. "The Effects of Ambient Air Pollution on School Absenteeism Due to Respiratory Illnesses." *Epidemiology* 12:43–54.

Gilliland, Frank D., Yu-Fen Li, Andrew Saxon, and David Diaz-Sanchez. 2004. "Effect of Glutathione-S-Transferase M1 and P1 Genotypes on Xenobiotic Enhancement of Allergic Responses: Randomised, Placebo-Controlled Crossover Stud." *Lancet* 363:119–125.

Goldstein, Michael. 1999. *Alternative Health Care: Medicine, Miracle, or Mirage?* Philadelphia: Temple University Press.

Greer, Linda, and Rena Steinzor. 2002. "Bad Science." *Environmental Forum* January/February, 28–43.

Klawiter, Maren. 1999. "Racing for the Cure, Walking Women, and Toxic Touring: Mapping Cultures of Action within the Bay Area Terrain of Breast Cancer." *Social Problems* 46:104–126.

Krieger, N., D. L. Rowley, A. A. Herman, B. Avery, and M. T. Phillips. 1993. "Racism, Sexism, and Social Class: Implications for Studies of Health, Disease, and Well-Being." *American Journal of Preventive Medicine* 9:82–122.

Krimsky, Sheldon. 2003. *Science in the Private Interest: Has the Lure of Profits Corrupted Biomedical Research?* Lanham, MD: Rowman & Littlefield.

Künzli, N., R. Kaiser, S. Medina, M. Studnicka, O. Chanel, P. Filliger, M. Herry, et al. 2000. "Public-Health Impact of Outdoor and Traffic-Related Air Pollution: A European Assessment." *Lancet* 356:795–801.

"Laxity in the Labs." 2001. *Boston Globe,* editorial, September 2, D6.

Le Roux, P., F. Toutain, and B. Le Luyer. 2002. "Asthma in Infants and Young Children. Prevention, Challenge of the 21st century?" *Archives of Pediatrics* 9:408s–414s.

Light, Donald. 2000. "Sociological Perspectives on Competition in Health Care." *Journal of Health Politics, Policy, and Law* 25:969–974.

McCally, Michael. 2002. Personal communication, November 15.

McConnell, R., K. Berhane, F. Gilliland, S. J. London, T. Islam, W. J. Gaudermann, E. Avol, H. G. Margolis, and J. M. Peters. 2002. "Asthma in Exercising Children Exposed to Ozone: A Cohort Study." *Lancet* 359: 386–391.

McCormick, Sabrina, Phil Brown, and Stephen Zavestoski. 2003. "The Personal Is Scientific, The Scientific Is Political: The Public Paradigm of the Environmental Breast Cancer Movement." *Sociological Forum* 18:545–576.

McKeown, Thomas. 1976. *The Modern Rise of Population.* New York: Academic Press.

Morello-Frosch, Rachel. 2002. "The Political Economy of Environmental Discrimination." *Environment and Planning. C, Government and Policy* 20:477–496.

Morello-Frosch, Rachel, Manuel Pastor, Carlos Porras, and James Sadd. 2002. "Environmental Justice and Regional Inequality in Southern California: Implications for Future Research." *Environmental Health Perspectives* 110 (Suppl. 2): 149–154.

Morgen, Sandra. 2002. *Into Our Own Hands: The Women's Health Movement in the United States, 1969–1990*. New Brunswick, NJ: Rutgers University Press.

Mortimer, K. M., L. M. Neas, D. W. Dockery, S. Redline, and I. B. Tager. 2002. "The Effect of Air Pollution on Inner-City Children with Asthma." *European Respiratory Journal* 19:899–705.

Ong, Elisa K., and Stanton A. Glantz. 2001. "Constructing Sound Science and Good Epidemiology: Tobacco, Lawyers, and Public Relations Firms." *American Journal of Public Health* 91:1749–1757.

Porter, Roy. 1997. *The Greatest Benefit to Mankind: A Medical History of Humanity*. New York: W. W. Norton.

Raffensperger, Carolyn, and Joel Tickner, eds. 1999. *Protecting Public Health and the Environment: Implementing the Precautionary Principle*. Washington, DC: Island Press.

Ries, L. A. G., M. P. Eisner, C. L. Kosary, B. F. Hankey, B. A. Miller, L. Clegg, and B. K. Edwards, eds. 2002. *SEER Cancer Statistics Review, 1973–1999*. Bethesda, MD: National Cancer Institute.

Rose, Geoffrey. 1985. "Sick Individuals and Sick Populations." *International Journal of Epidemiology* 14:32–38.

Rosenstock, L, and L. J. Lee. 1992. "Attacks on Science: The Risks to Evidence-Based Policy." *American Journal of Public Health* 92:14–18.

Rosner, David, and Gerald Markowitz. 1987. *Dying for Work: Workers' Safety and Health in Twentieth-Century America*. Bloomington: Indiana University Press.

Rudel, Rurhann A., David E. Camann, John D. Spengler, Leo R. Korn, and Julia G. Brody. 2003. "Phthalates, Alkylphenols, Pesticides, Polybrominated Diphenyl Ethers, and Other Endocrine Disrupting Compounds in Indoor Air and Dust." *Environmental Science and Technology* 37:4543–4553.

Ruzek, Sheryl Burt. 1978. *The Women's Health Movement: Feminist Alternatives to Medical Control*. New York: Praeger.

Ruzek, Sheryl Burt, Virginia L. Olesen, and Adele E. Clarke, eds. 1997. *Women's Health: Complexities and Differences*. Columbus: Ohio State University Press.

Schell, L., and A. Tarbell. 1998. "A Partnership Study of PCBs and the Health of Mohawk Youth: Lessons from Our Past and Guidelines for Our Future." *Environmental Health Perspectives* 106 (Suppl. 3): 833–840.

Schulte, P. A., and M. H. Sweeney. 1995. "Ethical Considerations, Confidentiality Issues, Rights of Human Subjects, and Uses of Monitoring Data in

Research and Regulation." *Environmental Health Perspectives* 103 (Suppl. 3): 69–74.

Schulte, P. A., D. Hunter, and N. Rothman. 1997. "Ethical and Social Issues in the Use of Biomarkers in Epidemiological Research, in Application of Biomarkers in Cancer Epidemiology." *International Agency for Research on Cancer* 1422:313–318.

Shapiro, Joseph. 1993. *No Pity: People with Disabilities Forging a New Civil Rights Movement.* New York: Random.

Shepard, Peggy M., Mary E. Northridge, Swati Prakash, and Gabriel Stover. 2002. Preface: Advancing Environmental Justice through Community-Based Participatory Research. *Environmental Health Perspectives* 110 (Suppl. 2): 139–140.

Steinbrook, Robert. 2002. "Protecting Research Subjects—The Crisis at Johns Hopkins." *New England Journal of Medicine* 346:716–720.

Szasz, Andrew. 1994. *Ecopopulism: Toxic Waste and the Movement for Environmental Justice.* Minneapolis: University of Minnesota Press.

Tesh, Sylvia. 2000. *Uncertain Hazards: Environmental Activists and Scientific Proof.* Ithaca, NY: Cornell University Press.

U.S. Environmental Protection Agency (EPA). 2003. *America's Children and the Environment: Measures of Contaminants, Body Burdens, and Illnesses.* 2nd ed. EPA 240-R-03-001. Available at www.epa.gov/envirohealth/children/ace_2003.pdf (accessed November 3, 2004).

Waitzkin, Howard. 2001. *At the Frontlines of Medicine: How the Health Care System Alienates Doctors and Mistreats Patients.* Lanham, MD: Rowman & Littlefield.

Weinberg, Alvin. 1972. "Science and Transcience." *Minerva* 10:209–222.

Weiss, Rick. 2004. "Peer Review Plan Draws Criticism under Bush Proposal, OMB Would Evaluate Science before New Rules Take Effect." *Washington Post*, January 15, A19.

10

Strategies for Alternative Science

BRIAN MARTIN

Professor Smith, to his class: This semester we've been looking at cultural contradictions of science, including contradictory popular images of science as liberator and science as oppressor, contradictory views of scientific research as autonomous and as socially determined, and contradictory conceptualizations of scientific practice as formal method and as localized craft activity. To conclude, I'd like to mention something we haven't covered: the idea of alternative science. Nicholas Maxwell, a philosopher of science, describes existing science as conforming to what he calls the "philosophy of knowledge." Knowledge is the goal, without any judgment about how that knowledge will be used. In other words, knowledge is seen as a good in itself, indeed almost an overriding good. Maxwell (1984, 1992) subscribes to an alternative that he calls the "philosophy of wisdom." In this vision, science would be oriented to solving pressing human problems, including hunger, inequality, environmental degradation, war, and oppression. Maxwell notes that lots of scientific research is driven by military and corporate funding, and that some of the world's most talented scientists and engineers devote their efforts to designing more ingenious fragmentation bombs or detergents that leave your dishes sparklingly clean. So here we have another contradiction: science, a system designed for creating objective knowledge actually ends up creating knowledge that is mainly useful to vested interests.

But there are visions of alternative science. The appropriate technology movement has pushed for technology designed for poor people in poor countries, such as efficient stoves and irrigation pumps, that can be locally produced and maintained (Darrow and Saxenian 1986). The alternative health movement has supported investigation of nonmedical approaches to health—for example, nutritional prevention and treatment of cancer (Johnston 2003). These examples are at the technology or applied end of the spectrum, but as we've seen, scientific research and technological development feed off and indeed help constitute each other in a process sometimes called *technoscience*. The point here is that the research that is actually done is only a portion of what could be done, and that powerful interest groups influence research agendas and outcomes. In this class we've studied science as it exists, but there is also a subterranean subject available for study, what might be called "undone science" (Woodhouse et al. 2002), into which much of Maxwell's philosophy of wisdom falls.

Existing research shows inklings of alternatives. There is some research on biological control of pests, though a lot more on pesticides. There is some research on cooperative enterprises, but a lot more on competitive markets. But in some areas, it's possible that we don't even know what alternatives are not being investigated, so dominant are conventional agendas.

Powerful groups pour massive amounts of money into scientific research to obtain products that serve their interests but also because of science's reputation as a source of objective knowledge. It is a central contradiction of science, then, that its image of objectivity is the source of the greatest threats to that objectivity. Alternative science is the embodiment of that contradiction.

CHRIS [*later*]: "Could I speak to you for a moment, Professor Smith?"

PROF: "Of course."

CHRIS: "I've really enjoyed taking your class 'The New Cultural Contradictions of Science.' It's given me a good understanding of the complexities of the science scene today and the way theory can be used to probe these complexities . . ."

PROF: "Thanks—"

CHRIS: "But there's one thing I'd like to ask. You probably know I'm involved in the local group 'Science Justice.' Right now we're trying to develop some plans for the next year, and we want to think about campaigns that can help transform science in a participatory, egalitarian direction. I'm just wondering what the theorists we've studied have to say about that."

PROF: "Well, they point to the role of structural factors in establishing the context for action. There's some insightful work on the way governments and corporations set agendas through media framing. That's certainly relevant to science . . ."

CHRIS: "But that's still more about what we're up against. What about insights about how we go about changing things?

PROF: "Have you looked at the social movement literature? Resource mobilization, political process theory, dynamics of contention? Some of the key sources are on the reading list."

CHRIS: "Yes, that was my first stop. But when I explained the key ideas to the others, they couldn't see their relevance. Everything was either too abstract or obvious to our experienced members."

PROF [*after a pause*]: "Let me think about this and check with a few others. I'll get back to you."

CHRIS: "Thanks."

This dialogue is fictional but points to a real issue: scholarly analysis of science, like most other scholarly work, has little useful to say directly to activists. In the past half century, the analysis of science has become increasingly sophisticated, with attention to complexities and contradictions, including the socially constructed nature of knowledge production, the roles of governments, corporations, and professional structures, and the impacts of globalization, regulation, and citizen action.

This analysis has many strengths. It undermines simplistic understandings of science common in the media and popular discourse as the truth, as neutral, and as an inevitable source of progress. It points to the role of social factors in all dimensions of science, opening the door to alternative conceptions of scientific knowledge, scientific practice, and institutions of science. It reveals, through many case studies, the intertwining of power and knowledge in science.

Yet this impressive evolving scholarly achievement has limitations, at least from the point of view of some who would like to use the ideas as a basis for action. Scholarly work is often difficult to understand, requiring considerable training and expertise. The audience of most scholarly papers is largely other scholars. Newcomers can have a difficult time making sense of the field. Furthermore, much of the analysis, in terms of relevance to practice, is not inspiring: it is analysis, after all, not a set of success stories or how-to manuals.

Scholars and teachers have devoted a lot of effort to debunking technological determinism (Smith and Marx 1994; Winner 1977), but this is largely superfluous for activists who go ahead and challenge technological developments without a thought about the theoretical issues involved. There is much sophisticated analysis of the social shaping of technology (MacKenzie and Wajcman 1999; Sørensen and Williams 2002), but it gives little guidance to those want to go out and help do the shaping. Michel Foucault is widely cited for his ideas about power and knowledge, but since the 1960s feminists and other activists have been debating and living issues of power and knowledge—epitomized by the slogan "the personal is political"—often without the slightest awareness of Foucault or other theorists. Indeed, it is tempting to suggest that scholars have picked up on a change in social consciousness, but rather than acknowledge social movements as the vanguard of this change they instead cite the scholars who best capture the same orientation in dense theory.

Over several decades, I've talked with numerous activists about theory and strategy, both in interviews and informally. It quickly becomes obvious that very few of them spend any time perusing academic writings. Many are completely caught up in current campaigns and look only at materials that are directly relevant. Jargon-ridden articles dealing with complexities and qualifications are not for them. There are some activists who have been deeply shaped by intellectual work—for example, women who have been inspired by prominent feminist writings. Also, there are some scholar-activists, in particular students and academics, who are involved in activist groups.

A large amount of activism is short-term, with immediate practical goals such as organizing a meeting or rally, circulating information, raising money, or dealing with internal differences. There's too much to do and not enough time or people to do it. Developing visions of alternatives and undertaking long-term planning are not often on the agenda.

So is it worthwhile for activists to comb the academic literature for useful material? Few scholars have much to say about how activists can be more effective in day-to-day campaigning: there's plenty of research on making factories and offices more efficient, even some on enriching work for the workers, but hardly anything focusing on activist groups. And those areas where activists can benefit most from intellectual input—alternatives and strategies—are precisely the areas given little attention by scholars.

Edward de Bono, most well known for the concept of lateral thinking, has developed many other tools for thinking (de Bono 1992). Among them are the "six thinking hats" that divide thinking into six categories: the white hat for dealing with information, the black hat for critical analysis, the red hat for emotional responses, the green hat for creative ideas, the yellow hat for optimism, and the blue hat for managing thinking (de Bono 1986). It is a simple observation that most academic work involves the white and black hats: information and critical analysis. I've given many a seminar presenting new ideas; the most common responses are requests for more information (white hat) and critical comments (black hat). Very seldom does anyone use the opportunity to suggest wilder ideas (green hat). My conclusion is that academics find safety in critique: critically analyzing the work of others minimizes the risk of counterattack, whereas presenting a new idea—which is almost bound to have limitations—is to open oneself to critique by others. If those others are referees, you may not be published. A black hat culture thus reproduces itself.

De Bono (1995) points out that critique as a method of obtaining truth—a black-hat approach—works if it cuts away at weak parts of evidence and argument, revealing a core of solid intellectual material. This might work well in "normal social science," analogous to Kuhn's normal science, but it is not the way to proceed if there are other, alternative constructions—in particular, when knowledge is created by design. It is worth noting that the postmodernist preoccupation with deconstruction elevates critique above all other intellectual tools.

In the spirit of de Bono's green and yellow hats, my plan here is to look at models and strategies. First I pick out a sample of visions of alternative science. Then I look at a selection of strategies to move toward these alternatives. Along the way I examine some different roles people can play in this process. Finally, I illustrate these ideas with a case study of defense technology.

There's a certain tension in writing about these matters in the company of some of the most committed critical thinkers in the field. My intent is not to criticize scholarly work; after all, I do plenty of it myself.

Rather, it is to point to some areas that both scholars and activists can explore. In keeping with my comments above about the limited accessibility of much academic writing, I attempt to write simply and clearly, knowing that readers of the draft will help me sharpen the argument. The black hat has its function! I aim to write clearly even though I'm aware of research (Armstrong 1980) showing that most readers of an abstract that was difficult to read thought the author to be higher in research competence than the author of a more readable abstract, even though the content of the abstracts was the same.

Visions of Alternative Science

To speak of alternative science may seem utopian, because current scientific institutions are so entrenched and difficult to change. It is useful to remember that science is dynamic and responsive to external pressures. Injections of funds can and do lead to shifts in research directions. What is entrenched about the system is the dominance of government, corporate, and professional influence over science. It is this that makes science for the people seem like wishful thinking.

Yet, on closer inspection, quite a bit of scientific research can be interpreted as "for the people." Much research is driven, directly or indirectly, by practical applications, and many applications are largely beneficial or innocuous, such as fabrics, toothbrushes, CDs, insulation, and bicycles. Much of the generic science associated with such products—such as kinetic theory or optics—seems not to be a source of problems. Things become more obviously "political" when applications are contentious: weapons, genetically engineered crops, surveillance equipment. The same holds for science that is driven by or incorporated in such technologies.

There are many possible visions of alternative science, so it may be more accurate to speak of alternative *sciences*. To illustrate possibilities, I present four examples.

1. *Science for the people (rational version).* Bernal (1939) had a vision of science at the service of society, in which an enlightened, rational government directed scientific research into areas of greatest benefit to society. This can be called a socialist model, if the concept of "socialism" can be divorced from its association with dictatorial regimes and used to denote a polity in which

government truly serves the needs of the people, at the same time subordinating special interests—corporations, churches, professions—that might shape science for their own ends.

This is a thoroughly technocratic vision: managers and experts have a great deal of power, which is assumed to be used entirely for the good of the population. Rather than pursuing research for profit and control, a rational government would put priority on human well-being in the widest sense. For example, manufacturing technology would be developed to produce practical products in safe and stimulating working conditions. Transport systems would be developed that balanced cost, equity of access, safety, environmental impact, and convenience.

2. *Science for the people (pluralist version).* In this vision, scientists undertake socially relevant research and development (R&D) due to oversight and pressure from citizens. Research agendas, rather than being dominated by corporate and government imperatives and thinking, are shaped by wider social priorities as articulated by individuals and groups that are in touch with genuine social needs. Social welfare is ensured not through rational assessment from the top but instead by numerous formal and informal channels of influence from the grassroots, including direct contact with scientists, citizen presence on advisory and funding bodies, citizen input into the training of scientists, widespread participation in public debates about social and scientific priorities conducted through the mass media and alternative media, and citizen involvement in formal processes of research planning and evaluation. The result is heavy citizen involvement in the social shaping of science, but without a single voice being guaranteed dominance in the public debate. Science would thus become responsive to a plurality of voices, resulting in diversity and flexibility.

3. *Science by the people.* Rather than research being done by professional scientists, in this vision citizens would themselves become scientists: many citizens would participate in research activities at some periods in their lives. In a utopian picture of science in China under the Cultural Revolution, Science for the People (1974) described how peasants and workers were involved in setting research problems and proposing solutions and how scientists were oriented to the problems of ordinary people. Though this image of Chinese science was undoubtedly unrealistic, it nonetheless offers a vision of self-managed science, namely science done *by* the people rather than *for* the people by professional

scientists. Science by the people implies a radical restructuring of education and scientific method, so that popular participation becomes both expected and much easier. An analogy would be the way that information searching has become democratized through the availability of the Internet.

4. *Science shaped by a citizen-created world.* If science is shaped by the society in which it is developed and applied (MacKenzie and Wajcman 1999), then a different world is likely to lead to a different science (Martin 1998). Scientific priorities would be quite different in a world in which workers and communities directly determine what is produced and how it is produced, in a world in which intellectual products are freely shared rather than owned, or in a world in which energy systems are built around local renewable sources, produced and maintained locally. In this vision of alternative science, the key is the way society is organized. Here, the form of organization is taken to be "citizen-created," at a local level, with widespread participation.

Though each of these visions is dramatically different from present-day science, nevertheless it is possible to see elements of each vision in science as it now exists. Aspects of vision 1, built around rational planning with the aim of serving the people, can be seen in some university research, some government research, and even some corporate research. Examples include research into energy efficiency, nutritional prevention of disease, human-centered manufacturing, and aids for people with disabilities. Indeed, a long list of socially relevant research could be compiled. Rationality and altruism in research are potentially compatible.

Vision 2, science for the people as a product of multiple influences on research, is most obvious in areas of contested policy, in which competing groups seek to influence research agendas (Primack and von Hippel 1974). When policies are challenged and debated, it is a sign that no single influence on research is hegemonic. Debates about climate change have led to a vast amount of research in the area, which in turn has fed into the ongoing debates. AIDS research has been stimulated and partially shaped by AIDS activists (Epstein 1996). There are also various means for more routine citizen input in science, such as science shops in the Netherlands and other countries (Farkas 1999), consensus conferences as used by the Danish Board of Technology and elsewhere (Fixdal 1997), and policy juries made up of randomly selected citizens (Carson and Martin 2002).

Vision 3 of science by the people can be found in some mainstream scientific areas such as astronomy, where amateurs play an important role (Ferris 2003). It is also found in community research, where groups of citizens undertake projects of direct concern to their lives, such as on local environmental issues (Murphy, Scammell, and Sclove 1997; Ui 1977). Science by the people is often defined out of existence by the boundary work of professional scientists, who have an interest in being the sole proprietors of what counts as science (Gieryn 1995).

Science shaped by a citizen-shaped world, vision 4, can be found wherever popular initiatives have changed the research agenda. Campaigns for occupational health and safety have led to altered research priorities, even though most of those campaigns focused on immediate issues, not research agendas. Just as important are areas where research has been reduced. The success of the movement against nuclear power has contributed to a reduction in nuclear research; the campaign that stopped supersonic transport aircraft led to the demise of much associated research.

Strategies

Each of the four visions can be related to wider strategies for creating alternative science. Out of many possible strategies, I consider just four. They have obvious connections with visions 1 to 4, respectively, but are not restricted to pursuing a single vision.

Strategy 1 is a state-led transformation of science. This is the traditional socialist approach, with control of the state achieved either by revolution or more gradually by election of a socialist party that implements policies bringing about a socialist society. This strategy, in both main variants, is widely recognized to have failed: most socialist states have collapsed, and most socialist parties elected to office have adapted to capitalism (Boggs 1986). Nevertheless, this strategy remains important in that much citizen effort is invested in supporting left-wing parties and promoting progressive policies within them.

It is possible to conceive of a different transformation of science from the top, led by capitalists. The idea would be for corporations to be colonized by managers who put a priority on the public interest. Very few activists even imagine such a strategy, much less put energy into it.

Large corporations are authoritarian in structure, with few formal openings for citizen or worker input, whereas systems of representative government have at least the facade of participation. If strategy 1 is recast as top-down transformation of science, then capitalist-led transformation can be labeled strategy 1C, state-led transformation as strategy 1S, and profession-led transformation as strategy 1P. Strategy 1C should not be dismissed too easily. There are some visionaries in the corporate sector who seek a transformation of capitalism into a more humane system (Soros 2002; Turnbull 1975).

Strategy 1 can also be adopted by individuals or attributed to them as an unconscious guide to behavior. Consider a scientist who works for a corporation or government, who is not linked with any outside groups, and who does not have explicit affinities with social causes. Such a scientist, in choosing research projects or undertaking evaluations of research, can choose to make decisions in the light of a belief that science should serve the public interest. For example, a biotechnologist might decide to explore genetically modified crops that need less rather than more pesticides; a weapons researcher might investigate designs that reduce long-term environmental impact; an automotive engineer might look at ways to reduce fatigue in drivers. Similarly, in evaluating the research of others, such scientists may give priority to options that are better for "the people" or those in greatest need rather than powerful and privileged groups. (See also the chapter by Woodhouse in this volume.)

Insider scientists need not draw attention to themselves in making these choices. Because of the interpretive flexibility available to researchers, choices can be justified on rational grounds, such as efficiency, cost, and simplicity, without having to argue in terms of human interest. If social welfare is explicitly accepted in an organization as a relevant criterion, then it is easier to justify choices that serve the public interest.

All this can happen in isolation from social movements and others who articulate alternatives to science in the service of power. When movements exist, then scientists are more likely to become aware of the down sides of their enterprises. Biotechnologists can become aware of the exploitation of genetic resources from indigenous peoples; military researchers can become aware of the devastating effects of weapons and war; automotive engineers can become aware of the environmental and human costs of automobilization. Scientists can choose to move out of damaging areas and into alternatives, though often such choices come at a major cost to one's career. If enough scientists push for change and enough policymakers and research managers are willing to accept or promote change, this is compatible with vision 2: the research agenda is responsive to wider social priorities.

Strategy 2 is pressure-group transformation of science. This includes, for example, campaigns against research on genetically modified crops and for development of cheap pharmaceutical drugs for common diseases in poor countries. Feminists, environmentalists, neighborhood groups, and many others play a role. This strategy relies on there being many pressure groups at the grassroots, sufficient to orient research to the public interest, though "the public interest" can be a multifaceted and changeable object due to differences between pressure groups. Citizen activism is now routine in many countries, but despite this there is relatively little direct citizen pressure on scientific research priorities. Strategy 2 would see a vast expansion of this pressure.

A limitation of this strategy is that pressure groups seldom provide a balanced representation of those in greatest need: the powerful have influential pressure groups, and the weak have few or none. In sporting metaphor, pressure group teams do not compete on a level playing field, and some teams are not even in the game. Another way to say the same thing is that pluralist politics operates within a polity with institutionalized bias.

Strategy 3 is living the alternative of participatory science. Rather than trying to bring about change from the top or by pressuring those at the top, this approach is direct: go ahead and start doing science by the people. Examples include amateur scientists in a few fields, such as astronomy and botany. The community research movement is closest to strategy 3: citizens, sometimes in collaboration with sympathetic scientists, undertake projects that are directly relevant to their concerns, such as dealing with environmental justice issues. (Sometimes these initiatives influence professional scientists, such as when disease sufferers promote a different conception of an illness than medical orthodoxy [Kroll-Smith and Floyd 1997]: this is a mixture of strategies 2 and 3.)

These initiatives serve several purposes. They are powerful learning experiences for participants, giving insight both into the process of scientific research and into the politics of science. They provide a demonstration effect, showing that nonprofessionals can make useful contributions to knowledge. In principle, they lay the basis for a gradual deprofessionalization of science. The limitation of this strategy is that the small initiatives in citizens' science are too easily marginalized. Professionals may ignore, tolerate, or denigrate and undermine amateurs in their fields, in any case not ceding significant prerogatives. Even highly successful citizen research usually has little effect on the multibillion-dollar professional research enterprise. One prominent exception is the open source movement, which is built on voluntary contributions to collective enterprise. Open source software has, in a short time, become a major challenge to proprietary software (Moody 2002), and the open source model is seen by some as an alternative method of production.

Strategy 4 is grassroots empowerment for social change that transforms the conditions under which science is done. This is vision 4 turned into a process. It involves a wide variety of social movements bringing about major change in personal relationships, working conditions, products produced, the energy system, and a host of other areas. These changes will inevitably have an effect on the content and practice of science.

Peace movements played a crucial role in bringing an end to most tests of nuclear weapons; this in turn led to a reduction in test-related nuclear research. Environmental movements put their concerns on agendas worldwide, with wide-ranging ramifications for research in many fields. This process could go much further. If peace movements were to be successful in abolishing nuclear weapons, then certain forms of nuclear knowledge could atrophy (MacKenzie and Spinardi 1995). If environmental movements could bring about substantial institutional reform—for example, by replacing industrial agriculture with organic farming or replacing automobile-centered transport systems with urban planning for pedestrians, cyclists, and public transport—then research agendas would be more drastically altered.

A shift to more cooperative and egalitarian interpersonal relationships—something sought by some feminists and others—would affect the ethos of science, undermining the research hierarchy in which a few elite scientists have a grossly disproportionate influence over research directions (Blissett 1972; Elias, Martins, and Whitley 1982).

The limitation of this strategy is that it relegates change in science to a secondary outcome: change elsewhere comes first and only afterward in science. This might not matter except that science plays a crucial material and ideological role in current social arrangements. R&D oriented to an alternative society is needed now to help aid the process of change.

These four strategies are hardly new: they encapsulate well-trodden political paths. Strategy 1 is the usual socialist and social democratic path. Strategy 2 is the familiar approach of instigating reform from below. Strategy 3, living the alternative, reflects the philosophy of anarchists, direct actionists, Gandhians, and some others. It is sometimes called "prefiguration": the alternative, practiced now, is modeled on a desired future and shows that it is possible.

Strategy 4 can be called "after the revolution." In traditional Marxist analysis and practice, class contradictions took priority, with other issues postponed until after the overthrow of capitalism. Feminists disagreed, arguing that patriarchy was not subsidiary to capitalist rule, and others similarly challenged the Marxist hierarchy of oppression. Strategy 4, as applied to science, assumes that change in science can be left until after change elsewhere.

Although it can be revealing to link the four strategies for change in science to wider strategies of social change, there is also a risk: many people would find it uncomfortable or offensive to be categorized as, or even associated with, socialism, reformism, anarchism, or after-the-revolutionism. These labels have all sorts of connotations that are potential distractions from the assessment of different forms of action. As discussed above, it is possible for scientists and nonscientists to work in a variety of different ways, and it may be counterproductive to introduce off-putting labels.

On the other hand, the value of pointing out connections between strategies for alternative science and wider strategies for social change is to draw attention to likely areas of strength and weakness. If a weakness of the strategy of social democracy is that party leaders become captives of the capitalist system, then it is worth looking for parallel weaknesses in the strategy of state-led transformation of science. But it would be unwise to dismiss this science strategy altogether on the grounds that social democracy has failed to live up to its initial expectations of replacing capitalist social relations.

Each strategy can be used as a guide for individuals who would like to help create alternative science. Scientists can act directly on the basis of their own judgments of the public interest or in response to popular movements. Citizens can apply pressure to scientists and science policy-makers. They can instigate or join community research projects themselves. And they can participate in movements that have the capacity to bring about social change that can shape research agendas.

My assumption here is that strategies need to be assessed in particular applications. In the jargon, strategies are "practical accomplishments" rather than theoretical conclusions. Looking at visions and model strategies can offer ideas to those trying to figure out what to do.

To see how this might work, I now turn to a case study: defense technology. This is an applied area with ties to more fundamental research. I choose an applied area because that is where social movements are more active, so that each of the models and strategies is potentially relevant. Activists whose targets are topology or nucleosynthesis are scarce on the ground. My aim is to pull out some insights that might apply to alternative science in general. Other examples in contested areas, such as climate change or biotechnology, would work just as well.

Defense Technology

A significant proportion of the world's scientists and engineers work in the military-industrial complex, with research covering nearly every field of natural and social science, such as oceanography, control engineering, and the psychology of groups (Mendelsohn, Smith, and Weingart 1988; Smith 1985). Military R&D therefore is a prime area for assessing visions and strategies. It is typical of R&D in applied areas, in which science and technology are assessed in relation to a purpose, in this case defense. On the other hand, military R&D is atypical to some extent in being driven by state imperatives, where cost is less of a consideration compared to areas of R&D driven by market factors.

In examining visions of alternative science, an immediate problem arises: there are competing visions of the future of defense technology. One is peace through strength, achieved with ever-improving technologies. Another is weapons design that minimizes civilian casualties. Another is designing weapons that are easy to use for defense but not for offense. Yet another is elimination of weapons altogether. For my purposes here, it is sufficient to specify a direction, rather than an end point: a humanitarian alternative science for defense should move in the direction of fewer casualties, lower environmental impact, greater orientation to defense (rather than offense), and greater orientation to nonmilitary means for achieving security and resolving conflicts. This provides a sufficient framework to assess each of the four visions of alternative science and the associated strategies.

Vision 1 is of science for the people implemented on a rational basis by policymakers and scientists. Given the continuation of massive R&D efforts to produce new weapons systems, this seems a forlorn hope. It is well known that militaries of leading powers seek to develop ever more powerful and effective tools to wage war; once such tools are available, other militaries follow suit to develop or acquire these weapons. Although the overall dynamic of military development seems in direct contradiction to vision 1, nevertheless there are elements of this vision in operation. Many government leaders have taken meaningful stands against military races. For example, quite a few governments have voluntarily refrained from developing nuclear weapons and have supported arms control treaties.

Strategy 1, a state-led transformation of military-related science, has some prospect of success in individual states, but this has been insufficient to redirect military R&D more widely. With the end of the Cold War in 1989 and the collapse of the Soviet Union two years later, there was much talk of a "peace dividend," namely a redirection of military spending toward civilian priorities. In practice, global military spending continued without drastic change, suggesting that it is driven internally more than by a rational examination of external threats.

Only three major industrialized countries adopted the alternative path of arming the population: Sweden, Switzerland, and Yugoslavia (Roberts 1976). However, this seems not to have led to major transformations of military R&D; the Yugoslav experiment ended in disastrous war making. A more radical alternative is to get rid of armies altogether. There are dozens of tiny countries without armies, of which Costa Rica is the most well known (Aas and Høivik 1986), and one-third of Swiss voters supported a referendum to abolish their army. But countries without armies have a negligible impact on global military-related R&D. Nor have any of them pioneered nonmilitary defense research programs.

Scientists and engineers are largely silent players in military R&D. Although many individuals refuse to be involved, militaries seem to have little trouble finding sufficient numbers of qualified people to do their bidding (Beyerchen 1977; Haberer 1969).

In summary, state-led strategies in the defense area have failed to restrain global military R&D and have done very little to support alternative models of defense.

Vision 2 is of defense R&D that is responsive to citizen input. Although some citizen groups push for more military spending, my focus here is on pressures in the opposite direction. There are many citizen movements, large and small, that can be cited. Anti-nuclear movements have been most prominent. There were major worldwide mobilizations against nuclear weapons in the late 1950s and early 1960s and then again in the 1980s, with some significant activism at other times, too. There have also been focused movements to oppose chemical and biological weapons, space weapons, and anti-personnel weapons, among others. Although there has been little formal assessment of the effectiveness of these movements, it is reasonable to conclude that they have played a significant role in stopping or restraining some types or deployments of weapons systems. But movement successes may not last. The 1972 treaty against anti-ballistic missiles was signed partly due to peace movement pressure, including the efforts of many scientists; in December 2001, the U.S. government quietly withdrew from the treaty.

Pressure groups have been far less successful in changing the agenda of defense R&D. There have been campaigns for what is called "peace conversion" or "economic conversion," namely the conversion of military production facilities to production for human needs, such as converting military vehicle production facilities to produce civilian vehicles (Cassidy and Bischak 1993; Melman 1988). Conversion activists have mustered powerful arguments and occasionally mobilized direct action, but they seem to have had relatively little impact. Some conversion efforts have focused on military research, with local successes but again apparently with relatively small impact overall (Reppy 1998; Schweitzer 1996).

The pressure group strategy has the advantage of tapping into potentially widespread citizen antagonism to war. The amount and range of peace activism is inspiring (Carter 1992). It includes mass rallies, vigils, strikes, boycotts, and blockades. Activism has been ably supported by intellectual work in collecting information, building arguments, writing articles, and producing documentaries. Although it is difficult to precisely trace cause and effect, it is plausible to argue that citizen efforts against war have been instrumental in deterring the use of nuclear weapons after 1945 and in helping bring a largely peaceful end to the

Cold War through the collapse of the Soviet bloc (Cortright 1993; Summy and Salla 1995).

On the other hand, peace movements have an erratic history, with periods of mass mobilization being less common than periods of relative quiescence. For example, anti-nuclear activism, after reaching a peak in the early 1980s, declined precipitously in the 1990s, even though arsenals of nuclear weapons remained largely unchanged. Overall, peace movements have failed to thwart the momentum of the military-industrial complex. Strategy 2, pressure-group politics, seems to have had only limited impacts on military R&D agendas.

Vision 3 is science by the people. What can this mean in the context of defense technology? It is possible to imagine self-managed teams of workers designing weapons or running factories to manufacture them, but this is hardly a vision of a society liberated from violence.

The most famous worker initiative for alternatives to military technology was the Lucas Aerospace workers' plan (Wainwright and Elliott 1982). In the 1970s at Lucas, a major British military contractor, workers were worried about loss of jobs and developed an alternative plan to produce nonmilitary products using their skills, in the process developing prototypes of road-rail vehicles and kidney machines. The Lucas Aerospace Shop Stewards' Committee initially sought ideas from experts but obtained a pitifully small response, so they turned instead to the workers, who produced a wealth of ideas. As well as being prolific innovators, the workers were community minded, putting priority on serving human needs rather than just the interests of the workers themselves. This experience provided encouragement that "technology by the people" would also be technology that served the broadest human interest.

The primary strategy of the Lucas workers was to propose their alternative to management, along the lines of strategy 2. Management, though, did everything possible to oppose the workers' initiatives, including turning down projects that promised profits. An obvious interpretation is that managerial control was more important than the prosperity of the enterprise, not to mention wider human welfare. The workers' development of prototypes can be considered an instance of strategy 3, living the alternative; this initiative captured imaginations in many countries.

One reason the Lucas workers' initiatives received so much attention is that they were highly unusual, as is any form of science by the people. Most peace conversion efforts have used pressure group approaches, with direct action by workers being rare. Most military workers are reluctant to jeopardize their jobs and wages by pushing for production of alternative products; taking initiatives can lead to reprisals.

There is another limitation to peace conversion: it does not provide a full alternative to military systems. In periods of low military threat, it can seem reasonable to convert some military production to civilian outputs, but not all of it. Therefore it is worthwhile considering entirely different alternatives. One such alternative is nonviolent defense: defense of a community through organized methods of nonviolent action including rallies, boycotts, strikes, sit-ins, and alternative institutions. Although at first glance this may sound impractical, actually there are many historical examples showing that popular nonviolent action can be effective against repressive regimes, such as the Iranian revolution of 1978–1979, the toppling of the Marcos dictatorship in the Philippines in 1986, the collapse of Eastern European communist regimes in 1989, the ending of apartheid in South Africa, and the end of Suharto's dictatorial rule in Indonesia in 1998. Inspired by such examples, theorists have proposed that with suitable preparation, nonviolent action could form the basis for a defense system. This is also called social defense, civilian-based defense, and defense by civil resistance (Burrowes 1996; Randle 1994).

Science and technology could play an important role in a system of nonviolent defense. For example, decentralized communication methods—including telephones and e-mail—are especially useful to opponents of aggression, whereas mass media are usually of more value to aggressors: in military coups, the first target is television stations. Decentralized energy sources, such as solar collectors and wind generators, are more suitable for defending a society nonviolently than large electricity-generating plants and large dams, which are vulnerable to both aggressors and terrorists. Proceeding through a range of systems—agriculture, water, manufacturing, housing—it is possible to come up with a wide-ranging agenda for science and technology suitable for nonviolent defense (Martin 1997, 2001).

In practice, initiatives have been taken in many of these areas. For example, there is a large amount of R&D on energy efficiency and renewable energy sources. This effort is inspired not by relevance to nonviolence defense but by other concerns, such as reducing environmental damage or promoting Third World development. It turns out that there is a strong compatibility between technology for nonviolent defense and what is commonly called appropriate technology. Appropriate technology—technology oriented to people's needs, often designed so that users can directly control and adapt it—lies close to the vision of science by the people (Boyle, Harper, and *Undercurrents* 1976; Illich 1973). Promotion of alternative technology by grassroots groups fits with strategy 3, living the alternative. So far, though, this approach has not been used as a means of supporting nonviolent defense. Indeed, nonviolent defense remains off the agenda even of most peace movements.

Vision 4 is of science shaped by a citizen-created world. Applied to defense technology, this means that defense policies would be determined by citizens in a participatory process, and that science and technology related to defense would then reflect those policies. What this might mean in practice depends on what policies would be chosen, for which there is no single answer, but some possibilities can be examined.

One obvious possibility is elimination of technologies that can kill or destroy on a large scale, that are targeted at civilians, or that are designed for repression of dissent. This includes everything from fuel-air explosives to land mines to thumb cuffs. Elimination of such technologies would shift R&D priorities. For example, missile research would be reduced whereas research into small-scale solar energy would not be affected. Another policy possibility is much greater emphasis on technologies designed for civilian purposes, for example research on how to keep people in civilian occupations healthy rather than how to keep soldiers alive and able to fight.

Looking more broadly, a citizen-created world might have a different political and economic system. Indeed, there might be a diversity of co-existing systems with the constraint that none is aggressive and domineering. To take an example, a community might introduce its own system of money and replace intellectual property laws with alternative systems to foster local innovation and creativity. With a considerable divergence from the economic and political conditions of research today, the result could well be research directions pursued in ways that differ considerably from present-day priorities and methods. This could include elements of vision 3, science by the people.

How might vision 4 be achieved? State-led strategies could in principle help bring about a citizen-created world, but in practice this approach has not led to a different direction for defense. Socialist governments, whether state socialist or social democratic, have largely followed the same road on defense as other governments, namely the usual form of military forces and weapons systems. Some socialist-oriented liberation struggles have adopted weapons and methods of struggle in the mode of guerrilla warfare. Once successful, though, they usually move to the conventional military model.

Pressure-group methods can help move toward a citizen-created world. In the defense area, though, most campaigns do not question basic assumptions about the need for (military) defense, much less the existence of the state, large-scale industry, and bureaucracy, all of which underpin today's military systems. Banning landmines by itself does not transform the military dynamic.

Strategies of empowerment, including science by the people, seem to have a greater potential to move toward a citizen-created world. The feminist movement has achieved many of its gains through changing people's thinking and behavior at the interpersonal level—an empowerment approach—rather than via a feminist state or via pressure-group influence on policymaking, though elements of both these strategies have played a role. Military systems, though, seem especially resistant to transformation: there are a few more women in western military forces than before, but without significant transformation of defense mandates or military R&D. Feminism has had a much greater influence on peace movements, helping to promote non-sexist behavior and egalitarian group dynamics (Brock-Utne 1985; Gnanadason, Kanyoro, and McSpadden 1996).

It is possible to argue that citizen action outside the peace movement has already had significant impact on defense-related research agendas. The worldwide movement against nuclear power has been largely successful in preventing nuclear power from becoming the lynchpin of energy systems (Falk 1982; Rüdig 1990). (In the United States, where rising costs were the proximate cause of the decline in nuclear power, citizen opposition was a key factor in forcing higher expectations of safety, with serious cost consequences. In most other countries, nuclear power was a state enterprise, not dircctly affected by cost considerations.) Although much citizen concern centered on nuclear reactor accidents and disposal of long-lived radioactive waste, there was also a strong link to proliferation of nuclear weapons. Furthermore, many campaigners were motivated by opposition to a nuclear future in which government repression would become essential to deal with the risk of criminal and terrorist activity in a "plutonium economy" (Patterson 1977). In other words, dependence on a potentially catastrophic and highly expensive energy system is associated with authoritarian politics. A plutonium economy would inspire nuclear-oriented R&D with a side menu of technologies for political control.

Citizen action has prevented this dystopian future, with elements of all four strategies playing a role. Grassroots activism, built around empowerment, has been the foundation of much of the movement. Developing energy alternatives, a facet of living the alternative, has been important in demonstrating that nuclear power is not necessary. Pressure-group politics has been important, and some governments have taken stands against nuclear power. The outcome has been the prevention of a plutonium economy, with consequent impacts on present-day research agendas—including defense research.

Because so many social movements have taken their inspiration from being *against* something, their achievements are better recognized by spelling out what the world might have become without them. This is certainly true in regard to defense. Without peace movements, it is possible that military technologies could be playing dominating roles in many more areas than at present: space weapons, biological weapons, technologies of political control, military models for education—the list is endless. Pessimists might say that we live in a military-dominated world, but compared to periods of total mobilization for war, civilian priorities are central to much of today's world.

Conclusion

I started out describing four visions of alternative science built around the concepts of science for the people and science by the people, and then I described four strategies for moving toward these visions. In the abstract, these visions and strategies may seem logical enough, but challenges arose when examining the specific case of defense technology. Several insights can be pulled from this case.

- To develop a vision of alternative *science*, it is first necessary to have a vision of alternative *society*. This may not be easy. In the case of defense technology, the key issue is choosing a vision for defense. It could be conventional military defense, non-offensive defense, security through promoting social justice, or nonviolent defense. In thinking about alternative science, there is a risk in not thinking creatively enough about alternative society.

- There is very little strategic thinking about how to achieve alternative science. Most social action is driven by immediate issues, with short time horizons. Thinking years ahead, or even decades ahead, is unusual (Schutt 2001). Instead, the usual focus is on stopping a new development or organizing the next rally or meeting. Short-term thinking fits most conveniently with pressure-group approaches, aiming at reform. These can be quite useful, but they leave wider changes to chance.

  Long-term strategic thinking need not be linked to central planning. It is also relevant to participatory approaches. The point of strategic thinking is to work out what to do now to help achieve long-term goals. It can be used by any group, indeed by individuals.
- Indirect approaches to alternative science—changing science by changing society—can be quite effective but have been neglected. Because science is so highly professionalized around an ideology of autonomy, direct citizen oversight of research seems to scientists like a threat, and science by the people sounds almost self-contradictory. On the other hand, social movements can change research agendas indirectly without an immediate threat to most scientists.

  To emphasize indirect approaches to alternative science—in other words, to change science by changing society—implies treating science as a "tough case," namely a part of society that is relatively resistant to change. That may or may not be correct. More experimentation with science by the people is needed in order to find out.

In looking at visions and strategies for alternative science, my main focus has been on implications for activists. Social action can only take place in the here and now, but it can be informed by a clear articulation of goals and methods. Activists usually know what they are against but less often have a well-developed picture of where they are going. Academic work in the typical mode of critique replicates this same imbalance.

Reflecting back on research, the implication is that there is much scope for scholarly work on visions and strategies. To be sure, there is some academic work in these areas, such as in the field of future studies. But as long as it remains in academic journals, it is unlikely to have much impact outside the academy. To be useful to activists, research needs to be different in method as well as content. An obvious candidate is participatory action research, which aims to bring about social change while developing knowledge (Whyte 1991).

Chris's comments got Professor Smith thinking. A few months later, the following discussion took place at a meeting of Science Justice:

CHRIS: Hi everyone. I'd like to introduce Professor Smith, who was my teacher last year.

PROF: Just call me Stef.

CHRIS: Okay. Perhaps you'd like to tell us some of your ideas for the group.

STEF: Actually, I'd rather hear about your activities and plans.

CHRIS: Sure, if you want. We need to review our plans anyway. Who'd like to start?

[*A lengthy discussion ensues.*]

CHRIS: Stef, after hearing all that, is there anything you'd like to say?

STEF: I've learned a lot from hearing about your successes and difficulties. You've talked about lots of possibilities for the next year. Have you thought about your goals for the next five or ten years?

[*No one responds for a moment.*]

CHRIS: I guess not. What are you thinking of?

STEF: In the group you have a really well developed understanding of your strengths and weaknesses and what you're up against, plus lots of ideas for campaigns. It might help to look at your long-term goals and then work backward to decide which campaigns should get top priority.

CHRIS [*tentatively*]: That sounds reasonable, just so long as we don't get into too much abstract theorizing. The current issues are really important and urgent, and we don't want to lose momentum.

STEF: Well, maybe a couple of you could work with me to develop some ideas to present to the group. I might be able to get one or two of my colleagues to help. But some of you would need to be involved to make it relevant.

CHRIS: I'm willing to try it. Anyone else? . . .

Leaving, Chris thinks, "I was worried that Stef might be too academic but it turned out all right." Stef thinks, "That was different! I wonder whether I'll be able to convince any of my colleagues to talk to these activists."

Acknowledgments

I thank Scott Frickel, Kelly Moore, Ned Woodhouse, and Elisabeth S. Clemens for valuable comments on drafts of this chapter.

References

Aas, Solveig, and Tord Høivik. 1986. "Demilitarization in Costa Rica: A Farewell to Arms?" In *Costa Rica: Politik, Gesellschaft und Kultur eines Staates mit Ständiger Aktiver und Unbewaffneter Neutralität,* ed. Andreas Maislinger, pp. 343–375. Innsbrück, Austria: Inn-Verlag.

Armstrong, J. Scott. 1980. "Unintelligible Management Research and Academic Prestige." *Interfaces* 10:80–86.

Bernal, J. D. 1939. *The Social Function of Science.* London: George Routledge & Sons.

Beyerchen, Alan D. 1977. *Scientists under Hitler: Politics and the Physics Community in the Third Reich.* New Haven, CT: Yale University Press.

Blissett, Marlan. 1972. *Politics in Science.* Boston: Little, Brown.

Boggs, Carl. 1986. *Social Movements and Political Power: Emerging Forms of Radicalism in the West.* Philadelphia: Temple University Press.

Boyle, Godfrey, Peter Harper, and the editors of *Undercurrents,* eds. 1976. *Radical Technology.* London: Wildwood House.

Brock-Utne, Birgit. 1985. *Educating for Peace: A Feminist Perspective.* New York: Pergamon.

Burrowes, Robert J. 1996. *The Strategy of Nonviolent Defense: A Gandhian Approach.* Albany: State University of New York Press.

Carson, Lyn, and Brian Martin. 2002. "Random Selection of Citizens for Technological Decision Making." *Science and Public Policy* 29:105–113.

Carter, April. 1992. *Peace Movements: International Protest and World Politics since 1945.* London: Longman.

Cassidy, Kevin J., and Gregory A. Bischak, eds. 1993. *Real Security: Converting the Defense Economy and Building Peace.* Albany: State University of New York Press.

Cortright, David. 1993. *Peace Works: The Citizen's Role in Ending the Cold War.* Boulder, CO: Westview.

Darrow, Ken, and Mike Saxenian, eds. 1986. *Appropriate Technology Sourcebook: A Guide to Practical Books for Village and Small Community Technology.* Stanford, CA: Volunteers in Asia.

de Bono, Edward. 1986. *Six Thinking Hats*. Boston: Little, Brown.
———. 1992. *Serious Creativity: Using the Power of Lateral Thinking to Create New Ideas*. London: HarperCollins.
———. 1995. *Parallel Thinking: From Socratic Thinking to de Bono Thinking*. Harmondsworth, UK: Penguin.
Elias, Norbert, Herminio Martins, and Richard Whitley, eds. 1982. *Scientific Establishments and Hierarchies*. Dordrecht, Germany: D. Reidel.
Epstein, Steven. 1996. *Impure Science: AIDS, Activism, and the Politics of Knowledge*. Berkeley: University of California Press.
Falk, Jim. 1982. *Global Fission: The Battle over Nuclear Power*. Melbourne: Oxford University Press.
Farkas, Nicole. 1999. "Dutch Science Shops: Matching Community Needs with University R&D." *Science Studies* 12(2): 33–47.
Ferris, Timothy. 2003. *Seeing in the Dark: How Amateur Astronomers Are Discovering the Wonders of the Universe*. New York: Simon and Schuster.
Fixdal, Jon. 1997. "Consensus Conferences as 'Extended Peer Groups.'" *Science and Public Policy* 24:366–376.
Gieryn, Thomas F. 1995. Boundaries of Science. In *Handbook of Science and Technology Studies*, ed. Sheila Jasanoff, Gerald E. Markle, James C. Petersen, and Trevor Pinch, pp. 393–443. Thousand Oaks, CA: Sage.
Gnanadason, Aruna, Musimbi Kanyoro, and Lucia Ann McSpadden, eds. 1996. *Women, Violence and Nonviolent Change*. Geneva: WCC.
Haberer, Joseph. 1969. *Politics and the Community of Science*. New York: Van Nostrand Reinhold.
Illich, Ivan. 1973. *Tools for Conviviality*. London: Calder and Boyars.
Johnston, Robert D., ed. 2003. *The Politics of Healing: A History of Alternative Medicine in Twentieth-Century North America*. New York: Routledge.
Kroll-Smith, Steve, and H. Hugh Floyd. 1997. *Bodies in Protest: Environmental Illness and the Struggle over Medical Knowledge*. New York: New York University Press.
MacKenzie, Donald, and Graham Spinardi. 1995. "Tacit Knowledge, Weapons Design, and the Uninvention of Nuclear Weapons." *American Journal of Sociology* 101:44–99.
MacKenzie, Donald, and Judy Wajcman, eds. *The Social Shaping of Technology*. 2nd ed. Buckingham, UK: Open University Press.
Martin, Brian. 1997. "Science, Technology and Nonviolent Action: The Case for a Utopian Dimension in the Social Analysis of Science and Technology." *Social Studies of Science* 27:439–463.
———. 1998. "Technology in Different Worlds." *Bulletin of Science, Technology and Society* 18:333–339.
———. 2001. *Technology for Nonviolent Struggle*. London: War Resisters' International.

Maxwell, Nicholas. 1984. *From Knowledge to Wisdom: A Revolution in the Aims and Methods of Science.* Oxford: Basil Blackwell.

———. 1992. "What Kind of Inquiry Can Best Help Us Create a Good World?" *Science, Technology, & Human Values* 17:205–227.

Melman, Seymour. 1988. *The Demilitarized Society: Disarmament and Conversion.* Montreal: Harvest House.

Mendelsohn, Everett H., Merritt Roe Smith, and Peter Weingart, eds. 1988. *Science, Technology and the Military.* Dordrecht, Germany: Kluwer.

Moody, Glyn. 2002. *Rebel Code: Linux and the Open Source Revolution.* New York: Perseus.

Murphy, Danny, Madeleine Scammell, and Richard Sclove, eds. 1997. *Doing Community-Based Research: A Reader.* Amherst, MA: Loka Institute.

Patterson, Walter C. 1977. *The Fissile Society.* London: Earth Resources Research.

Primack, Joel, and Frank von Hippel. 1974. *Advice and Dissent: Scientists in the Political Arena.* New York: Basic Books.

Randle, Michael. 1994. *Civil Resistance.* London: Fontana.

Reppy, Judith, ed. 1998. *Conversion of Military R&D.* Basingstoke, UK: Macmillan.

Roberts, Adam. 1976. *Nations in Arms: The Theory and Practice of Territorial Defence.* London: Chatto and Windus.

Rüdig, Wolfgang. 1990. *Anti-Nuclear Movements: A World Survey of Opposition to Nuclear Energy.* Harlow, UK: Longman.

Schutt, Randy. 2001. *Inciting Democracy: A Practical Proposal for Creating a Good Society.* Cleveland: Spring Forward Press.

Schweitzer, Glenn E. 1996. *Moscow DMZ: The Story of the International Effort to Convert Russian Weapons Science to Peaceful Purposes.* Armonk, NY: M. E. Sharpe.

Science for the People. 1974. *China: Science Walks on Two Legs.* New York: Avon.

Smith, Merritt Roe, ed. 1985. *Military Enterprise and Technological Change: Perspectives on the American Experience.* Cambridge, MA: MIT Press.

Smith, Merritt Roe, and Leo Marx, eds. 1994. *Does Technology Drive History? The Dilemma of Technological Determinism.* Cambridge, MA: MIT Press.

Sørensen, Knut H., and Robin Williams, eds. 2002. *Shaping Technology, Guiding Policy: Concepts, Spaces and Tools.* Cheltenham, UK: Edward Elgar.

Soros, George. 2002. *George Soros on Globalization.* New York: PublicAffairs.

Summy, Ralph, and Michael E. Salla, eds. 1995. *Why the Cold War Ended: A Range of Interpretations.* Westport, CT: Greenwood.

Turnbull, Shann. 1975. *Democratising the Wealth of Nations from New Money Sources and Profit Motives.* Sydney: Company Directors Association of Australia.

Ui, Jun. 1977. "The Interdisciplinary Study of Environmental Problems." *Kogai—The Newsletter from Polluted Japan* 5(2): 12–24.

Wainwright, Hilary, and Dave Elliott. 1982. *The Lucas Plan: A New Trade Unionism in the Making?* London: Allison & Busby.

Whyte, William Foote. 1991. *Participatory Action Research.* Newbury Park, CA: Sage.

Winner, Langdon. 1977. *Autonomous Technology: Technics-out-of-Control as a Theme in Political Thought.* Cambridge, MA: MIT Press.

Woodhouse, Edward, David Hess, Steve Breyman, and Brian Martin. 2002. "Science Studies and Activism: Possibilities and Problems for Reconstructivist Agendas." *Social Studies of Science* 32:297–319.

11

Powered by the People

Scientific Authority in Participatory Science

KELLY MOORE

Over the past three decades, nonscientists have become more involved in the design, production, and use of science. To be sure, professional scientists still create the vast majority of scientific knowledge. Yet there is an unmistakable increase in the types and levels of nonscientist participation in scientific knowledge production and science policy decisions. Thus, coalitions of scientists and amateurs work on projects ranging from the restoration of ecosystems to bird surveys; individuals at risk of or experiencing illness use direct action to demand that questions and methods that they deem important are included in research studies; and networks of scientists, often supported by government agencies such as the Environmental Protection Agency and the National Institute of Environmental Health Sciences, now regularly work with citizens to design and carry out studies. Citizens participate in science policy decisions in town meetings on technoscientific issues (Sclove 1997); consensus conferences (Guston 1999; Fischer 2000); government-sponsored "participatory design" programs (Laird 1993; Futrell 2003); and public hearings, surveys, and citizen review meetings (Fiorino 1990).[1] The varied forms of scientist-citizen interaction have been examined under the rubric of *participatory science,* often with great enthusiasm for the ways in which they are thought to develop useful knowledge for multiple stakeholders and to expand democratic practice into science.

In this chapter, I turn attention toward another feature of participatory science: its effects on the political authority of science in public debate. By authority I mean the ability to command power and influence. The sources of scientific authority have traditionally included the idea that science ultimately benefits all people ("progress"); that competence in science requires years of specialized training; that it is a unified social activity based on common methodological and theoretical bases; and that scientific knowledge is, after vetting by scientists over time, ultimately objective, independent of political, moral, and social influences. Participatory science both supports and challenges these ideas, but it does not do so in a uniform fashion. That is because participatory science refers to a very wide variety of practices, initiated by different groups of people for very different reasons. "Participatory" conjures up an image of equality and balance, a harmonious project in which well-intentioned parties come to mutual agreement. Such an image obscures the fact that in general, some person or group has to initiate it for a specific purpose. Initiators set agendas, the terms of debate, and define the resources, languages, and venues for discussion and adjudication available. By more closely examining initiators, it is possible to see how participatory science undermines and reinforces the political authority of science.[2]

Drawing on studies of and original research on participatory science on subjects including health, ornithology, and ecological restoration, I compare three kinds of participatory science: that which is initiated by *activists, professionals,* and *amateurs* distinguishing them according to how knowledge is generated and used, in which disciplines it is most commonly found, the relationships between professional and nonprofessional actors, and the particular challenges that each type poses to the authority of science. These are heuristics only; in reality of course, amateurs, activists, and professionals may coinitiate projects, and indeed, they may have ongoing collaborations, as Morello-Frosch and her coauthors point out in chapter 9. The analytic value of examining participatory science based on single initiators is to build theoretical ideas that can be investigated in other settings.

Origins of Participatory Science in the United States

In the United States, participatory research has been facilitated by three institutional changes: the undermining of some features of the authority of science in the 1960s and early 1970s as a result of the student and anti–Vietnam war movements, the development of a new kind of knowledge-based health social movement in the 1970s (Morello-Frosch et al. this volume), and the growth of the regulatory state.

Social movements of the 1960s and 1970s shaped the rise of participatory science in several ways. Younger activists and some intellectuals involved in the student and antiwar movements were highly critical of authority in general. They challenged the idea that experts or other leaders knew what was best for American society, and sought to include a wider range of people in formal political and other kinds of public decisions. Activists closely scrutinized the work of practically anyone in authority, from religious leaders, to artists, to professionals. Among those under the critical eye of activists were scientists. Activists argued that scientists' claim to serve the public good was contradicted by their involvement in the chemical and weapons industries and the Department of Defense and by their use of the idea of objectivity to mask political decisions (Moore forthcoming). At the same time they were targets of social movement activists, scientists were constitutive of them as well (Moore and Hala 2002; Hoffman 1989; Frickel 2004). These scientists agreed that science ought to be more responsive to a broader range of human needs. Under these conditions, nonprofessionals began to assert their right to challenge the uses to which science was put, and some scientists began to find ways to work closely with community groups on issues of mutual concern.

For some other some scientists, these motivations were complemented by the knowledge that gaining the cooperation of people who were subjects of research and learning from nonexperts who had experience with agriculture and health issues could lead to better scientific knowledge. One of the influences on the development of networks of scientists seeking practical benefits from nonprofessionals' involvement came from agricultural programs in newly independent or developing countries in Africa, Asia, and Latin America in the early 1960s (Hall 1993; Fals Borda 1987; Masters 1995). Agricultural productivity was critical to the survival of these countries, and so they looked for inexpensive and reliable methods of increasing yields and controlling pests. Farmers shared their knowledge with government agents who could then distribute best practices to other farmers, and government agents would encourage farmers to engage in agriculture experiments with their crops or try new practices. Visitors to these countries, or those who knew scientists in these countries through professional or personal networks, learned of these practices, and brought them back to the United States (Hall 1993:xiii). These political networks of reformers, academics, and schools (e.g., the Highlander Folk School) promoted "action research" that was centered in debates over Jürgen Habermas's ideas about forms of knowledge and their emancipatory potential (Habermas 1984–1987). For many of those involved in action research, the goal was to treat the making of technologies as a vehicle for social justice. Thus, one of the distinct features of action research was that the *process* of participation was intended to be liberatory in the sense of demonstrating and enacting the power of ordinary people to use their skills and ideas for their own betterment (Park, Brydon-Miller, Hall, and Jackson 1993). The best political analogy to this type of research is participatory democracy, which is intended to allow stakeholders to participate as equals in decision making rather than appointing representatives or proxies (Polletta 2003).

The activism of the 1960s and the development of participatory and action research at an international level was complemented by the development of community-based health activism in the U.S. and abroad in the 1970s. Women and African Americans were at the forefront of these new challenges, arguing that the ways that doctors had treated them and framed their illnesses and injuries were incorrect and unjust. They began to use direct action (Hoffman 1989) and publications (e.g., *Our Bodies, Ourselves* [Boston Women's Health Book Collective 1973]), and to force college and university health centers to provide information about contraception and abortion (Morgen 2002). Other movement groups rejected scientifically-prescribed treatments for mental health (Crossley 1998; Brown 1984) and categorizations of disability and homosexuality as medical problems (D'Emilio 1988; Shapiro 1993). In the 1970s and 1980s, health-based activism continued to grow, and so did the movement for environmental justice, organized around the health implications of pollution (Szasz 1994). Rising education levels also helped to increase the ability of movement groups to challenge the claims of scientists and medical researchers.

At the same time as these social movements developed, the growing regulation of production in industry set in place new laws that allowed nonscientists to participate in certain kinds of sociotechnical policy decisions. Before the second half of the twentieth century, most industrial regulation concerned distribution practices, such as transport and sales. The 1960s saw the growth of the regulation of production. Environmental legislation was one of the most important forms of such regulation, and one that set the stage for the involvement of noncredentialed experts in debates about sociotechnical issues. In 1969, President Nixon established the Citizen's Committee for Environmental Quality, formally mandated to participate in the executive branch's newly formed Council on Environmental Quality. The act also mandated that the Environmental Protection Agency "make available to states, counties, municipalities, institutions, and individuals, advice and information useful in restoring, maintaining, and enhancing the quality of the environment" (NEPA 1969:sec. 102g). Perhaps as important, the act mandated that different constituents, including those with economic, health, and other interests, ought to be served by the legislation, opening the door for lawsuits. The EPA followed the establishment of many state environmental protection acts, thereby providing legitimacy for citizen involvement and for the claims made by citizens about the harms to and by the environment (Dunlap and Mertig 1991; Hays 2000; Rome 2003). Environmental regulations were followed by greater regulation of drugs and medical treatments and the rise of professionalism in areas that were previously considered outside the realm of medicine. These, too, often contained rules that made it possible for the public to access scientific information and file lawsuits based on this information.

A related change that helped give rise to participatory science was the institutionalization of research ethics into rule and law. Professionals' use of humans as subjects used to be based on the supposition that humans were objects whose individual rights were overridden by the larger public good presumed to come from almost any scientific study. In the eighteenth, nineteenth, and early-twentieth centuries, scientists informally struggled to weigh the harms done to research subjects and the potential benefits, but it was not until later in the twentieth century that this kind of debate was institutionalized in rules (Halpern 2004). One of the most important influences on the development of these rules was World War II, through the work of the inter-Allied Scientific Commission on Medical War Crimes (Weindling 2001). The group was organized to examine the cruel and deadly experimentation to which the Nazis subjected their prisoners. Later, the deliberate exposure of American soldiers to radiation, the tragedy of the Tuskegee syphilis experiments, plutonium experiments, birth control experiments on Puerto Rican women, and other studies in which harm was done to human subjects led scientists, politicians, and laypeople to develop protocols for informed consent (Faden and Beauchert 1986; Welsome 1999). The idea of informed consent was one of the most dramatic changes in science in the twentieth century. Moving humans out of the category of objects that were equivalent to nonliving things permitted individuals to make greater legal and moral claims on scientists and those who employed scientists.

Finally, nonprofit organizations and government agencies have played an important role in generating participatory research. In 1992, a total of 178 nations adopted Agenda 21, a comprehensive plan of action for environmentally sensitive sustainable development, overseen by the newly formed United Nations Commission on Sustainable Development (United Nations 2004). One of the key components of Agenda 21 was funding and plans for local, national, and international community-based research. Over the last decade, the agency has sponsored hundreds of community-based environmental research projects around the world, most of them with a capacity-building component to ensure the perpetuation of similar research projects. Similarly, participatory research has become a required part of many public health research grants in the United States, according to scholar and participatory research practitioner Meredith Minkler, because philanthropists and government funders have "become discouraged by the often modest to disappointing research of more traditional research and intervention efforts in many low-income communities of color" (Minkler, Blackwell, Thompson, and Tamir 2003:1210). Backed by a handful of foundations, including the Aspen Institute and the W. K. Kellogg Foundation as well as by the National Institutes of Health and the Environmental Protection Agency, participatory research is now a legitimated and fundable research method.

In general, the origins of participatory research thus come from the needs of scientists and citizens to acquire knowledge that can solve practical problems, and from political theories that see collective participation in all kinds of activities as a source of liberation and human fulfillment. In practice, these goals are not always met, for reasons I examine below.

PROGRESS FOR WHOM? ACTIVIST-INITIATED PARTICIPATORY SCIENCE

Activist-based participatory science is usually initiated by people who are concerned about a specific and immediate problem for which they seek to gather information to be used in an adversarial political process. These include anti-toxics/anti-pollution campaigns (Brown 1992; Bullard 1994; Lichterman 1996; Allen 2003) and health social movement groups (Morello-Frosch et al. this volume) that seek to stop the source of pollution or toxicity or to force scientists to examine new questions or use new methods of research and implementation (Brown, Kroll-Smith, and Gunter 2000). As Phil Brown describes popular epidemiology, it is "the process by which laypersons gather scientific data and other information, and also direct and marshal the knowledge of other experts in order to understand the epidemiology of a disease" (Brown 1992:269). In this type of participatory science, scientists are usually asked to support the community's struggle by providing scientific information and legitimacy or by collecting and analyzing information with the assistance of the community.

What "assistance" means varies considerably, however. Sometimes community-based science works well, but at other times, scientists and others who wish to involve community members in research run into several problems. What people mean by community and their expectations and desires for participation can vary. One of the key features of this type of participatory science is that scientists are invited in by the community, not chosen as subjects by scientists first, even though in the past they may have been subjects of research. Reardon (this volume) argues that one of the problems in participatory science is that scientists and those who are subjects of research sometimes do not see subjects in the same way. Second, scientists and community members may have different ideas about what *community* means (Lichterman 1996). Whereas scientists may be comfortable working in groups that are based on the mutual interests of all participants, community members might emphasize other bases for participation, such as having a long history of association with other members.

West Harlem Environmental Action (WE ACT), studied by Phil Brown and his collaborators, provides a good case of a successful scientist-movement collaboration (Brown 1997; Brown and Ferguson 1995; Morello-Frosch, this volume). In 1986 the city of New York built a sewage treatment plant next to the Hudson River at the western edge of Harlem. Women residents began to organize to draw attention to the health problems that residents reported had grown since the plant began running. Having no scientific data on the relationship between community health and the plant, they relied on their own local, often personal knowledge about the plants' effects on them and their families (Brown and Ferguson 1995). The complaints were at first dismissed by city and state agencies, but WE ACT quickly gained support from city and state representatives from Harlem who took part in demonstrations and prodded city and state agencies to deal with WE ACT's complaints. The attention was undoubtedly the result of the fact that one of WE ACT's founders, Peggy Shepard, was a member of the New York City Council. WE ACT, with the help of political leaders, raised money to hire the Center for Natural and Biological Systems at Queens College, run by longtime environmental activist Barry Commoner, to produce a report on pollutants from the sewage plant. The report documented increased levels of many kinds of pollutants as compared with other parts of the city. WE ACT, along with the Natural Resources Defense Council, used these findings to successfully sue the city and the state using the pro bono services of the prestigious law firm Paul, Weiss, Rifkind, Wharton and Garrison (WE ACT 2005).

The one-million-dollar settlement has been used to build a larger infrastructure of environmental justice programs in Harlem, including several partnerships with Columbia University School of Public Health, to examine health problems associated with pollutants in northern Manhattan.The problems include the levels of exhaust coming from a large bus depot in Washington Heights, and the development, in collaboration with the Cornell University Department of City and Regional Planning, of an easily used GIS-mapping system that will make health and environmental data easily accessible to communities (www.weact.org).

The routinization of activist-initiated research, as in the case of WE ACT's transition to a stabilized group, can also be seen in the inclusion of gay men in deciding on treatments for and research on HIV and AIDS and the establishment of Indian environmental health centers on many Native American reservations. The institutionalization of these programs makes nonscientists' experiences and evidence more important in problem solving and in the creation of general scientific knowledge.

Yet, as Epstein (1996) and numerous others point out, gaining a seat at the table of science may also mean that nonscientists lose the leverage that their specifically political and moral claims once allowed them. Moral pressure is the process by which activists publicize the contradiction between scientists' professed goals of helping people and their inability or unwillingness to respond to the needs of activists. Public demonstrations, including marches and vigils, are one of the most common ways of bringing attention to this contradiction, and so are media campaigns and lobbying those groups who control the funding of the scientists in question. When activist-initiated participatory research works well, there is no need for these tactics; but when it does not, moral pressure is one form of leverage that activists can use.

Activists may also gain leverage by developing their own studies. This strategy works best when the group in question has the organizational skills and resources, and the credibility, to carry out such work. It may be particularly effective when the results of activist-initiated research solves problems or puzzles of interest to scientists and when activists come from relatively high status groups.

By identifying these sources of leverage, I do not mean to assert that all activist-initiated participatory research will be an overtly contentious struggle between scientists and activists; clearly the case of groups like WE ACT and others demonstrates that such collaborations can be fruitful for all involved. Rather, my claim is that the power that activists have is based in part on their ability to contest one of the major sources of the legitimacy of science, and that is that science ultimately serves all people. Surely few people believe that science always leads to benefits, but what activist-initiated research does is to put front and center the ways in which social structural relationships provide some groups with more benefits from science than others. Second, by raising new questions and providing new evidence, activist-initiated science undermines the idea that scientists ought to monopolize knowledge making. The kind of specialized knowledge that scientists have may not be enough to answer questions of importance to all people. For some sociologists of science, these might be familiar arguments.

Less obvious, however, is that the terms we use to describe these participants suggests that despite our recognition of how they challenge the authority of science as a political tool, the procrustean bed of liberal politics cannot easily accommodate these challengers. "Lay-experts," "citizen-scientists," "citizen-participants," "community-based lay researchers," and other terms suggest an uncomfortableness with what, precisely, these people have to contribute and how to conceive of their role in the politics of science. There is no such elaborated set of terms for scientists involved in participatory research, which implies that either their participation is understood to be seamlessly in line with the idea of scientists as servants of the people, or that we take for granted that they are not transformed or otherwise challenged by the experience of participation. A more elaborated vocabulary to describe scientists who engage in participatory research ("activist scientist" does not capture the range of reasons scientists have for participation) would flesh out the power of participatory science to change scientists' professional and political roles.

PROFESSIONAL-INITIATED PARTICIPATORY SCIENCE

The second type of participatory research is that which is initiated by professional researchers. Since the 1980s, the inclusion of (presumed) beneficiaries in the design, execution, and application of research has grown as a result of legal mandates. For reasons of space, I focus here on one area, environmentalism. During the Reagan and Bush I administrations, fewer laws were passed to include ways in which citizens could become involved in the creation of scientific knowledge regarding the environment. The Clinton administration, spurred in part by the claims of environmental justice groups (Topper 2004), put in place a series of laws and executive orders that encouraged such participation by granting funds and by mandating citizen participation. In 1992, the Environmental Protection Agency developed a series of community-based programs to address environmental justice issues, especially those confronting Native American communities. In 1993, within the EPA the Office of Environmental Justice was established, charged with finding ways to assist citizen groups in making legal claims about environmental issues and in gathering the evidence that they might need to make such claims. In February 1994, President Clinton expanded that mandate by signing Executive Order 12898, which directed federal agencies to address environmental justice in minority and low-income populations. The order directed agencies to identify and address "as appropriate" adverse human and environmental effects of programs, policies, and activities (Hays 2000:78). The EPA currently sponsors several programs designed to engage communities in the process of designing, implementing, and using research.

Even as the EPA changes rules that permit activities that minority and poor groups are unlikely to benefit from, such as preventing companies from having to comply with clean-air regulations when they refit existing infrastructure, the agency is also sponsoring relatively innovative projects to link scientists and other professionals from the EPA and universities with people in communities that suffer from or that might suffer from health and environmental hazards. For example, the agency provides small grants to environmental justice partnerships, usually between universities and community groups. The grant must go to a 501(c)(3) group, effectively cutting out groups without grant-writing capacity or stable nonprofit status, likely to be the poorest groups. Mainly such grants are provided to university–environmental justice group partnerships—one of the most common ways in which professionals and nonprofessionals are linked in professionally sponsored participatory research (EPA, Community Assistance Program 2003).

A good example of university–local group collaboration is the Dickinson University Alliance for Aquatic Research Monitoring (ALLARM) Program. Founded in 1986 to study local water quality, in 1996 the program expanded its focus to work in collaboration with local groups who wished to monitor the water quality in the area. The Technical Support Team, made up of students under the supervision of the program director, helps to coordinate water collection and analysis projects at the behest of local groups. The students engage in the laboratory work, while the local groups choose sites and engage in routine monitoring (www.dickinson.edu/storg/allarm). There are hundreds of such projects across the United States, centered in small colleges like Dickinson, at large state universities, and at elite universities. Like the institutionalized version of WE ACT, these projects routinize professional-nonprofessional interactions in research projects to produce data and practical action that is mutually beneficial.

Yet it would be a mistake to treat such collaborations as routinely successful. Professional-initiated participatory research activity runs the constant risk of reproducing power relations that ultimately undermine the trust of those whom professionals believe they are assisting. For example, the EPA is carrying out pilot studies designed as methods through which the EPA can learn best (and worst) practices in order to organize future community-based research projects. In these studies, the EPA uses a technical team to help collect, organize, and analyze the data, and the presents data in formats that are thought to be easily understood. Yet these projects have turned out not to be particularly successful, because participants were too often treated as audiences and became disappointed because results were contrary to what they expected and only identified problems without providing solutions for them. While the potential beneficiaries were told about the project and regularly consulted, they were not involved in carrying out the research itself. Rather than abandoning the cause, the EPA continues to fund partnerships and is developing a manual that can be used by community organizers who wish to know more about how to organize and act to solve environmental pollution problems in poor and minority areas (EPA, Interagency Working Group 2004). By not involving nonprofessionals in the design and execution of research and instead treating community consent as all that is necessary for research to be participatory, such projects continue the early "information model," popularized by scientists active in the Ban the Bomb movement in the 1950s and 1960s. In the information model, citizens are told about dangers or promises that have been studied by scientists. The expectation—and one that has not been shown to have much merit—is that nonprofessionals will use "information" to evaluate their own or others' behavior and will modify it according to what is most rational and reasonable given the information (Moore 1996).

Participatory science initiated by professionals might, at one level, be interpreted to have no effect on the authority of science. After all, scientists are serving the public good by helping those in need, and they still produce "good science" that can be used for their professional advancement and circulated for use by other scientists. But the participation of nonscientists raises an important question: why isn't all science engaged with nonscientist stakeholders? In Martin's (this volume), terms, this would be "science by a citizen-shaped world." The existence of some kinds of participation, in a limited number of fields at that, draws attention to the lack of such participation in other fields. The claim that this is because subjects like health or the environment are easier to understand, while physics, chemistry, and other physical sciences are too abstract for most people to comprehend, will not work, because we know that citizens understood the dangers of nuclear weapons quite well in the public education campaigns undertaken by scientists and citizen groups after World War II (Smith 1965). I argue, then, that the existence of participatory science in some areas but not others—as well as the existence of many failures in professional-initiated participatory science—suggests that the authority of science based on its ability to serve the interests of all people is placed on rockier footing than it would be otherwise.

AMATEURS IN PARTICIPATORY SCIENCE

There is a third set of initiators of participatory science who have received less attention than scientist participants. Because they are not professionally accredited scientists and they lack the knowledge and skill bases that scientists rely upon, amateurs often enter the research process in ways that are very different than how activists enter, and they rely on different resources and forms of power to effect change. Amateurs have long been contributors to ornithology and astronomy and in recent years have become more important in ecological research; they will likely become more important in the future, especially in ecological sciences (Ainley 1979/1980; Gross 2003; Ferris 2002). Their involvement in scientific research is based on a vocational interest in the process, not concerns about outcomes.

I distinguish between two kinds of amateur science. One is what I call "marginal amateur" science, based on Collins and Evans's (2002) idea of "marginal experts," or people who have knowledge that is not part of a well-respected knowledge regime. Some marginal amateur science is religiously based. While this approach is seemingly an oxymoron, creationists seek to use scientific methods to make religious doctrine consistent with science by undermining scientific claims about human origins or the origins of the earth. Creationist science involves people who are not credentialed experts in a scientific field. There is no credible, widely accepted scientific field that studies the role of God in forming and developing life, and involves people who use aspects of scientific method to examine questions that have some resemblance to mainstream scientific question.

The efforts of creationists to counter evolutionary theory began with the 1920 Scopes Trial. Until 1958, however, creationists were relatively quiet on the subject, in part because public schools did not teach evolutionary theory in any clear and rigorous fashion and certainly did not depict it as a challenge to Christian interpretations (Binder 2002:43). The 1968 *Epperson vs. Arkansas* case ruled that no state could prohibit the inclusion of evolution in science curriculum (44); this ruling pushed people who believed in creationism to begin organizing to have creationism returned to school curricula. In the 1980s, the Republican Party began to help organize fundamentalist Christians as voters and as political agenda setters. Emboldened by their newfound political power, they embarked on the expansion of creationist science and efforts to have evolution taught as a "theory," not a fact, in public schools. By theory, they mean a Scottish commonsense realist idea of knowledge as that acquired through direct sense experience, not the positivist understanding of theory as a predictive set of propositions supported by evidence or potentially testable using evidence. Another main point of difference is that creation scientists conclude that any disagreements about a subject means that it is not a fact. Thus, people agree with something because it is a fact. In contrast, social studies of science scholars argue that something is a fact because people agree that it is.

Anti-evolutionists go by various names, including "Young Earth" proponents, anti-evolutionists, and proponents of intelligent design (see the Institute for Creation Research, www.icr.org). Some of the leading proponents of these theories have degrees in chemistry and mathematics (but few in biology). One of the techniques that they use is typical in science, especially for those on the margins of a field: they find shreds of evidence to cast doubt on the idea that all members of the scientific community agree on all findings related to a particular subject. By taking apart scientific arguments piece by piece, creationist scientists hope to undermine the foundations of evolutionary theory. For example, at a poster session at the 2003 American Geophysical Union conference, physicist Russell Humphrys, formerly of Sandia Laboratories, presented a paper that challenged the age of a particular kind of granite, based on the amount of helium still left in the rock. He argued that his findings were consistent with an earth age of 6000 years (Humphreys 2003). There are thousands of such papers in existence, produced by scientists who have credentials in a specific scientific field and who choose to extend their expertise to the area of creationism. According to Schindler (2003) noncreationist scientists see many of the creationist scientists as scientists "gone bad," because they fail to use the same standards in their creationist research as they do in other work. Some creationist scientists thus remain credible in non-creationist science fields, but not as investigators of a subject that has no standing in mainstream science.

A second strategy used by creationist scientists is to build upon existing scientific theory, thereby linking their claims to already established ideas. For example, one writer, an assistant professor of psychology at Bowling Green University, seized upon Albert Syzent-Györgi's concept of syntropy (the idea that in addition to the tendency toward entropy in the universe, there is a tendency toward order and the building of higher- and higher-level organisms) and Syzent-Györgi's concern with why nonliving things seem to stop combining at low levels of complexity, while living things can become very complex. The psychologist argued that Syzent-Györgi's puzzle can easily be understood by positing a supernatural origin for life (Bergman 1977). The logic of this argument is, of course, flawed; one could posit all sorts of reasons for differences in the complexity of living and nonliving things, none of which have to do with the supernatural. In order to be considered scientific, the claim would have to be supported by evidence or a sophisticated theory consistent with existing knowledge. By attempting to support claims by drawing on existing scientific theory, and by equating scientific disagreement with a complete lack of evidence, amateurs use many of the tools of science but fail to provide the evidence or logic that other scientists consider necessary for creationist theory to be accepted as a science. Moreover, few creationist scientists begin with the null hypothesis, nor do they seek or accept disconfirming evidence. If the sine qua non of success in scientific research is having one's ideas accepted by one's peers, then clearly those developing alternatives to evolution and a fifteen-billion-year-old earth in their efforts to understand human origins have been unsuccessful. While anti-evolutionists may enroll other scientists in their claim making, the vast majority of scientists do not do so. A similar dynamic occurs in the debates over extraterrestrial life; despite proponents' uses of scientific methods of observation and theorizing, few scientists are convinced, based on empirical evidence, that there is extraterrestrial life except perhaps in the tiniest forms on nearby planets.

A second major type of amateur participatory science has a long history in the United States and a distinctly different relationship to professional scientists. The "marginal" amateurs such as the anti-evolutionists and extraterrestrial seekers have little credibility with geologists, astronomers, and biologists, but other amateurs play an important role in developing scientific knowledge. In what I call "vocational" amateur science, scientific knowledge is produced by people who are not paid to produce knowledge, nor do their lives or health depend on the knowledge that they produce. Sometimes this knowledge is respected by and used by academic researchers. Bird-watchers, for example, have a long history of assisting professionals in identifying species and subspecies of birds and helping to conserve areas frequented by birds. At present, there are at least 650 "opportunities for birders" to participate in volunteer activities listed by the American Birding Association (2004). Much of the work is organized around restoring bird habitats and monitoring bird behavior and migration. Birders can help for a day or a week or a season and can engage in activity in the United States, Canada, or internationally. "Citizen scientists," as the American Birding Association calls them, collect data that is submitted electronically or on paper to scientists, who use the data for their own purposes.

The idea that amateurs could be used as reliable data collectors for scientific ornithology studies was popularized in Canada and the United States by Erica Dunn, a wildlife biologist who was working at a private bird observatory in Canada. In 1976, her husband, a British-born biologist, suggested that Dunn consider having nonscientists participate in surveys, based on the model that was established by the British Trust for Ornithology in 1932 in which volunteers provided systematically collected data to scientists. Dunn set to work organizing data collection and analysis. Her method of coordinating people and organizing data was successful and soon spread to many other bird groups and Canadian and U.S. agencies. Dunn's innovation was not the involvement of amateurs, but the systematization of data collection and analysis so that it could be used in many settings (Dunn 2004; Ainley 1978). Her methods are now in use in many government, university, and private settings.

Another example of the involvement of vocational researchers comes from the field of native habitat restoration. Natural area restoration projects, once mainly the province of professionals, now attract amateurs as well. The Society for Ecological Restoration (SER), formed in 1987, is the most prominent restoration group in the world, with members drawn from thirty-seven countries. Not merely a professional association, the society's active members include scientists, amateurs, volunteers, policymakers, indigenous peoples, and others (Higgs 2003). The society encourages collaborations between scientists and amateurs, with the latter participating in annual meetings and SER-sponsored projects in many of the same ways that scientists do (SER 1988–2004). In a growing number of settings, they undertake projects on their own (Gobster and Hull 2000). Gross (2003) uses the term "public experiments" for these new efforts by amateurs to restore and create healthy biosystems. More often, amateurs become involved with such projects through publicly funded restoration projects that rely on community members engaged in the labor involved in restoration. In activities organized as group works, dozens of people may pull "weeds" (any kind of plant that is deemed to be outside its appropriate boundaries), plant grasses and trees, or set fires, among other activities (Moore 2004). In San Francisco, for example, the Presidio, recently decommissioned as a military site, is now being transformed into a recreational and educational site. Amateurs play important leadership roles by developing plans for specific projects, supervising volunteers, gathering data on the progress of flora and fauna, and generating reports (Holloran 2000). In other restoration projects, amateurs begin the projects on their own and only later come into contact with scientists who have mutual interests in their projects (Gobster and Hull 2000).

Vocational amateurs are sources of information about what works and what does not in restoration projects. In a study of restoration projects in Illinois and California, Gross (2003) found that the "knowledge of volunteers can often be an addition to, and even a very fundamental substitute for, more academic oriented science" (151). He found that amateurs, some of whom began as volunteers under the tutelage of scientists, came up with independent and useful ideas about the dynamics of soil, forests, and grasses while working on their own. An amateur with thirty years' experience in restoration predicted even more involvement in the future: "I think citizen scientists, who are doing science for the love of science, for the love of exploration . . . will change the way science works. They will increase interest in standard science and make important additions to it. They will have off-beat ideas, that you would never get a grant for" (quoted in Gross 2003:152).

The implications of amateur science for the authority of science must be parsed according to the goals and methods that are used. In vocational science, despite their lack of professional training and supervision, amateurs are able to contribute to the development of knowledge production precisely because they are unbound by the constraints of funding and strict question formulation and often have the time to explore multiple directions at once rather than being constrained by the dictates of well-planned research studies. In doing so, they challenge the ideas that strict attention to objectivity and careful research plans are the ideal avenues for the production of useful knowledge. Amateurs typically express feelings of great affection or love for their subjects—normally off-limits, at least consciously, for scientists, who are normatively required to demonstrate at least a modicum of "distance" from their subjects.

Marginal experts such as creationists, however, pose a different set of challenges to the authority of science. Although they do not typically contribute to the pool of knowledge, they are using some of the tools of science to investigate subjects that are at the edges of what is considered legitimate in scientific research. If such activities had no effect on the authority of scientists, they might well ignore these activities. But they do not. Scientists, as well as public officials, spend time criticizing the creationists because they call into question what constitutes theory and evidence in science. The creationists are not entirely successful, of course, since scientists have not adopted commonsense realism as the main way of knowing about nature. Marginal experts, however, do reveal an important feature of science that first came to be a public issue in the 1950s in nuclear-testing debates: experts disagree. If science is supposed to be a coherent whole based on common practices and theories, how is it that controversies emerge, and what are we, as nonscientists, to make of them? Marginal science, at least of the types that I have examined, call into question the degree to which science is a coherent whole, thereby allowing nonscientists to intervene in debates, however unsuccessfully or successfully.

Intersections of Types, and the Future of Participatory Science

Taken together, the three types of participatory research come from the desire by nonprofessionals to have their needs and interests represented in scientific discourse and practice, the motivations of professionals to use their work for social justice concerns, and legal changes that have led to the de-legitimation of the idea that humans are mere objects of scientific research. At the most general level, it would be difficult to imagine the involvement of nonprofessionals in scientific research projects without recognizing that respect for authorities has declined significantly since the 1970s, opening up the possibility that experts may not know best.

The structure of the American political system has made it possible for different combinations of people, activities, subjects, and practices to come together. As compared to most other democratic nations, the United States has a decentralized government that allows multiple points of access and encourages interest group participation, and it has a history of developing policy and practice from local, social movement, and state-level experiments (Amenta 1998; Skrentny 1996; Clemens 1997). Moreover, professions operate relatively independently of the state, governing themselves through associations, permitting professionals to engage in new kinds of action without direct government oversight. These structures make it possible for many types of participatory activity to flourish. One would expect far fewer of these in nation-states that are more centralized and hierarchical, or those in which populist political traditions are less visible.

Contentious science and amateur participatory science call into question whether professionals are the unique source of scientific knowledge and the sources of standards of systematization and evidence. Amateur science, and to some extent activist-initiated participatory science, asks what counts as systematic evidence and controlled experimentation. In particular, public experiments in restoration ecology bear significant resemblance to AIDS activists' calls for "impure subjects," (Epstein 1996) for they both seek practical benefits with less regard for the protocols and needs of professional researchers. Both contentious science and professionally initiated science call critical attention to the idea of science-as-progress by directing attention to how scientific practices may harm people, and the role that nonprofessionals can play in directing science toward more useful ends.

Ideal types can draw attention to analytic points, but they cannot do justice to the complexity of participatory science in practice. Most of the participatory research in the United States relies on the collaboration of professionals and nonprofessionals, not the independent work of amateurs and activists. Although I have categorized types in terms of initiators, decision points, as institutionalist theory has shown, are not always clear-cut. Practices and ideas are likely to emerge from interactions between groups, not from a highly rational, linear application process (see Owen-Smith this volume). What are becoming more institutionalized in the United States are collaborations between universities and local groups, some of which may be ongoing, and some of which may be initiated by local groups that want assistance solving a sociotechnical problem, and short-term research studies that incorporate the ideas and practices of communities into design, execution, and application of research.

These two activities are dependent for their routinization on the willingness of scientists to engage in such activities and the availability of political sites of influence for nonprofessionals. Despite the trend toward commercialization in universities, they remain places where scientists may find funds and support for participatory research projects, especially if they can include students and raise funds. Given the call for more hands-on science teaching, and given the increase in funds for cost-effective, practical problem-solving research from foundations and government agencies, university-community participatory projects are likely to grow. In the 1970s, many professional science associations formed subsections that were devoted to making their science more relevant to the public (Moore 1996); while many of the subunits mainly provide information for the public, they have also spawned networks of scientists who engage in participatory research. Finally, in the past ten years scientific information has become more easily available through electronic resources. Given the continued increase in education levels in the United States, we can expect that more nonprofessionals will have access to ideas that they can use to counter professional expertise and to resist the efforts of professionals who may wish to marginalize their inclusion by treating subjects as mere receptacles for information.

The continuation of participatory research requires that the participation of nonprofessionals is legitimated and made legally possible. Participatory research is uncommon in the physical sciences and only appears in a small number of biological sciences, suggesting that its legitimacy is not well established. The trend in government funding for scientific research may be making it more legitimate, for in a growing number of subjects, including nanotechnology, public participation in the development of projects and applications are required. This does not, of course, necessarily suggest that such research will come close to allowing nonprofessionals much involvement—it may turn out that again involvement may mean the provision of information to nonprofessionals—but it does open up the possibility that it will encourage more scientists to seek ways to include nonprofessionals.

Acknowledgments

Many thanks to Elisabeth S. Clemens, Scott Frickel, Brian Martin, and Rhys H. Williams for their comments on earlier drafts.

Endnotes

1. See Latour (2004), Callon (1999), Ezrahi (1990), Kleinman (2000), Woodhouse (1991), Woodhouse and Nieumsa (2001), Rip (2003), Irwin (1995), and Kitcher (2003) for more theoretical discussions of the benefits and drawbacks of institutionalizing citizen participation in science policy and research.

2. For analytic purposes, I consider only U.S.-based participatory science. Because the United States has a decentralized political system that offers multiple points of access for citizens (although access clearly does not translate into success), and because scientists are less controlled by the state than they are in other countries, examining the United States enables us to see a wide variety of interactions between scientists and nonscientists in knowledge production. Decentralization, however, makes it very difficult to identify how widespread participatory science projects and programs are.

References

Ainley, Marianne Gosztonyi. 1979/1980. "The Contribution of the Amateur to North American Ornithology: A Historical Perspective." *Living Bird* 18: 161–177.

Allen, Barbara. 2003. *Uneasy Alchemy: Citizens and Experts in Louisiana's Chemical Corridor*. Cambridge, MA.: MIT Press.

Amenta, Edwin. 1998. *Bold Relief: Institutional Politics and the Origins of Modern American Social Policy*. Princeton, NJ: Princeton University Press.

American Birding Association, in cooperation with the Bureau of Land Management, U.S. Forest Service, U.S. Fish and Wildlife Service, and the National Park Service. 2004. *Opportunities for Birders*. Colorado Springs, CO: American Birding Association. Available at http://americanbirding.org/opps/voldinam.htm (accessed May 25, 2005).

Bergman, Albert. 1977. "Albert Syzent-Györgi's Theory of Syntropy and Creationism." *Impact* 54:12–22.

Binder, Amy. 2002. *Contentious Curricula: Afrocentrism and Creationism in American Public Schools*. Princeton, NJ: Princeton University Press.

Boston Women's Health Book Collective. 1973. *Our Bodies, Ourselves*. New York: Simon and Schuster.

Brown, Phil. 1984. "The Right to Refuse Treatment and the Movement for Mental Health Reform." *Journal of Health Policy, Politics, and Law* 9:291–313.

———. 1992. "Popular Epidemiology and Toxic Waste Contamination: Lay and Professional Ways of Knowing." *Journal of Health and Social Behavior* 33: 267–281.

———. 1997. "Popular Epidemiology Revisited." *Current Sociology* 45:137–156.

Brown, Phil, and Faith T. Ferguson. 1995. "Making a Big Stink: Women's Work, Women's Relationships and Toxic Waste Activism." *Gender and Society* 9:142–174.

Brown, Phil, Steve Kroll-Smith, and Valerie J. Gunter. 2000. "Knowledge, Citizens, and Organizations: An Overview of Environments, Disease, and Social Conflict." In *Illness and the Environment: A Reader in Contested Medicine*, ed. S. Kroll-Smith, P. Brown, and V. J. Gunter, pp. 9–25. New York: New York University Press.

Bullard, Robert D. 1994. *Dumping in Dixie: Race, Class, and Environmental Quality.* Boulder, CO: Westview Press.

Callon, Michel. 1999. "The Role of Lay People in the Production and Dissemination of Scientific Knowledge." *Science, Technology and Society* 4:81–94.

Clemens, Elisabeth S. 1997. *The People's Lobby: Organizational Innovation and the Rise of Interest Group Politics in America, 1890–1920.* Chicago: University of Chicago Press.

Collins, Harry M., and Robert Evans. 2002. "The Third Wave of Science Studies: Studies of Expertise and Experience." *Social Studies of Science* 32: 235–296.

Crossley, Nick. 1998. "R. D. Laing and the British Anti-Psychiatry Movement." *Social Science and Medicine* 47:877–889.

D'Emilio, John. 1998. *Sexual Politics, Sexual Communities: The Making of a Homosexual Minority in the United States, 1940–1970.* Chicago: University of Chicago Press.

Dunlap, Riley, and Angela G. Mertig. 1991. "The Evolution of the U.S. Environmental Movement from 1970–1990." *Society and Natural Resources* 4: 209–218.

Dunn, Erica. 2004. Telephone interview with the author February 22; e-mail correspondence with the author, February 23.

Environmental Protection Agency (EPA). Community Assistance Program, 2003. "Lessons Learned from the Baltimore Community Environmental Partnership." www.epa.gov/oppt/cahp/cattlessons.html (accessed November 22, 2003).

———. Interagency Working Group on Environmental Justice. 2004. www.epa.gov/compliance/environmentaljustice/interagency/index.html (accessed November 22, 2003).

———. Office of Environmental Justice. 2003. *Environmental Justice Collaborative Problem-Solving Grants Request for Applications.* Author's files, June 6.

Epstein, Steven. 1996. *Impure Science: AIDS, Activism, and the Politics of Knowledge.* Berkeley: University of California Press.

Ezrahi, Yaron. 1990. *The Descent of Icarus: Science and the Transformation of American Democracy.* Cambridge, MA: Harvard University Press.

Faden, Ruth R., and Tom L. Beauchart, in collaboration with Nancy M. P. King. 1986. *A History and Theory of Informed Consent.* New York: Oxford.

Fals Borda, Orlando. 1987. "The Application of Participatory-Action Research in Latin America." *International Sociology* 1:329–347.

Ferris, Timothy. 2002. *Seeing in the Dark: How Backyard Stargazers Are Probing Deep Space and Guarding Earth From Interplanetary Peril.* New York: Simon and Schuster.

Fiorino, Daniel J. 1990. "Environmental Risk: A Survey of Institutional Mechanisms." *Science, Technology, & Human Values* 15:226–243.

Fischer, Frank. 2000. *Citizens, Experts, and the Environment: The Politics of Local Knowledge*. Durham, NC: Duke University Press.

Frickel, Scott. 2004. *Chemical Consequences: Environmental Mutagens, Scientist Activism, and the Rise of Genetic Toxicology*. New Brunswick, NJ: Rutgers University Press.

Futrell, Robert. 2003. "Technical Adversarialism and Participatory Collaboration in the U.S. Chemical Weapons Disposal Program." *Science, Technology, & Human Values* 28:451–482.

Gobster, Paul H., and R. Bruce Hull, eds. 2000. *Restoring Nature: Perspectives from the Social Sciences and Humanities*. Washington, DC: Island Press.

Gross, Matthias. 2003. *Inventing Nature: Ecological Restoration by Public Experiments*. Lanham, MD: Lexington Books.

Guston, David H. 1999. "Evaluating the First U.S. Consensus Conference: The Impact of the Citizens' Panel on Telecommunications and the Future of Democarcy." *Science, Technology, & Human Values* 24:451–482.

Habermas, Jürgen. c1984–1987. *Theory of Communicative Action*. Trans. Thomas McCarthy. Boston: Beacon Press.

Hall, Budd. 1993. Introduction to *Voices of Change: Participatory Research in the United States*, ed. Peter Park, Mary Brydon-Miller, Budd Hall, and Ted Jackson, pp. xiii–xxii. Westport, CT.: Bergen and Garvey.

Halpern, Sydney. 2004. *Lesser Harms: The Morality of Risk in Medical Research*. Chicago: University of Chicago Press.

Hays, Samuel P. 2000. *History of Environmental Policy since 1945*. Pittsburgh: University of Pittsburgh Press.

Higgs, Eric. 2003. *Nature by Design: People, Natural Process, and Ecological Restoration*. Cambridge, MA: MIT Press.

Hoffman, Lily. 1989. *Politics of Knowledge: Activist Movements in Medicine and Planning*. Albany: State University of New York Press.

Holloran, Pete. 2000. "Seeing the Trees through the Forest: Oaks and History in the Presidio." In *Reclaiming San Francisco: History, Politics, Culture*, ed. James Brook, Chris Carlsson, and Nancy J. Peters, pp. 353–352. San Francisco: City Lights Books.

Humphreys, Russell. 2003. "Recently Measured Helium Diffusion Rate for Zircon Suggests Inconsistency with U-Pb Age for Fenton Hill Granodiorite." Poster presented at the Annual Meeting of the American Geophysical Union, San Francisco, CA.

Irwin, Alan. 1995. *Citizen Science: A Study of People, Expertise, and Sustainable Development*. New York: Routledge.

Kitcher, Philip. 2003. *Science, Truth and Democracy*. New York: Oxford University Press.

Kleinman, Daniel L., ed. 2000. *Science, Technology, and Democracy*. Albany: State University of New York Press.

Laird, Frank N. 1993. "Participatory Analysis, Democracy, and Technical Decision Making." *Science, Technology, & Human Values* 18:341–361.

Latour, Bruno. 2004. *Politics of Nature: How to Bring the Sciences Into Democracy.* Cambridge, MA: Harvard University Press.

Lichterman, Paul. 1996. *The Search for Political Community: Activists Reinvent Commitment.* Cambridge, MA: Cambridge University Press.

Masters, Janet. 1995. "A History of Action Research." In *Action Research Electronic Reader,* ed. I. Hughes. University of Sydney, www.scu.edu.au/schools/gcm/ar/arr/arow/rmasters.html (accessed May 26, 2005).

Minkler, Meredith, Angela Glover Blackwell, Mildred Thompson, and Heather Tamir. 2003. "Community-Based Participatory Research: Implications for Public Health Funding." *American Journal of Public Health* 8: 1210–1214.

Moore, Kelly. 1996. "Doing Good While Doing Science: American Science and the Creation of Public Interest Science Organizations, 1955–1975." *American Journal of Sociology* 101:1121–1149.

———. 2004. "DeNaturalizing Nature: Native Habitat Restoration in San Francisco, 1998–2003." Paper presented at the Annual Meeting of the American Sociological Association, San Francisco, August.

———. Forthcoming. *Disruptive Science: Social Movements, American Scientists, and the Politics of the Military, 1945–1975.* Princeton: Princeton University Press.

Moore, Kelly, and Nicole Hala. 2002. "Organizing Identity: The Creation of *Science for the People.*" *Research in the Sociology of Organizations* 19:309–335.

Morgen, Sandra. 2002. *Into Our Own Hands: The Women's Health Movement in the United States, 1969–1990.* New Brunswick, NJ: Rutgers University Press.

National Environmental Policy Act of 1969. United States Congress.

Park, Peter, Mary Brydon-Miller, Budd Hall, and Ted Jackson, eds. 1993. *Voices of Change: Participatory Research in the United States.* Westport, CT: Bergen and Garvey.

Piven, Frances Fox and Richard A. Cloward. 1979. *Poor People's Movements: How They Succeed, Why They Fail.* New York: Vintage.

Polletta, Francesca. 2003. *Freedom Is an Endless Meeting: Democracy in American Social Movements.* Chicago: University of Chicago Press.

Rip, Arie. 2003. "Constructing Expertise: In a Third Wave of Science Studies?" *Social Studies of Science* 33:419–434.

Rome, Adam. 2003. "'Give Earth a Chance': The Environmental Movement and the Sixties." *Journal of American History* 90:525–555.

Schindler, Amy. 2003. "How Creationists Use the Mantle of Science." Paper presented at the Annual Meeting of American Sociological Association, Atlanta, August 16–19.

Sclove, Richard. 1997. "Research by the People, for the People." *Futures* 29: 541–551.

Skrentny, John. 1996. *The Ironies of Affirmative Action: Politics, Justice, and Culture in America.* Chicago: University of Chicago Press.

Smith, Alice Kimball. 1965. *A Peril and a Hope: The Scientists' Movement in America, 1945–1947.* Chicago: University of Chicago Press.

Society for Ecological Restoration (SER). 1988–2004. Annual meeting programs.

Szasz, Andrew. 1994. *Ecopopulism: Toxic Waste and the Movement for Environmental Justice.* Minneapolis: University of Minnesota Press.

Topper, Hank, Environmental Protection Agency. 2004. Telephone interview by the author, January 17, 2004.

United Nations Division of Economic and Social Affairs Division of Sustainable Development. www.un.org/esa/sustdev/documents/agenda21/english/agenda21toc.htm (accessed June 13, 2005).

Weindling, Paul. 2001. "The Origins of Informed Consent: The International Scientific Commission on Medical War Crimes, and the Nuremberg Code." *Bulletin of the History of Medicine* 75:37–71.

Welsome, Nancy. 1999. *The Plutonium Files: America's Secret Medical Experiment in the Cold War.* New York: Delacourt Press.

West Harlem Environmental Action. N.D. www.weact.org/history.html (accessed May 25, 2005).

Woodhouse, Edward J. 1991. "The Turn toward Society? Social *Re*construction of Science." *Science, Technology, & Human Values* 3:390–404.

———, and Dean A. Nieumsa. 2001. "Democratic Expertise: Integrating Knowledge, Power, and Participation." In *Knowledge, Power and Participation in Environmental Policy Analysis,* eds. Matthis Hisschemoller, Rob Hoppe, William N. Dunn, and Jerry R. Ravetz, pp. 73–96. New Brunswick, NJ: Transaction.

3

Science and the Regulatory State

12

Institutionalizing the New Politics of Difference in U.S. Biomedical Research

Thinking across the Science/State/Society Divides

STEVEN EPSTEIN

In this chapter I describe a recent wave of reforms in biomedical knowledge production and pharmaceutical drug development in the United States. The reforms affect how researchers go about designing clinical studies, with downstream consequences for the medical care that all of us receive. More abstractly, they influence how matters such as citizenship, identity, and difference are understood in the United States, as well as how the politics of race and gender are played out. Given their implications, it would be worthwhile for me to specify these changes in detail, account for their origins, assess their significance, and explore their ramifications, but that is not my agenda here (see Epstein 2003a, 2003b, 2004a, 2004b, 2004c). Instead, in these pages, my use of the case study is somewhat more opportunistic. I provide a brief sketch of the case, describing the reform coalition and tracing the pathways of change. But my goal is to extract a series of conceptual points that, though grounded in the specific example, hold the potential for wider applicability. In particular, I pay attention to the *politics of categorization* and their role in the institutionalizing of what I call a *biopolitical paradigm.*

The concepts that I develop here have a hybrid character: they are meant to demonstrate the potential convergence of analyses in political sociology, science and technology studies, and social theory.[1] My goal is to suggest (and in these pages I cannot do more than suggest) that this cross-disciplinary fusion is useful for understanding the nexus of science/state/society relations. Because my empirical case is situated on the terrains of "the state," "medical science," and "social movements" and yet confounds attempts to distinguish clear borders between these theoretical entities, it is especially well suited for the task.[2]

The New Biomedical Politics of Inclusion

Since the mid-1980s, an eclectic assortment of reformers has argued that expert knowledge about human health is dangerously flawed. According to this critique, biomedical research has presumed a "standard human" and has been conducted without adequate regard for human variation. Those leveling the charge (including health activists, clinicians, scientists, and political leaders in the United States[3]) have pointed to numerous culprits in the general failure to attend to biomedical difference, but to the extent that advocates have brought about change, they have done so primarily by targeting the state. Reformers have trained their attention on the U.S. Department of Health and Human Services (DHHS) and especially two of its component agencies: the National Institutes of Health (NIH), the world's largest funder of biomedical research; and the Food and Drug Administration (FDA), the gatekeeper for the licensing of new therapies for sale.

Under pressure from within and without, these federal agencies have ratified a new consensus that biomedical research must become routinely sensitive to human differences of various sorts, especially sex and gender, race and ethnicity, and age.[4] Academic researchers receiving federal funds and pharmaceutical manufacturers hoping to win regulatory approval for their company's products are now enjoined to include women, racial and ethnic minorities, children, and the elderly as research subjects in many forms of clinical research; measure whether research findings apply equally well to research subjects regardless of their categorical identities; and question the presumption that findings derived from the study of any single group, such as middle-aged white men, might be generalized to other populations (see Table 1). Surprisingly,

Table 1. The "Inclusion-and-Difference" Paradigm: Guidelines, Rules, and Statutes (partial list)

1986	NIH introduces policy urging inclusion of women.
1987	NIH introduces policy urging inclusion of minorities.
1988	FDA guideline calls for analyses of subpopulation differences (gender, age, race/ethnicity) in new drug applications.
1989	FDA guideline calls for inclusion of elderly patients in clinical trials.
1993	FDA lifts 1977 restriction on inclusion of women.
	Congress passes the NIH Revitalization Act, which mandates inclusion of women and minorities and the study of subgroup differences in federally funded research.
1994	NIH publishes guidelines implementing the NIH Revitalization Act.
1995	FDA calls for presentation of demographic data (age, gender, race) for all new drug applications.
1997	Congress passes FDA Modernization Act, providing extension of patent exclusivity to drug manufacturers who test approved drugs in pediatric populations.
	FDA final rule requires labeling information on geriatric use of drugs.
1998	FDA final rule requires tabulation of data by demographic subgroup.
	NIH announces policy on inclusion of children in clinical research.
2000	FDA publishes final rule giving the agency power to impose a "clinical hold" on new drug applications that fail to provide data by demographic group.
2003	Congress passes the Pediatric Research Equity Act, granting the FDA the authority to require testing of drugs in pediatric populations.

despite attention to the move to include greater numbers of *women* in biomedical research (Auerbach and Figert 1995; Narrigan et al. 1997; Weisman 1998; Baird 1999; Eckenwiler 1999; Weisman 2000; Corrigan 2002), social scientists have had little to say about the more broad-scale, identity-centered redefinition of U.S. biomedical research practice that encompasses multiple social categories.[5]

How were these changes propelled? Advocates of reform made a series of interlinked arguments—or, to use a helpful concept from the scholarship on social movements, they put forth into the public arena a series of collective action frames (Snow et al. 1986; Benford and Snow 2000). Much like those promoting affirmative action in the workplace and in higher education, reformers sought to demonstrate that women, people of color, and other groups had been numerically underrepresented in the domain of health research in the past. They complained that attempts to keep "vulnerable populations" out of harm's way by excluding them from research amounted to misguided protection that only resulted in the inadequate study of those groups. They characterized the overemphasis on the study of white men as an instance of false universalism—treating the experience of the dominant group in society (such as white men) as the universal experience, the norm, or the standard, rather as the particular experience of that group alone. They called attention to health disparities, suggesting that a more diverse representation of groups in research would help level those inequalities. Finally, they argued that underrepresentation matters because of medical differences across groups: it was crucially important to include women, people of color, children, and the elderly as subjects in research because the findings from studies of middle-aged white men simply could not be assumed to be generalizable.

In essence, what reformers were demanding went well beyond the demographics of research populations: what they sought was full citizenship. Biomedical inclusion was not just a matter of counting up bodies; it also was a broader indicator of "who counted." Thus, as biomedical research increasingly came to be viewed as a domain in which social justice could be pursued, questions of scientific method became fused with questions of cultural citizenship—of the ways in which groups express their belonging to the social whole. These were high-stakes claims, but ones that resonated with many people who might not otherwise have concerned themselves much with the fine details of research methodology. My case study thereby joins a rich and growing body of scholarship that demonstrates how science and technology in general, and biomedicine in particular, have become increasingly central to the modern constitution of citizenship and public identity (Akrich 1992; Rabinow 1996; Rose 2001; Petersen and Bunton 2002; Petryna 2002; Rabeharisoa and Callon 2002; Briggs and Mantini-Briggs 2003; Heath, Rapp, and Taussig 2004). This work suggests that as the biosciences ever more thoroughly remake our bodies and selves, they also redefine our relationship to the state and civil society.

Who, then, were these reformers? This broad-ranging critique of biomedical practice was propelled by a diverse set of individuals and groups, including "policy entrepreneurs" (Kingdon 1984; Oliver and Paul-Shaheen 1997; Weisman 1998), politicians, advocates, interest groups, and social movements. By and large, "women's health" was the driving wedge—not surprisingly so. Women as a class are the largest social category invoked in these debates, and any failure on the part of biomedicine to attend satisfactorily to more than half the U.S. population seems especially egregious. But in addition, feminist movements of the 1970s and 1980s helped create new possibilities for social change in this arena. Indeed, different political "lines" within feminism had different effects, all contributing, directly or indirectly, to the concern with research on women's health.[6]

On the more left-leaning end of the movement, radical feminists and socialist feminists had promoted a thoroughgoing critique of patriarchal practices and assumptions. Activists within the feminist women's health movement had adapted this critique to address sexism within the medical profession and the health care industry specifically. The legacy of the feminist women's health movement was a deep skepticism toward the mainstream medical profession, a critique of many of its characteristic practices, and a strong emphasis on women's personal autonomy and control over their bodies. These sensibilities were important not only for their direct impact on women who grew to care about the politics of biomedical research, but also because they were absorbed by many activists who became involved in organizing around HIV/AIDS or breast cancer, and whose concern with these medical conditions led them to focus on research politics and practices.

On the more moderate end of the broader women's movement, liberal feminists had pushed for the mainstreaming of women within all branches of U.S. society, with results that are consequential for the developments I describe here: because of the relative successes of this project of mainstreaming, women—at least in limited numbers—had risen into positions of prominence in government, the medical profession, and the world of scientific research. And some of these women, influenced by feminist ideals, were especially inclined to use their positions to press for reforms of biomedicine.

Once women put forward their critiques, they opened up a space of possibility that others could occupy. Racial and ethnic minorities, for example, followed with arguments that they, too, were underserved by modern medicine and underrepresented in study populations. These advocates drew strength from traditions of health activism within those minority communities as well as from advocacy organizations within the women's health movement that represented the interests of women of color. In addition, here as well, the recent successes of the challenging group in gaining entry into the medical profession made a difference, as did their political organization. In the case of African American physicians, for example, their representative organization, the National Medical Association, had grown from 5,000 members in 1969 to 22,000 members in 1977 (Watson 1999:154).

Thus, many of the key actors promoting reform represented women's health, either directly or indirectly: female members of Congress and their staffs (especially the Congressional Caucus for Women's Issues); female "insiders" at agencies such as the NIH; health professionals and scientists sympathetic to women's health issues; activist lawyers; the broader women's health movement; activists concerned with the effects of diseases such as AIDS and breast cancer on women; and new, professionalized women's health advocacy groups that arose in the early 1990s, such as the Society for the Advancement of Women's Health Research (SAWHR). But, over the course of the 1990s, other actors also played important roles in expanding the critique, including the American Academy of Pediatrics, geriatricians, politicians concerned about minority health issues (including members of the Congressional Black Caucus), physicians working primarily with racial and ethnic minority patients, and pharmaceutical companies interested in diversifying their markets or developing niche markets.

In thinking about the implications of this case for science studies and political sociology, it is useful to examine the characteristics of this set of actors who brought this reform wave into being. A consideration of the composition of what could be termed a tacit coalition should cause us to reflect critically on the use of common conceptual oppositions in sociological analysis. This is not a simple story of how challengers from "outside" the state moved in, disrupted the status quo, and transformed the state's operations. Rather, the reform coalition spilled across the normally recognized divides between state and society, as well as those between experts and the laity, science and politics, and the powerful and the disenfranchised. Nor was this movement simply in the business of attacking existing institutions. Instead, here as elsewhere, "social movements [played] a double-edged role: they [de-institutionalized] existing beliefs, norms, and values embodied in extant forms, and [established] new forms that instantiate new beliefs, norms, and values" (Rao, Morrill, and Zald 2000:238).

Perhaps most importantly, the case demonstrates the "fuzzy and permeable boundary between institutionalized and non-institutionalized politics" (Goldstone 2003) and underscores the risk "[of assuming that] social movements are discrete entities that exist *outside* of government" (Skrentny 2002:5; see also Jenness 1999). As Mark Wolfson notes in his analysis of a similar case, the tobacco control movement, too often analysts of social movements tend to see the state simply as a movement's "target," "sponsor," or "facilitator," or as the provider or denier of "opportunities" for activism. But in many cases, "fractions of the state are often allied with the movement in efforts to change the policies of other fractions." In such cases—for which he proposes the label "interpenetration"—"it is hard to know where the movement ends and the state begins" (Wolfson 2001:7, 144–145).[7]

"Categorical Alignment Work"

The successful installation of these reforms was by no means guaranteed. In fact, this movement for inclusion encountered resistance on multiple fronts. Defenders of scientific autonomy opposed the politicizing of research and argued that it should be up to scientists, not policymakers, to determine the best ways to conduct medical experiments. Conservatives decried the intrusion of "affirmative action," "quotas," and "political correctness" into medical research. Ethicists and health activists expressed concern about the risks of subjecting certain groups, such as children, to the risks of medical experimentation in large numbers. Furthermore, many proponents of medical universalism argued that biological differences are less medically relevant than fundamental human similarities: when it comes right down to it, people are people. The ranks of dissenters and critics included politicians, representatives of the pharmaceutical industry, NIH-funded researchers, some prominent DHHS officials, and conservative writers and critics, as well as a small but influential group of statisticians who were experts on clinical trial methodology.

What accounted for the victory of proponents of reform over those who were suspicious of or hostile to these efforts? I cannot do the topic justice here; but I would like to focus on one important part of the answer, which concerns the reformers' successful engagement with the politics of classification. Of course, classification is a critical point of intersection between science and technology studies, political sociology, and social theory. On the one hand, classification, including the classification of human groupings, is central to scientific practices of description and generalization (Dupré 1993:17–84; Bowker and Star 1999). On the other hand, as Paul Starr has observed, official classifications "are sewn into the fabric of the economy, society and the state" and provide incentives for action of all sorts (Starr 1992:154, 160). In modern, formally democratic polities, governments are often placed in the business of deciding which categories will count as legitimate, and they often remake the structure of political opportunities by providing benefits on the basis of categorical membership (160–161; see also Espiritu 1992; Nagel 1995; Porter 1995).

Even more emphatically, Pierre Bourdieu referred to systems of social categorization as "the stakes, par excellence, of political struggle" (Bourdieu 1985:729); and he emphasized the tremendous power of the modern state "to produce and impose . . . categories of thought that we spontaneously apply to all things of the social world," and to "[impose] common principles of vision and division" (Bourdieu 1998:35, 45). However, no state can fully monopolize the power to classify, and as Rogers Brubaker and Frederick Cooper remind us, "the literature on social movements . . . is rich in evidence on how movement leaders challenge official identifications and propose alternative ones" (Brubaker and Cooper 2000:16). Indeed, work on modern, identity-oriented social movements has suggested that self-naming and self-classifying are among their most consistent preoccupations (Melucci 1989). Thus a wide range of actors and institutions may participate in projects of human classification.

Appropriately, much scholarly attention has focused on the ways in which classifiers *compete* to establish their preferred classification scheme as legitimate and to institutionalize it across social domains. Whose categories will rule? Which will travel beyond their worlds of origin? When will official categories provoke resistance from those who perceive a lack of fit between classification systems and the identities of everyday life (Bowker and Star 1999:197–210)? Bourdieu, in particular, assumed that all parties will seek a "monopoly of legitimate *naming*," employing as symbolic capital in that struggle "all the power they possess over the instituted taxonomies" (Bourdieu 1985:731–732, emphasis in the original). Bourdieu argued, moreover, that it is the state that will tend to be victorious in such struggles (Bourdieu 1985, 1998:35–63).

It is important, however, to consider an alternative possibility, one that the case described here exemplifies. Instead of using their symbolic capital to impose a preferred classification scheme as legitimate and to displace the classifications of others, individuals may perform what I call "categorical alignment work" (Epstein 2004c), causing classification schemata that already are roughly similar to become superimposed or aligned with one another. In such situations, categorical terms (such as *black, male,* or *child*) can function as what Leigh Star and James Griesemer have termed "boundary objects"—objects "which are both plastic enough to adapt to local needs and the constraints of the several parties employing them, yet robust enough to maintain a common identity across sites" (Star and Griesemer 1989:93).[8] Here the groups designated both by state and biomedical authorities correspond roughly with mobilized collective identities, or with what Michèle Lamont and Virág Molnár call the "phenomenology of group classification": how ordinary individuals "think of themselves as equivalent and similar to . . . others [and] how they 'perform' their differences and similarities" (Lamont and Molnár 2002:187). In such cases, bureaucratic and scientific classifications come to be treated as functionally equivalent both to one another and to the categories of everyday life.

The reform coalition opposing the "standard human"[9] was consolidated by means of this sort of categorical alignment: the successful superimposition of modes of authoritative social categorization prevalent in different domains of state and society, in order to build bridges connecting those domains. That is, reformers acted as if it were self-evident that the mobilization categories of identity politics, the biological categories of medical research, and the social classifications of state bureaucratic administration were all one and the same system of categorization. For example, through the use of a single set of racial and ethnic descriptors that bridged domains of "science," "society," and "the state," advocates solidified the notion that the political desire to improve minority health coincided with biomedical researchers' interests in describing racial and ethnic subpopulation differences, and that efforts to accomplish these goals could be administered and monitored using the authorized list of racial and ethnic categories identified on the U.S. census. In this way, categorical terms for race and ethnicity functioned as boundary objects, moving across and stitching together diverse social worlds.[10]

Such an accomplishment should not be taken for granted. In a recent analysis of a failed scientific endeavor, the Human Genome Diversity Project, Jennifer Reardon (this volume, 2001, 2004) has shown how that project—similarly concerned with inclusion and difference—came to be doomed when geneticists' desires to sample the diversity of the human genome in an inclusive way clashed with the desires of indigenous groups to represent themselves. In that case, categories simply failed to travel across social worlds. By contrast, in the case I analyze, reformers consistently framed their arguments in ways that facilitated categorical alignment. For example, advocates placed particular emphasis on the argument that underrepresentation in clinical trials was harmful not only for reasons of justice and equity (that is, the idea that everyone deserves "equal time" from biomedicine) but also because of medical differences across groups—differences generally understood in biological terms. Here advocates positioned themselves as representing the latest in medical knowledge. For example, numerous studies have analyzed the unequal distribution by sex and race or ethnicity of genetic variants of the "cytochrome P450" enzymes that metabolize many pharmaceutical drugs, and researchers have argued that such differences cause a standard dose of these medications to have different effects in different groups. Researchers also became aware of striking differences in the efficacy of antihypertensive drugs across racial groups in the United States, and consequently the idea that African American patients respond better to diuretics than to beta-blockers became an established medical maxim.

The potent conjunction of political and ethical arguments with scientific ones helped accomplish what social movement scholars term "frame bridging": "the linkage of two or more ideologically congruent but structurally unconnected frames regarding a particular issue or problem" (Snow et al. 1986:67). And this bridging of normative and technical concerns—and the related fusing of sociopolitical with scientific categories that I term categorical alignment—facilitated a convergence of interests among diverse actors and their agreement on specific remedies.

The marker of successful categorical alignment work is that it becomes invisible in hindsight: the superimposition of political classifications with scientific ones seems natural and inevitable. But there are many bases by which claims of inequality could plausibly be put forward in the domain of health research, including by social class and geographic region; and likewise there are many ways of representing the dispersion of biological or genetic differences within the human population. So it was not foreordained that medical and political categories would come to be aligned in this case—indeed, some of the opponents of the proposed new inclusionary policies did their best to disrupt or refute the logic of categorical alignment.

In understanding the success of categorical alignment in this case, it is vital to attend to the work of state officials. Government health officials within DHHS agencies played a crucial part in cementing categorical alignment via their specification of categorical terms. In particular, they helped determine the standard set of categories that are authorized under the call for research inclusiveness and for the study of biomedical differences. These recognized categories principally include "male," "female," various gradations of age, and an official list of racial and ethnic terms, though occasionally other categories are invoked. Such specifications have power: it is worth noting that this categorization schema leaves certain identities, such as "working class," largely outside the rubric of the new concern with inclusion and difference.

In its employment and enforcement of this classification scheme, DHHS practice has been shaped by what political sociologists call "policy legacies" (Weir, Orloff, and Skocpol 1988; Skrentny 2002) and cultural sociologists and scholars of social movements call "transpositions" (Sewell 1992:17 n.9): officials have imported standards and classifications from other domains, including other government agencies that have dealt with issues of inclusion and affirmative action. For example, as I have already indicated, DHHS agencies approach the question of racial and ethnic differences in health by means of the census categories promulgated by the Office of Management and the Budget. By virtue of simply adopting categories already employed elsewhere in state administration, the DHHS has been able to finesse the question of the precise correspondence between political and scientific classifications (that is, they have avoided calling attention to their own categorical alignment work).

Precisely by bridging manifestly "scientific" and "political" arguments, reformers framed their arguments in ways that promoted an alignment of classificatory projects in three domains: the self-nominating practices of identity-based social movements, the segmenting practices of biomedical science, and the counting and labeling practices of the state. On the basis of the alignment, proponents of inclusion were able to *act as if* the social movement identity labels, the biomedical terms, and the state-sanctioned categories were all one and the same set of classifications—that is, that the politically salient categories were simultaneously the scientifically relevant categories. And it followed from this presumption that political and biomedical remedies could be pursued simultaneously through a single project of reform.

"Biopolitical Paradigms"

The new expectations governing medical research are codified in a series of federal laws, policies, and guidelines issued between 1986 and the present, which require or encourage research inclusiveness and the measurement of difference (refer back to Table 1). The reform wave also can be traced through the creation of bureaucratic offices, whether by congressional mandate or by internal DHHS action (Table 2). For example, by 1994, offices for women's health existed within the NIH, the FDA, and the Centers for Disease Control and Prevention (CDC), as well as higher up, at the departmental level under the assistant secretary for health.

"On the ground," biomedical researchers are most likely to encounter the new inclusionary mandate when applying for NIH funds to conduct clinical studies. The standard grant application form "PHS 398" now has a chart on which investigators must enter their study recruitment targets by sex/gender, ethnicity, and race. If investigators do not intend to include both women and men and a range of racial and ethnic groups, then they must explain the rationale for exclusion as part of their grant application text (U.S. Public Health Service 2004). All applications to the NIH are coded by review panels to indicate whether inclusion by sex/gender, race, ethnicity, and age is deemed acceptable or unacceptable (Table 3); if unacceptable, then a "bar-to-funding" is imposed on the proposal until the issue of inclusion has been satisfactorily addressed (Hayunga and Pinn 1996). If a proposal is funded, the investigator must submit regular reports on accrual of subjects, demonstrating that the actual demographics of the study are consistent with the inclusion plan that was proposed. To comply with congressional reporting requirements, the NIH maintains a database on the overall demographics of research subjects. Pharmaceutical companies submitting "new drug applications" to the FDA face somewhat similar, if less specific, reporting requirements.

Clearly, the new emphasis on inclusion and difference encompasses much more than a set of categories: it is undergirded by an infrastructure

Table 2. The "Inclusion-and-Difference" Paradigm: Offices (partial list)

1985	DHHS Office of Minority Health created.
1988	CDC Office of the Associate Director for Minority Health created.
1990	NIH Office of Research on Women's Health created.
	NIH Office of Research on Minority Health created.
1991	DHHS Office on Women's Health created.
1994	CDC Office of Women's Health created.
	FDA Office of Women's Health created.
2000	Congress passes bill to convert NIH Office of Research on Minority Health into National Center on Minority Health and Health Disparities
	Legislation introduced to create DHHS Office of Men's Health
	Assistant Secretary for Health appoints DHHS interagency committee to consider possibility of an Office of Lesbian and Gay Health.
2001	Agency for Healthcare Research and Quality establishes an Office of Priority Populations Research.

of standard operating procedures, encoded in regulations, and enforced and overseen by new bureaucratic offices. How might we best conceptualize the institutional structure that has developed in response to the reform wave that I have described? While it bears affinities with what some analysts of technoscience have called a "bandwagon" (Fujimura 1988) and others have called a "platform" (Keating and Cambrosio 2000), I am attracted to Peter Hall's concept of the "policy paradigm," a term that he adapts from the work of Thomas Kuhn. As Hall describes it, a policy paradigm is "a framework of ideas and standards that specifies not only the goals of policy and the kinds of instruments that can be used to attain them, but also the very nature of the problems they are meant to be addressing. Like a *Gestalt,* this framework is embedded in the very terminology through which policymakers communicate about their work, and it is influential precisely because so much of it is taken for granted and unamenable to scrutiny as a whole" (Hall 1993:279).[11]

In the modern world of what Foucault (1980) called "biopolitics"—in which new forms of biological knowledge become constitutive of new modes of governance and administration—it would behoove us to broaden Hall's concept beyond his understanding of policy. I believe we might speak usefully of "biopolitical paradigms" that traverse the

Table 3. "Summary Card of Inclusion Codes for Scientific Review Group Reviewers" (revised October 1998)

Gender Code	Minority Code	Children Code
First character = G	*First character* = M	*First character* = C
Second character:	*Second character:*	*Second character:*
1 = Both genders	1 = Minority and nonminority	1 = Children and adults
2 = Only women	2 = Only minority	2 = Only children
3 = Only men	3 = Only nonminority	3 = No children included
4 = Gender unknown	4 = Minority representation unknown	4 = Representation of children is unknown
Third character:	*Third character:*	*Third character:*
A= Scientifically acceptable	A = Scientifically acceptable	A = Scientifically acceptable
U = Scientifically unacceptable	U = Scientifically unacceptable	U = Scientifically unacceptable

Source: "Review and Award Codes for the NIH Inclusion of Children Policy," NIH Office of Extramural Research, 26 March 1999.

Note: Examples:
G1A = Both genders, scientifically acceptable
M3U = Only nonminorities, scientifically unacceptable
C2A = Only children, scientifically acceptable

boundaries between the life sciences and state policymaking, simultaneously specifying goals, methods, and procedures for each. To characterize the particular biopolitical paradigm that is at the core of the present case study, I use the term "inclusion-and-difference paradigm." The name reflects the two substantive goals: the inclusion of members of various groups generally considered to have been underrepresented previously as subjects in clinical studies, and the measurement of differences across groups with regard to treatment effects or biological disease processes.

By adopting the language of paradigms, I mean to underscore the considerable degree of inertia that can inhere in these regimes. As Geof Bowker and Leigh Star (1999:14) observe in their study of standards and institutional infrastructures, because successful standards "have significant inertia," changing them or ignoring them can be difficult, time-consuming, and costly. Similarly, the political scientist John Kingdon has noted that "establishing a principle is so important because people become accustomed to a new way of doing things and build the new policies into their standard operating procedures. Then inertia sets in, and it becomes difficult to divert the system from its new direction" (Kingdon 1984:201). Numerous political sociologists, using terms such as "policy legacies" and "policy feedback," have described the "path-dependent" character of policymaking and how "decisions at one point in time can restrict future possibilities by sending policy off onto particular tracks" (Weir 1992:9; see also Skocpol and Amenta 1986; Weir, Orloff, and Skocpol 1988; Burstein 1991; Dobbin 1994; Skrentny 2002).

At the same time, this inertia has the effect of stabilizing policies in ways that may permit their expansion or migration. Analysts of "domain expansion" have described how policies deemed successful will often migrate from one policy domain to another, or how policies such as affirmative action (Skrentny 2002) or hate crime legislation (Jenness 1995) will come to be extended to more and more social groups and categories over time. These tendencies toward the standardization, migration, and transposition of policy also have their correlates in the spread of social problems from one arena to another (Hilgartner and Bosk 1988:72) and in the diffusion of collective action frames and tactics from one social movement to the next (Meyer and Whittier 1994:277; McAdam 1995:219; Benford and Snow 2000:627–628). An excellent example from my case study (which I discuss at length in Epstein 2003b) is the partially successful attempt by lesbian and gay health advocates, during the final years of the Clinton administration, to incorporate the attribute of sexual orientation under the rubric of the inclusion-and-difference paradigm.

The inclusion-and-difference paradigm infuses the life sciences with new political meaning. It takes two different areas of scientific and governmental concern—the status of socially subordinated groups and the meaning of biological difference—and weaves them together by articulating a distinctive way of asking and answering questions about them. Perhaps most centrally, the paradigm asserts new ways of standardizing humanity and classifying human groups. While some might see the inclusion-and-difference paradigm as an example of how biomedicine (for better or for worse) gets politicized, it might just as well be taken as evidence of the converse—how, in the present period, governing gets "biomedicalized." Medical research thereby becomes reconceived as a domain in which a host of political problems can get worked out—the nature of social justice, the limits of citizenship, and the meanings of equality at the biological as well as social levels.

A Final Note: Elective Affinities and Modernity

In its extended use of a single example and its synoptic treatment of complex theoretical debates, this essay can be nothing more than a promissory note toward a more satisfactory analysis. Nevertheless, I do believe that the conceptual apparatus that I have outlined here can fruitfully be employed in a wide range of cases. It is plausible to postulate an elective affinity between the enhanced classificatory practices of modern democratic states, the assertive self-naming practices of identity-based social movements, and the new segmenting techniques of the biosciences. If so, then analyses of the relation between institutional change, the politics of classification, and the emergence of new biopolitical configurations take on considerable importance. To conduct them, we need new interdisciplinary approaches that bridge the gaps, but also build on the unappreciated affinities between science and technology studies, political sociology, and social theory, among other fields.

Acknowledgments

This chapter is part of a larger research project and thus indirectly benefits from the advice of many more people than I can acknowledge here. Neither am I able to reference in the bibliography the many works that have influenced my thinking in this project. However, I am particularly grateful to the editors of this volume, Scott Frickel and Kelly Moore, for their helpful comments on a previous draft as well as their insights into how my work contributes to the project they articulate in this volume. Elisabeth Clemens also provided important suggestions. This material is based upon work supported by the National Science Foundation under Grant No. SRB-9710423. Any opinions, findings, and conclusions or recommendations expressed in this material are those of the author and do not necessarily reflect the views of the National Science Foundation. The work on which this chapter was based was also supported by an Investigator Award in Health Policy Research from the Robert Wood Johnson Foundation. Finally, this work was supported by funding from the University of California, San Diego.

Endnotes

1. In my approach to the study of state policymaking, I draw on *neo-institutional analyses of state (and non-state) actors and organizations* (Weir, Orloff, and Skocpol 1988; Burstein 1991; Weir 1992; Dobbin 1994; Skocpol 1995; Skrentny 1996; Moore 1999; Skrentny 2002); *Foucauldian studies of governmentality* (Burchell,

Gordon, and Miller 1991; Mitchell 1999; Rose 2001); and *Bourdieu's work on states and symbolic violence* (Bourdieu 1985, 1998).

My goal, then, is to bring the above literatures together with recent work on *biomedical knowledge, practice, and technologies* (Conrad and Gabe 1999; Keating and Cambrosio 2000; Lock, Young, and Cambrosio 2000; Mol 2002; Petersen and Bunton 2002; Clarke et al. 2003; Franklin and Lock 2003; Timmermans and Berg 2003), as well as with the growing bodies of scholarship that seeks to reveal the *centrality of gender, race, and sexuality both to the organization of the modern state* (Omi and Winant 1986; Connell 1990; Duster 1990; Brown 1992; Luker 1998; Goldberg 2002) *and to the workings of science* (Wailoo 1997; Schiebinger 1999; Tapper 1999; Shim 2000; Montoya 2001; Braun 2002; Briggs and Mantini-Briggs 2003; Duster 2003; Oudshoorn 2003; Kahn 2004; Reardon 2004).

2. Data for this study have been obtained from 72 semi-structured, in-person interviews in and around Boston, New Haven, New York, Baltimore, Washington, Atlanta, Ann Arbor, Chicago, Denver, Boulder, San Francisco, Los Angeles, and San Diego. Those interviewed included past and present NIH, FDA, and DHHS officials; clinical researchers; pharmacology researchers; biostatisticians; medical journal editors; drug company scientists; women's health advocates and activists; bioethicists; members of Congress; congressional aides; lawyers; representatives of pharmaceutical company trade associations; experts in public health; and social scientists. Additional primary data sources include documents and reports from the NIH, the FDA, the CDC, the DHHS, and the U.S. Congress; archival materials from health advocacy organizations; materials from pharmaceutical companies and their trade organizations; articles, letters, editorials, and news reports published in medical, scientific, and public health journals; and articles, editorials, letters, and reports appearing in the mass media.

3. My focus in this chapter is the United States. I am not aware of any other countries except Canada that have issued formal policies on research inclusion. Clearly, cross-national comparisons and a more global perspective not only would deepen the analysis presented here but also would be consistent with the kind of theoretical bridging work that this chapter is meant to encourage.

4. The categorical terms used in this chapter are meant to represent the terms employed by the actors I studied, in all the ambiguity of everyday usage. On the history of the biomedical reliance on age, sex, and race categories, see also Hanson 1997.

5. However, the focus by the DHHS on specific social identities has attracted scrutiny from opponents of these emphases, such as Satel (2000), who criticize what they term the intrusion of "political correctness" into medicine.

6. The following discussion borrows the accepted understandings of distinctions between radical feminism, socialist feminism, and liberal feminism in the 1970s and afterward. See, for example, Echols 1989.

7. For another example of an "interpenetrated" movement, see my analysis of AIDS treatment activism in Epstein 1996. Analyses of interpenetrated movements that take up questions related to science and technology point us in the direction of a larger theoretical and empirical project. On one side, scholars have suggested models for understanding the mutual constitution of state and society and for abandoning reified and holistic notions of "the state" (Abrams 1988; Barkey and Parikh 1991; Mitchell 1999; Steinmetz 1999). On the other side, a central project of science and technology studies has been to theorize the mutual constitution of the state and science, and to describe the tight historical intertwining of technoscientific development and modern state formation (Shapin and Schaffer 1985; Latour 1987; Hacking 1990; Jasanoff 1990; Mukerji 1994; Porter 1995; Carroll 1998; Desrosières 1998). Case studies of interpenetrated technoscientific movements demand that we find ways of bringing these literatures together to understand the ever-evolving relations among, and the redefinitions of the boundaries between, science, society, and the state.

8. Categorical alignment might also usefully be seen as a subtype of what Geof Bowker and Leigh Star call "convergence": the "double process by which information artifacts and social worlds are fitted to each other and come together" (Bowker and Star 1999:49, 82). My emphasis on alignment also builds on Joan Fujimura's (1987) analysis of the alignment of work as an important component of scientific activity.

9. On the politics of the "standard human," see Epstein 2004a.

10. There are important affinities here between categorical alignment and the social and political practices of commensuration described by Espeland and Stevens (1998).

11. Of course, Kuhn's concept of paradigms has been subjected to many critiques over the years, and it is not my purpose here to resurrect the concept in toto. For example, in proposing the concept of "platforms," Keating and Cambrosio reject the Kuhnian language of paradigms as inferior because of its presumption of shared understandings among actors (Keating and Cambrosio 2000:47). But as the above quotation from Hall makes explicit, it is entirely possible to speak of paradigms without assuming consciously shared understandings.

References

Abrams, Philip. 1988. "Notes on the Difficulty of Studying the State (1977)." *Journal of Historical Sociology* 1:58–89.

Akrich, Madeleine. 1992. "The De-Scription of Technical Objects." In *Shaping Technology/Building Society: Studies in Sociotechnical Change,* ed. W. E. Bijker and J. Law, pp. 205–224. Cambridge, MA: MIT Press.

Auerbach, Judith D., and Anne E. Figert. 1995. "Women's Health Research: Public Policy and Sociology." *Journal of Health and Social Behavior* 35 (extra issue):115–131.

Baird, Karen L. 1999. "The New NIH and FDA Medical Research Policies: Targeting Gender, Promoting Justice." *Journal of Health Politics, Policy and Law* 24:531–565.

Barkey, Karen, and Sunita Parikh. 1991. "Comparative Perspectives on the State." *Annual Review of Sociology* 17:523–549.

Benford, Robert D., and David A. Snow. 2000. "Framing Processes and Social Movements: An Overview and Assessment." *Annual Review of Sociology* 26: 611–639.

Bourdieu, Pierre. 1985. "The Social Space and the Genesis of Groups." *Theory and Society* 14:723–744.

———. 1998. *Practical Reason: On the Theory of Action.* Stanford, CA: Stanford University Press.

Bowker, Geoffrey C., and Susan Leigh Star. 1999. *Sorting Things Out: Classification and Its Consequences.* Cambridge, MA: MIT Press.

Braun, Lundy. 2002. "Race, Ethnicity, and Health: Can Genetics Explain Disparities?" *Perspectives in Biology and Medicine* 45:159–174.

Briggs, Charles L., and Clara Mantini-Briggs. 2003. *Stories in Times of Cholera: Racial Profiling during a Medical Nightmare.* Berkeley: University of California Press.

Brown, Wendy. 1992. "Finding the Man in the State." *Feminist Studies* 18:7–34.

Brubaker, Rogers, and Frederick Cooper. 2000. "Beyond 'Identity.'" *Theory and Society* 29:1–47.

Burchell, Graham, Colin Gordon, and Peter Miller, eds. 1991. *The Foucault Effect: Studies in Governmentality.* Chicago: University of Chicago Press.

Burstein, Paul. 1991. "Policy Domains: Organization, Culture, and Policy Outcomes." *Annual Review of Sociology* 17:327–350.

Carroll, Patrick Eamonn. 1998. "Engineering Ireland: The Material Constitution of the Technoscientific State." PhD diss., Sociology (Science Studies), University of California, San Diego, La Jolla.

Clarke, Adele E., Janet K. Shim, Laura Mamo, Jennifer Ruth Fosket, and Jennifer R. Fishman. 2003. "Biomedicalization: Technoscientific Transformations of Health, Illness, and U.S. Biomedicine." *American Sociological Review* 68:161–194.

Connell, R. W. 1990. "The State, Gender, and Sexual Politics." *Theory and Society* 19:507–544.

Conrad, Peter, and Jonathan Gabe, eds. 1999. *Sociological Perspectives on the New Genetics.* Oxford: Blackwell.

Corrigan, Oonagh P. 2002. "'First in Man': The Politics and Ethics of Women in Clinical Drug Trials." *Feminist Review* 72:40–52.

Desrosières, Alain. 1998. *The Politics of Large Numbers: A History of Statistical Reasoning.* Cambridge, MA: Harvard University Press.

Dobbin, Frank. 1994. *Forging Industrial Policy: The United States, Britain, and France in the Railway Age.* Cambridge: Cambridge University Press.

Dupré, John. 1993. *The Disorder of Things: Metaphysical Foundations of the Disunity of Science.* Cambridge, MA: Harvard University Press.

Duster, Troy. 1990. *Backdoor to Eugenics.* New York: Routledge.

———. 2003. "Buried Alive: The Concept of Race in Science." In *Genetic Nature/Culture: Anthropology and Science beyond the Two-Culture Divide,* ed. A. H. Goodman, D. Heath, and M. S. Lindee, pp. 258–277. Berkeley: University of California Press.

Echols, Alice. 1989. *Daring to Be Bad: Radical Feminism in America, 1967–1975.* Minneapolis: University of Minnesota Press.

Eckenwiler, Lisa. 1999. "Pursuing Reform in Clinical Research: Lessons from Women's Experience." *Journal of Law, Medicine and Ethics* 27:158–188.

Epstein, Steven. 1996. *Impure Science: AIDS, Activism, and the Politics of Knowledge.* Berkeley: University of California Press.

———. 2003a. "Inclusion, Diversity, and Biomedical Knowledge Making: The Multiple Politics of Representation." In *How Users Matter: The Co-Construction of Users and Technology,* ed. N. Oudshoorn and T. Pinch, pp. 173–190. Cambridge, MA: MIT Press.

———. 2003b. "Sexualizing Governance and Medicalizing Identities: The Emergence of 'State-Centered' LGBT Health Politics in the United States." *Sexualities* 6:131–171.

———. 2004a. "Bodily Differences and Collective Identities: Representation, Generalizability, and the Politics of Gender and Race in Biomedical Research in the United States." *Body and Society* 10:183–203.

———. 2004b. "Beyond the Standard Human?" In *Reckoning with Standards,* ed. S. L. Star and M. Lampland. Manuscript submitted for publication.

———. 2004c. "'One Size Does Not Fit All': Standards, Categories, and the Inclusion-and-Difference Paradigm in U.S. Biomedical Research." Manuscript submitted for publication.

Espeland, Wendy Nelson, and Mitchell L. Stevens. 1998. "Commensuration as a Social Process." *Annual Review of Sociology* 24:313–343.

Espiritu, Yen Le. 1992. *Asian American Panethnicity: Bridging Institutions and Identities.* Philadelphia: Temple University Press.

Foucault, Michel. 1980. *The History of Sexuality.* Vol.1: *An Introduction.* Trans. R. Hurley. New York: Vintage Books.

Franklin, Sarah, and Margaret Lock, eds. 2003. *Remaking Life and Death: Toward an Anthropology of the Biosciences.* Santa Fe, NM: School of American Research Press.

Fujimura, Joan. 1987. "Constructing 'Do-Able' Problems in Cancer Research: Articulating Alignments." *Social Studies of Science* 17:257–293.

———. 1988. "The Molecular Biological Bandwagon in Cancer Research: Where Social Worlds Meet." *Social Problems* 35:261–283.

Goldberg, David Theo. 2002. *The Racial State.* Malden, MA: Blackwell.

Goldstone, Jack A. 2003. "Introduction: Bridging Institutionalized and Non-institutionalized Politics." In *States, Parties, and Social Movements,* ed. J. A. Goldstone, pp. 1–24. Cambridge: Cambridge University Press.

Hacking, Ian. 1990. *The Taming of Chance.* Cambridge: Cambridge University Press.

Hall, Peter A. 1993. "Policy Paradigms, Social Learning, and the State: The Case of Economic Policymaking in Britain." *Comparative Politics* 25:275–296.

Hanson, Barbara. 1997. *Social Assumptions, Medical Categories.* Greenwich, CT: JAI Press.

Hayunga, Eugene G., and Vivian W. Pinn. 1996. "Implementing the 1994 NIH Guidelines." *Applied Clinical Trials,* October, 35–40.

Heath, Deborah, Rayna Rapp, and Karen-Sue Taussig. 2004. "Genetic Citizenship." In *A Companion to the Anthropology of Politics,* ed. D. Nugent and J. Vincent, pp. 152–167. London: Blackwell.

Hilgartner, Stephen, and Charles L. Bosk. 1988. "The Rise and Fall of Social Problems: A Public Arenas Model." *American Journal of Sociology* 94:53–78.

Jasanoff, Sheila. 1990. *The Fifth Branch: Science Advisers as Policymakers.* Cambridge, MA: Harvard University Press.

Jenness, Valerie. 1995. "Social Movement Growth, Domain Expansion, and Framing Processes: The Case of Violence against Gays and Lesbians as a Social Problem." *Social Problems* 42:145–170.

———. 1999. "Managing Differences and Making Legislation: Social Movements and the Racialization, Sexualization, and Gendering of Federal Hate Crime Law in the U.S., 1985–1998." *Social Problems* 46:548–571.

Kahn, Jonathan. 2004. "How a Drug Becomes 'Ethnic': Law, Commerce, and the Production of Racial Categories in Medicine." *Yale Journal of Health Policy, Law and Ethics* 4:1–46.

Keating, Peter, and Alberto Cambrosio. 2000. "Biomedical Platforms." *Configurations* 8:337–387.

Kingdon, John W. 1984. *Agendas, Alternatives, and Public Policies.* Boston: Little, Brown.

Lamont, Michèle, and Virág Molnár. 2002. "The Study of Boundaries in the Social Sciences." *Annual Review of Sociology* 28:167–195.

Latour, Bruno. 1987. *Science in Action: How to Follow Scientists and Engineers through Society.* Cambridge, MA: Harvard University Press.

Lock, Margaret, Allan Young, and Alberto Cambrosio, eds. 2000. *Living and Working with the New Medical Technologies: Intersections of Inquiry.* Cambridge: Cambridge University Press.

Luker, Kristin. 1998. "Sex, Social Hygiene, and the State: The Double-Edged Sword of Social Reform." *Theory and Society* 27:601–634.

McAdam, Doug. 1995. "'Initiator' and 'Spin-Off' Movements: Diffusion Processes in Protest Cycles." In *Repertoires and Cycles of Collective Action,* ed. M. Traugott, pp. 217–239. Durham, NC: Duke University Press.

Melucci, Alberto. 1989. *Nomads of the Present: Social Movements and Individual Needs in Contemporary Society*. Philadelphia: Temple University Press.

Meyer, David S., and Nancy Whittier. 1994. "Social Movement Spillover." *Social Problems* 41:277–298.

Mitchell, Timothy. 1999. "Society, Economy, and the State Effect." In *Culture: State-Formation after the Cultural Turn,* ed. G. Steinmetz. Ithaca, NY: Cornell University Press.

Mol, Annemarie. 2002. *The Body Multiple: Ontology in Medical Practice*. Durham, NC: Duke University Press.

Montoya, Michael J. 2001. "Bioethnic Conscripts: Biological Capital and the Genetics of Type 2 Diabetes." Paper presented at the annual meeting of the American Sociological Association, Anaheim, CA, August.

Moore, Kelly. 1999. "Political Protest and Institutional Change: The Anti-Vietnam War Movement and American Science." In *How Social Movements Matter,* ed. M. Giugni, D. McAdam, and C. Tilly, pp. 97–118. Minneapolis: University of Minnesota Press.

Mukerji, Chandra. 1994. "Toward a Sociology of Material Culture: Science Studies, Cultural Studies and the Meanings of Things." In *The Sociology of Culture: Emerging Theoretical Perspectives,* ed. D. Crane, pp. 143–162. Oxford: Blackwell.

Nagel, Joane. 1995. "American Indian Ethnic Renewal: Politics and the Resurgence of Identity." *American Sociological Review* 60:947–965.

Narrigan, Deborah, Jane Sprague Zones, Nancy Worcester, and Maxine Jo Grad. 1997. "Research to Improve Women's Health: An Agenda for Equity." In *Women's Health: Complexities and Differences,* ed. S. B. Ruzek, V. L. Olesen, and A. E. Clarke, pp. 551–579. Columbus: Ohio State University Press.

Oliver, Thomas R., and Pamela Paul-Shaheen. 1997. "Translating Ideas into Actions: Entrepreneurial Leadership in State Health Care Reforms." *Journal of Health Politics, Policy and Law* 22:721–788.

Omi, Michael, and Howard Winant. 1986. *Racial Formation in the United States: From the 1960's to the 1980's*. New York: Routledge & Kegan Paul.

Oudshoorn, Nelly. 2003. *The Male Pill: A Biography of a Technology in the Making.* Durham, NC: Duke University Press.

Petersen, Alan, and Robin Bunton. 2002. *The New Genetics and the Public Health.* London: Routledge.

Petryna, Adriana. 2002. *Life Exposed: Biological Citizens after Chernobyl.* Princeton, NJ: Princeton University Press.

Porter, Theodore M. 1995. *Trust in Numbers: The Pursuit of Objectivity in Science and Public Life*. Princeton, NJ: Princeton University Press.

Rabeharisoa, Vololona, and Michel Callon. 2002. "The Involvement of Patients' Associations in Research." *International Social Science Journal* 54:57–65.
Rabinow, Paul. 1996. *Essays on the Anthropology of Reason*. Princeton, NJ: Princeton University Press.
Rao, Hayagreeva, Calvin Morrill, and Mayer N. Zald. 2000. "Power Plays: How Social Movements and Collective Action Create New Organizational Forms." *Research in Organizational Behavior* 22:237–281.
Reardon, Jennifer. 2001. "The Human Genome Diversity Project: A Case Study in Coproduction." *Social Studies of Science* 31:357–388.
———. 2004. *Race to the Finish: Identity and Governance in an Age of Genomics*. Princeton, NJ: Princeton University Press.
Rose, Nikolas. 2001. "The Politics of Life Itself." *Theory, Culture and Society* 18: 1–30.
Satel, Sally. 2000. *P.C., M.D.: How Political Correctness Is Corrupting Medicine*. New York: Basic Books.
Schiebinger, Londa. 1999. *Has Feminism Changed Science?* Cambridge, MA: Harvard University Press.
Sewell, William H. 1992. "A Theory of Structure: Duality, Agency, and Transformation." *American Journal of Sociology* 98:1–29.
Shapin, Steven, and Simon Schaffer. 1985. *Leviathan and the Air-Pump: Hobbes, Boyle and the Experimental Life*. Princeton, NJ: Princeton University Press.
Shim, Janet K. 2000. "Bio-Power and Racial, Class, and Gender Formation in Biomedical Knowledge Production." *Research in the Sociology of Health Care* 17: 173–195.
Skocpol, Theda. 1995. *Social Policy in the United States: Future Possibilities in Historical Perspective*. Princeton, NJ: Princeton University Press.
Skocpol, Theda, and Edwin Amenta. 1986. "States and Social Policies." *Annual Review of Sociology* 12:131–157.
Skrentny, John David. 1996. *The Ironies of Affirmative Action: Politics, Culture, and Justice in America*. Chicago: University of Chicago Press.
———. 2002. *The Minority Rights Revolution*. Cambridge, MA: Harvard University Press.
Snow, David A., E. Burke Rochford Jr., Steven K. Worden, and Robert D. Benford. 1986. "Frame Alignment Processes: Micromobilization and Movement Participation." *American Sociological Review* 51:464–481.
Star, Susan Leigh, and James R. Griesemer. 1989. "Institutional Ecology, 'Translations' and Boundary Objects: Amateurs and Professionals in Berkeley's Museum of Vertebrate Zoology, 1907–39." *Social Studies of Science* 19:387–420.
Starr, Paul. 1992. "Social Categories and Claims in the Liberal State." In *How Classification Works: Nelson Goodman among the Social Sciences*, ed. M. Douglas and D. Hull, pp. 159–174. Edinburgh: Edinburgh University Press.

Steinmetz, George. 1999. "Introduction: Culture and the State." In *Culture: State-Formation after the Cultural Turn*, ed. G. Steinmetz, pp. 1–49. Ithaca, NY: Cornell University Press.

Tapper, Melbourne. 1999. *In the Blood: Sickle Cell Anemia and the Politics of Race.* Philadelphia: University of Pennsylvania Press.

Timmermans, Stefan, and Marc Berg. 2003. *The Gold Standard: The Challenge of Evidence-Based Medicine and Standardization in Health Care.* Philadelphia: Temple University Press.

U.S. Public Health Service. 2004. Grant Application Instructions (PHS 398). National Institutes of Health 1998. Revised September 2004. www.nih.gov/grants/funding/phs398/phs398.html (accessed June 9, 2005).

Wailoo, Keith. 1997. *Drawing Blood: Technology and Disease Identity in Twentieth-Century America.* Baltimore: Johns Hopkins University Press.

Watson, Wilbur H. 1999. *Blacks in the Profession of Medicine in the United States: Against the Odds.* New Brunswick, NJ: Transaction.

Weir, Margaret. 1992. *Politics and Jobs: The Boundaries of Employment Policy in the United States.* Princeton, NJ: Princeton University Press.

Weir, Margaret, Ann Shola Orloff, and Theda Skocpol. 1988. "Introduction: Understanding American Social Politics." In *The Politics of Social Policy in the United States,* ed. M. Weir, A. S. Orloff, and T. Skocpol, pp. 3–27. Princeton, NJ: Princeton University Press.

Weisman, Carol S. 1998. *Women's Health Care: Activist Traditions and Institutional Change.* Baltimore: Johns Hopkins University Press.

———. 2000. "Breast Cancer Policymaking." In *Breast Cancer: Society Shapes an Epidemic,* ed. A. S. Kasper and S. J. Ferguson, pp. 213–243. New York: St. Martin's Press.

Wolfson, Mark. 2001. *The Fight against Big Tobacco: The Movement, the State, and the Public's Health.* New York: Aldine de Gruyter.

13

Creating Participatory Subjects

Science, Race, and Democracy in a Genomic Age

JENNY REARDON

In April of 1993, Kenneth M. Weiss, molecular anthropologist and incoming chair of the North American Regional Committee (NAmC) of the Human Genome Diversity Project, explained in a letter to the U.S. Congress:

> The National Institutes *mandates affirmative action* in all of its research grants, to ensure that all of our nation's people are served by the research and clinical establishment. How is it that we can allocate vast sums of money to studying human genes [the Human Genome Project] in a way that specifically *excludes* consideration of relevant diversity? (Weiss 1993: 44; emphasis added)

The Diversity Project, Weiss argued, would correct for this exclusionary "Eurocentric" tendency of the Human Genome Project in two ways (Bowcock and Cavalli-Sforza 1991; Human Genome Organization 1993). First, it would diversify the populations from which scientists collected DNA—in addition to "Caucasians," it would sample indigenous populations and "major ethnic groups" in the United States (Weiss 1993,: 44).[1] Second, it would seek to "partner" with the diverse populations they sought to sample, including them in the project's design and regulation. In short, organizers envisioned an "affirmative action" project in the fullest sense: one that sought to include all races and ethnicities as both objects and subjects.

From one perspective, this episode might count as a success story for democratization and might illustrate American scientists' changing relationship with citizens. Rather than maintaining a wall that divides science and civil society, Diversity Project organizers' efforts to include and "partner" with diverse populations might represent a positive step toward equalizing power relations between scientists and laypeople and toward building research based on the democratic values of participation and inclusion. However, the story of the Diversity Project troubles this hopeful narrative. Rather than embracing project organizers' efforts to promote their participation, many potential subjects resisted their efforts. This chapter illustrates why. In so doing, it seeks to reframe debates about participation and science by highlighting the fundamental questions about the constitution of subjects and the operation of power at stake in even the best-intentioned efforts to "democratize" science.

To date, scientists, policymakers, and academics alike have largely overlooked these consequential questions. Instead, in their accounts, participation figures as an a priori good—a practice designed to promote the inclusion of subjects in processes of knowledge production from which they have been wrongly excluded (Sclove 1995; Kitcher 2001). In Weiss's statement these subjects are members of "major ethnic groups." In the philosopher of science Phillip Kitcher's call for "well-ordered science," a science that is consistent with democratic values, these subjects are "women, children, members of minorities, and people in developing countries" (Kitcher 2001:117–135). It is the exclusion of these subjects that primarily concern Weiss, Kitcher, and other scientists and scholars of science.

In this chapter I argue that while a focus on inclusion of subjects is laudable, those committed to social justice could gain much by beginning their critique one step earlier, at the point of subject formation—at the point of defining who, for example, is a member of a "major ethnic group" like "African American," and who can legitimately define and speak for such a subject's interest. As the theorist of power Michel Foucault and political theorists inspired by his work have ably demonstrated, forms of power particular to liberalism tie one's formation as a subject who can act and participate in society (i.e., subjectivity) to institutionalized forms of subjection (or control by another) (Foucault 1977; Butler 1992; Cruikshank 1999:21). Consequently, these social theorists demonstrate, participation is not an innocent act. Instead, it is an institutionalized governmental practice that expresses some values and interests while excluding others.[2]

This is particularly evident in cases like the Diversity Project, where researchers and potential research subjects do not share the same cultural values and structural positions in society. To date, most literature on lay participation in research has focused on the activities of members of disease groups, such as breast cancer and AIDS activists (Epstein 1996; Klawiter 2000). These activists, for the most part, are white, middle to upper class, and well educated, and are likely to share values and interests with researchers who are also predominantly white, middle to upper class, and well educated (Epstein 1991; Klawiter 2000). In these cases, the manner in which efforts to "democratize" science might express particular values and interests is harder to discern. This is not true in the case of environmental health and human genetic diversity research, where those asked to participate in research (by virtue of their status as potential research subjects or subjects of environmental pollution) are often members of socially and economically disadvantaged racial and ethnic groups. In these instances where researchers and research subjects occupy different positions within structures of power, the manner in which efforts to create participatory subjects express particular institutionalized forms of power cannot be missed.

Diversity Project organizers' efforts to create "partnerships" with research subjects provide an exemplary case. As in past efforts guided by the logic of affirmative action, the Diversity Project's efforts to include previously excluded groups in an institution sanctioned by dominant society (this time genetics) provoked many courted for inclusion to ask: Whose interests would inclusion serve? Would participation in the design and conduct of genetic research advance the interests of the disempowered groups the project hoped to sample, or would it merely serve to legitimate the project through creating "exhibitions of representations" (Amit-Talai and Knowles 1996)?

In this chapter I focus on the organization and conduct of meetings as a key site where these questions about the material reality and effects of participation come into clear view. I demonstrate that despite the well-intentioned efforts of Diversity Project organizers, the latent conceptual and material structures embedded in organizers' notions of "partnership" acted to exclude indigenous groups and minorities in the United States from the meeting spaces where research agendas for the Project took shape, and the rules for governing human genetic diversity research formed. Indeed analogous to the effects of engaging farmers as active participants in the agrofood innovation process, described in this volume by Steven Wolf, Diversity Project organizers' efforts to create active, participatory subjects (or "partners") proved far from empowering. To the contrary, by defining subjects in a manner that undermined their rights to self-determination and demeaned their character, efforts to create participatory subjects created the conditions for these subjects' further subjection. I conclude the chapter with some modest recommendations for how to address this paradoxical predicament.

Discourses of Participation

From the very start, issues of research subject participation concerned organizers of the Diversity Project. Participation, after all, proved basic to their goals: if people refused to take part, the project would not move forward. However, understandings of what this participation would mean, and how it would be sought, changed over time. These changes begin to make evident the connections between ideas of participation and particular sets of values, interests, and relations of power.

At first, an instrumental goal guided the researchers' decisions: they wished to gain access to the largest number of "isolated human populations"—populations deemed to have the most "informative genetic records" (Cavalli-Sforza et al. 1991:490). To reach these people, organizers did not in the first instance seek the direct participation of these populations, but rather that of the "anthropologists, medical researchers, and local scientists who already have access to the more isolated groups" (Roberts 1991:1617). At this point in the evolution of the project, proposers did not anticipate that they would directly interact with populations. Rather, most expected that those who already had the necessary "working relationships" would facilitate DNA collection from "recommended populations" (1617). Thus, not surprisingly, organizers focused most of their efforts on enrolling these professionals, not the actual populations.[3] Any thought devoted to figuring out how to encourage the participation of "recommended populations" drew upon past examples of providing basic health care. As *Science* reported: "Already, says [Mary-Claire] King, the group is thinking of what it can offer to the populations in return, such as medical supplies" (Roberts 1991:1617).

However, following charges by physical anthropologists in 1992 and early 1993 that the Diversity Project might reignite biological racism and exacerbate the oppression of indigenous peoples, organizers' ideas about who should be included and how began to change.[4] In particular, some began to recognize the necessity of addressing broader issues of racism and the politics of inclusion. They acknowledged that those who originally proposed the project were all white Europeans and Americans, and if they did not want the Diversity Project to be viewed as a project organized by "colonialists," the composition of its leadership would have to change (Interview A4).[5]

The new goal of generating nonwhite and non-Western support and leadership for the project prompted supporters to organize an international planning meeting for the fall of 1993. As one member of the Diversity Project's North American Regional Committee explained:

> [W]e had decided at the Washington [Bethesda, ELSI] meeting that there should be a world meeting and that people from developing countries should be invited and get a say as to what should go on if there was going to be such a project. So that was the purpose of the Sardinia meeting—to get people from Japan, from Africa, from India, places where some sampling was likely to go on . . . to have those people involved in establishing a structure that might operate the Project for the world. (Interview B)

This international meeting that gathered together scientists from around the world interested in the study of human genetic diversity took place seven months later in Sardinia, Italy.[6] The new understanding of participation that shaped it marked the beginning of the effort to "align" the project to a different discourse of participation—one shaped by policies of inclusion in the United States, most notably, affirmative action.[7]

As noted above, organizers' efforts to create a participatory "affirmative action" initiative took many forms. Not only did organizers seek to expand the kind of objects included in the project (not just "Caucasian DNA," but the DNA of indigenous, racial, and ethnic groups); they also sought to promote the inclusion of populations the project hoped to sample in the design, conduct, and regulation of Diversity Project research. My focus here is on this latter effort—in particular, the effort to transform research subjects into subjects who stood alongside researchers as participants in every step of the research process.

The Historical and Political Context of Partnerships

The inspiration for the Diversity Project's proposal to form "partnerships" with research subjects came from a talk given by the biological anthropologist Michael Blakey at a November 1993 Wenner-Gren Foundation meeting on the Diversity Project.[8] At this meeting, Blakey shared his experiences fostering "public engagement" in the African Burial Ground Project (ABGP). Blakey served as the scientific director of the ABGP, a project that sought to analyze the African Burial Ground, the main cemetery for Africans in colonial New York City, which had been "rediscovered" in 1991 during the course of breaking ground on a new building. As part of the public engagement part of the project, those working on the ABGP distributed copies of the 130-page research design to all interested persons and held hearings in downtown Manhattan and Harlem to receive feedback. The goal was to make sure that the ABGP reflected the interests of the "descendant or culturally-affiliated community" by asking questions that were of interest to the community and also "appropriate for a scientific program" (Blakey 1998:400). African Americans were also included on the epistemological grounds that they would "bring a perspective that might help rectify the distorting effects of Eurocentric denial of the scope and conditions of African participation in the building of the nation" (400). According to Blakey, through this process members of the "African-American public" came to own the project.[9] He concluded, echoing the famous finale of Lincoln's Gettysburg Address, "Ours is a study of the people, by the people, and for the people principally concerned" (400).

Diversity Project organizers hoped something similar could be achieved in their initiative. Following the November 1993 Wenner-Gren meeting, members of the North American Regional Committee (NAmC) proposed that populations become "partners" in the research process. Not only would they provide DNA samples, but they would also formulate research questions, design sampling strategies, and collect samples. This new understanding of research subjects found formal articulation in the "Partnerships with Participating Population" section of the Diversity Project's Model Ethical Protocol.[10] As the Protocol explains:

> The key word is "with." We believe that, ideally, researchers involved in collecting samples for the HGD Project should be closely connected

> with the populations that provide those samples, connected not as "scientists" and "subject," but as partners. They might even be the same people, as communities may undertake to sample themselves in order to participate in the Project. (NAmC 1997:1468)

Just as Michael Blakey, an African American, had been enlisted to study the remains of African Americans, so Diversity Project organizers argued that members of different racial and ethnic groups should study their own genes:

> Indians work with India. We have no American Natives . . . and we don't have a Black person working on Africa, but we have . . . Asians working on Asia, for instance—not only Indians, but also Chinese and so on. (Interview E)

Another organizer defended the sincerity of this proposal:

> [We] would not just be sneering at these people. I don't think we are pretending—I think we would love to have a Kenyan in our lab, or a Yanomama in our lab if we could get a qualified person, or a scientist from Brazil in our lab. Or help set up a lab in Brazil. (Interview A2)

And indeed, organizers of the Diversity Project did make efforts to both enroll and train members of populations they identified as of interest for sampling.[11]

Blakey's notion of "descendant or culturally-affiliated communities" in the ABGP and the protocol's "partnership" provision differed in at least one crucial way.[12] Blakey's collaboration with African American communities in New York City derived from the belief that ethically it was the right thing to do. The Model Ethical Protocol, on the other hand, proposed "partnerships" both because organizers believed that it was the right thing to do and because they believed that partnerships would help generate support for the project.[13] For example, the protocol explicitly noted that "partnerships" could overcome the problem of real or perceived exploitation:

> Sampling for the HGDP will involve numerous tasks. . . . We recommend that, whenever possible, the researchers use local people in performing these tasks. These activities will allow the local population to be more engaged in the research. Their involvement in the collection process, like their involvement in the planning, *should also prevent the reality or perception that the researchers are exploiting or abusing the population.* (NAmC 1997:1471; emphasis added)

This passage gains significance if we recall that at this time organizers regarded the perception of exploitation as a major cause of opposition to their project.

In addition to involvement in the planning and collection process, the protocol argued that populations should be involved in the results of the research. Again, many of the reasons given were instrumental:

> Too often, participants in research projects have no ongoing involvement after donating their samples and time. This might generate resentment; it will inevitably waste an opportunity to get the public more involved in, and knowledgeable about, science. (Ibid.:1472)

In this passage, as well as many others, the authors of the Model Ethical Protocol focus on why partnerships are good for researchers and for science: they could help reduce resentment and enroll "the public" in learning about science.[14] They bring into view the ties between efforts to create participatory subjects and particular values and interests. These ties, apparent in the text of the Model Ethical Protocol, would soon become the subject of debate as Diversity Project organizers began to present the notion of "partnerships" to potential research subjects.

Participation in Whose Interest?

Challenges to the belief that "partnerships" would benefit studied populations arose even before "partnership" gained a formal articulation in the Model Ethical Protocol, first published in October 1995. These challenges would first emerge at meetings the MacArthur Foundation supported to promote dialogue between members of the Diversity Project's NAmC and American Indian tribes in the United States and First Nations in Canada.[15] At the first of these meetings held in San Francisco in March 1994, representatives of the NAmC introduced Blakey's model of including members of affected communities in the planning of research. These Diversity Project advocates assumed that, as in the case of the ABGP, researchers and communities could discover questions that would interest them both. Thus, they approached the meetings with the expectation and goal of discovering common ground. However, it quickly became evident that this assumption grew out of a particular history of race and racism and a particular set of social structures and values that were not shared by all.

Recall that the notion of "partnership" grew out of Blakey's experiences working with African American communities in New York City. In this case, one could reasonably argue that scientific questions about genealogical "roots" would be of interest to researchers and "people" alike. The popular success of Alex Haley's *Roots* spoke volumes about the desire of many African Americans to recover ancestral stories erased by the experience of slavery (Haley 1976). However, as the Diversity Project debates would make clear, one could not make a similar claim about members of tribes in the United States or First Nations in Canada. No analogous demand by "the people" for origins research existed among these groups. To the contrary, many expressed an explicit lack of interest in this kind of research. As one participant at a meeting designed to promote dialogue between Diversity Project organizers and indigenous groups in the Americas explained:

> We are not interested in you proving the land bridge because we have stories already that tell us that it is not true. We question the validity of certain scientific procedures like carbon dating and can prove to you on your own terms that is not accurate. We don't need linguistic studies because we know who talks like us, and so if it is all those things, then we don't want it. (Interview F)

The Out-of-Africa theory, the scientific theory that all humans originated in Africa, did potentially form a bridge to one tenet of the indigenous rights movement—that all indigenous people were connected (Interview M; Akwasasne Notes 1978). But beyond this, genetic research into human origins did not interest those asked to represent the interests of "Native Americans" at meetings with Diversity Project organizers. Questions about the relationships between the indigenous peoples of the Americas, they argued, had been answered long ago.[16]

Some even found insulting the attempts by organizers' to forge links to their stories about origins. At the second MacArthur Foundation–supported meeting held at Stanford University in January of 1998, one Diversity Project representative reportedly told participants that researchers were discovering what they already knew: we are all one (Interview G). Far from building support for the project, these statements only aggravated concerns. For many, they denoted a certain lack of awareness of the role that previous scientific claims about a "common stock" had played in legitimating "civilizing" projects that had stripped American Indian tribes in the United States of their languages and cultures (Weaver 1997:16).

Participation at Whose Cost?

Those asked to represent "Native Americans" in meetings with Diversity Project organizers argued that participation, in addition to not serving the interests of their communities, could have significant material costs. In particular, some worried that genetics threatened to drain valuable resources that could be used to address more immediate problems. As an attendee of the 1998 Stanford meeting explained:

> Just in December, elders froze in South Dakota because of substandard housing. And then you look at the millions and millions of dollars that are being spent to research indigenous peoples' DNA with absolutely no regard for our lives, and maybe even no intent to ensure our survival as peoples. I honestly don't think that there is much of an interest really to make sure that we survive as who we are. (Interview J)

Sampling indigenous DNA might answer questions of interest to scientists, but would it contribute to the survival of an oppressed people? Would genetic research lead to greater awareness of and attention to the health needs of research subjects, or would it merely drain resources from basic health care? Those asked to review the project asked organizers to answer these basic questions.

Some also noted that the work required placed a burden on the groups targeted for sampling. As one indigenous rights leader explained, there were many demands on his time. Whose time and money, he asked, would be spent raising awareness about these issues? (Interview K). And as the late Hopi geneticist Frank Dukepoo told the National Bioethics Advisory Commission: "Indian people have turned to me to do a lot of interpreting of what's going on here. We don't have the person power to deal with these issues" (Dukepoo 1998). In short, many asked to participate in the dialogue about the Diversity Project argued that it placed an undue burden on them to learn about and raise awareness about research that did not support their interests.[17] Some clarified that they were learning about the project not because they thought it could help communities, but because they feared it would endanger them:

> The reason why I work on it now is because I understand from walking in two worlds what one world is trying to do and the potential dangers of that to the other world. So I am compelled morally to work on this stuff now. And it has become a serious threat. In my estimation, this could happen. (Interview F)

For many years, this point proved difficult for project organizers to understand. They believed with sincerity that the Diversity Project would interest and benefit all people.[18] Many members of the populations they wished to sample remained unconvinced.

Members of indigenous groups targeted for sampling worried not only that participation would drain communities of valuable resources, but that it might also threaten their rights to territory, resources, and self-determination. As a primer on genetics published by the Indigenous Peoples Council on Biocolonialism (IPCB) later explained:

> Scientists expect to reconstruct the history of the world's populations by studying genetic variation to determine patterns of human migration. In North America, this research is focused on validation of the Bering Strait theory. It is possible these new "scientific findings" concerning our origins can be used to challenge aboriginal rights to territory, resources and self-determination. Indeed, many governments have sanctioned the use of genomic archetypes to help resolve land conflicts and ancestral ownership claims among Tibetans and Chinese, Azeris and Armenians, and Serbs and Croats, as well as those in Poland, Russia, and the Ukraine who claim German citizenship on the grounds that they are ethnic Germans. (Indigenous Peoples Council on Biocolonialism 2000:24)[19]

These fears that indigenous rights might be undermined were not based in fantasy. For example, some tribes in the United States grew very concerned about a bill introduced into the Vermont state legislature (Vermont H. 809) proposing that the state "establish standards and procedures for DNA-HLA testing to determine the identity of an individual as a Native American, at the request and expense of the individual" (Yona 2000).[20]

Finally, others worried that in addition to eroding legal rights, genetic studies of origins might threaten practices that sustain and create tribal identities, such as writing and telling stories. As the Kiowa novelist Scott Momaday explains, telling stories is

> imaginative and creative in nature. It is an act by which man strives to realize his capacity for wonder, meaning and delight. It is also a process in which man invests and preserves himself in the context of ideas. Man tells stories in order to understand his experience, whatever it may be. The possibilities of storytelling are precisely those of understanding the human experience. (Vizenor 1978:4)

Some asked by Diversity Project organizers to comment on the initiative from a "Native American" perspective worried that human genetic diversity research threatened to place the power to tell these life-sustaining stories in the hands of scientific and technical elites who were almost exclusively non-Native. This shift in power, some argued, might in a very real way amount to genocide.

Worries that the Diversity Project would indeed generate destructive representations of Native peoples would be sparked by the very text of the project's Model Ethical Protocol. As a Native attorney in health policy explained, the image he got from the entire Model Ethical Protocol (and not just from the Partnerships section) was not one of empowered subjects, but of reactive objects:

> It makes tribal people look absolutely unsophisticated. I get images of someone who really has not had much contact. It gave me the image of tribal people as object . . . it did not give me much image of them as people. They were objects. They were reactive objects, however. And the purpose of this—one of the things the scientists would have to do—is to elicit the proper reaction. (Interview F)

Indeed, the words "scientifically unsophisticated" do appear on the very first page of the Model Ethical Protocol. Specifically, it states: "[This Protocol] deals expressly with the ethical and legal issues that are raised when a project seeks DNA explicitly from populations, not individuals, especially when those populations maybe be *scientifically unsophisticated* and politically vulnerable" (NAmC 1997:1; emphasis added). Drafters of the proposed protocol assumed that populations, and not researchers, lacked necessary knowledge and sophistication. They argued that members of groups would see the value of the research and would be motivated to formulate their own research questions, if given the opportunity to learn about the project and gain some "sophistication" on genetic matters. There is no suggestion that members of populations might also be in a position to educate researchers.

As Steven Epstein has noted in his work on the democratic potential of AIDS activism, these models of knowledge and power that seek to "simply [disseminate] scientific knowledge in a 'downward' direction—creating a community-based expertise—seem potentially naïve, or at a minimum, insufficient. In the worst-case scenario, such a strategy transforms the recipient of knowledge into an object of power" (Epstein 1991: 55). This was precisely the concern of many asked to review the Diversity

Project: far from permitting them to function as active subjects, their construction as participatory subjects threatened to further objectify and disempower them.

Given these many potential problems with participation in genetic origins research, and the Diversity Project in particular, many potential research subjects had difficulty understanding why organizers thought the project would be of interest to their communities. Not only could they not understand this from their own perspectives, but they could not understand it from the perspective of what organizers had stated about the Diversity Project's goals and interests. For example, documents distributed before one meeting in the spring of 1994 presented research questions defined by the scientists who had proposed the initiative. How, one MacArthur meeting participant reportedly asked, did organizers reconcile this difference between the claim that the project would answer community questions and the reality that the only questions articulated so far had been those of scientists? NAmC representatives responded by arguing that the project was changing and now sought to address issues of interest to populations targeted for sampling. They cited health-related issues as the primary example. One member of the NAmC, for example, talked about the improvements that population sampling could bring to studies of diabetes. Others discussed the importance of assessing natural variation for studies of disease. And, indeed, some at this meeting did express interests in these studies and would continue to be interested in the possible health benefits. But the question remained: would the research proposed by Diversity Project organizers actually lead to such benefits?

Future experiences would make clear that many had reason to be suspicious. In 1997, a researcher contacted a tribe about the possibility of participating in a diabetes study. Further inquiries into the request revealed that no such study existed (Interview G). Instead, researchers sought to collect tissue samples for a biological diversity preservation project. In light of such experiences, some Native Americans began to ask whether researchers were interested in their health, or whether the researchers simply wanted the DNA for their own biological preservation projects. These Native Americans began to refer to these promises of health benefits as "false hopes" that had been tacked onto the Diversity Project goals in order to attract populations to the Project (Interview G; Montana field notes, October 10, 1998).[21]

Indeed, even some organizers of the Diversity Project recognized that disease research had been added to entice groups to participate, and that the promise of health benefits had perhaps been disingenuous. As one organizer explained the problem:

> At that point [after the Bethesda ethics meeting], you see, the issue really started to come up, a serious issue: Is it an evolutionary study, or in the long run would it have something to do with medicine? . . . If you want to study evolution you don't have to study families . . . sample sizes are smaller, you are not interested in association between diseases and markers, so you don't need so many data points. So I would have to say frankly that there was a conflict within the Project about that. Because you are trying to sell it on its political utility, but the sample sizes you could eventually afford to pay for weren't going to be big enough to do genotype/phenotype [studies]. (Interview A4)

In other words, although organizers thought disease research would be important because it would give something back to communities, they also later recognized that it moved beyond the scope of their proposed research. Such research would require "genotype/phenotype" information that the project did not plan to collect, and money it never expected to secure.[22]

As this case of health research illustrates, although project organizers might sincerely have wanted to ask questions of interest to the populations it proposed to sample, it remained a reality that these populations had not proposed the Diversity Project; population geneticists and molecular evolutionary biologists had. Thus, not surprisingly, the questions the project could support reflected the interests of the originating scientists.

Material Practice: Meetings

To change human genetic diversity research so that it reflected the interests of populations as well as researchers, many Diversity Project organizers stated on numerous occasions that members of populations should become involved in the discussions and meetings where research agendas took shape. However, in practice, meetings on human genetic diversity research continued to embody the values of scientists and the ethicists who facilitated their projects. Rather than opening up opportunities for change, they threatened to merely reproduce patterns of exclusion and privilege. The following account of three meetings held on human genetic diversity research between the fall of 1996 and the fall of 1998 illustrates this problem.

OPEN DOORS FOR WHOM?

On 6 September 1996, the First International Conference on DNA Sampling was convened at the Delta Hotel in Montreal, Quebec. As an organizer of the meeting later recalled:

> The purpose of the meeting was to find out on DNA sampling who is doing what. Who is sampling for epidemiology? Who is sampling for newborns? Who is sampling for diversity? (Interview H)

According to the organizers of the Montreal conference, the goal was to gather "the facts" about who was doing DNA sampling and for what purposes. Meetings discussing "issues" (such as commercialization) would follow.[23] All were "welcome to attend," but it would be the scientists—those who allegedly knew "the facts" about human genetic diversity research—who would speak.

This, however, was not the experience of the members of indigenous and public-interest groups who attempted to attend the conference. Coincidentally, on the day that the DNA sampling conference began, a meeting of the scientific and technical advisory committee to the United Nations Convention on Biological Diversity (CBD) adjourned just a few blocks down the street. Members of indigenous and public-interest groups had been asked to speak at this CBD meeting, but the same invitation had not been extended at the conference up the street on DNA sampling—an issue of equal or even greater concern to indigenous people. Although not explicitly invited, members of these NGOs did try to attend the DNA conference in order to voice their concerns.[24] However, security guards blocked their entrance. Once rebuffed at the door by not only hotel security guards but the Royal Canadian Mounted Police and the Montreal police, members of the Assembly of First Nations, Cultural Survival Canada, the Asian Indigenous Peoples Pact, the Asian Indigenous Women's Network, RAFI (Rural Advancement Foundation International), and numerous other indigenous and public-interest groups set up a protest outside of the Delta Hotel challenging conference participants to "'check their ethics at the door'" (Benjamin 1996).

The conference organizer reportedly defended her decision not to invite indigenous people to speak on the grounds that the issues surrounding DNA sampling were no more relevant to "any one population than another," and it simply was not possible to invite all the potentially interested groups (Interview H). She did, however, insist that all had been welcome to attend and could have exercised their "democratic right" to speak from the conference floor (Benjamin 1996). To make participation easier, organizers offered a special reduced student conference rate of $35 (Interview H).[25]

This defense did not ring true with the CBD representatives who experienced something very different on the site. First, not only did they believe that they had been denied their "democratic right" to speak from the floor, but they reported that they had been denied their "much less democratic right" to pay admission to attend the conference, as they had been removed by the police before they could even reach the registration desk (Benjamin 1996). Second, to claim that representatives of indigenous groups had not been invited to speak at the meeting because it was a "fact-finding" meeting—and so would not address "issues"—represented a blatant rewriting of history. If "commercialization" was one of these "issues"—as the organizer claimed—then the DNA sampling meeting did address "issues." Indeed, commercialization was the topic of Plenary 6, "Commercialization and Patents" (First International Congress on DNA sampling 1996).

In short, representatives of indigenous and public-interest groups who attempted to attend the DNA sampling meeting in Montreal found the organizers' defenses of the meeting structure to be weak and only demonstrative of how indigenous people were being excluded from the debate about human DNA sampling and human genetic diversity research. As their Open Letter to the meeting charged:

> This conference could have provided an opportunity for the open debate and public scrutiny that is so desperately needed. Instead, the indigenous peoples who are the primary targets of DNA sampling have been largely shut out of the conference and *participation* restricted to the proponents of mass sampling and commercialization of human genes. As a consequence, the "First International Conference on DNA Sampling" appears likely to stand as merely the most recent example of the genomic industry's callous disregard for the rights and safety of the peoples it targets. (Cultural Survival Canada et al. 1996; emphasis added)

Although claiming in words to be participatory, critics argued that organizers of conferences on human genetic diversity research had acted to exclude the very groups who would be the objects of research.

WHO SETS THE AGENDA?

Charges that the debate about the human genetic diversity research—in particular, the debate about the Diversity Project—excluded Native and indigenous voices reached a crescendo in January 1998. In that month, Diversity Project organizers invited members of American Indian tribes in the United States and Canadian First Nations to Stanford to attend a second MacArthur Foundation–supported meeting designed to promote dialogue between project organizers and Native peoples.[26] The meeting ended in deadlock, however, as once again scientists and the ethicists set the agenda, and no clear goals or benefits for Native peoples emerged.

Following this meeting, some in attendance decided that they needed to organize their own meeting on their own territory to discuss the Diversity Project. Organizers of this follow-up meeting, which took place in October 1998 on the Flathead Reservation in Montana, invited members of the Diversity Project's NAmC to attend. However, the NAmC refused the invitation on the grounds that the meeting was "biased." In particular, the chair objected to the use of the words "colonialism" and "biopiracy" in the title of the meeting: "Genetic Research and Native Peoples: Colonialism through Biopiracy." Judy Gobert, dean of Math and Sciences at Salish Kootenai College, co-organizer of the meeting, and an attendee at the Stanford meeting, responded:

> As to the conference theme "Colonialism through Biopiracy," I can do nothing to change it. I strongly suggested to Tribal Leadership that we choose a different theme. They were very adamant about retaining this theme. Their position has been and continues to be that HGDP, the Env. [Environmental] Genome Project and others must provide more than a "for the good of the world" or the "pursuit of knowledge" as a rationale for Tribal people to participate in such studies. Considering these are the very people you want to collect samples from, I wouldn't argue semantics with them. Tribes feel very strongly they have been exploited from the time of the first landings on this shore. Everyone who knows of these projects sees them as benefiting the researchers,

> not Tribal peoples. If you are afraid to argue your points before Tribal Leaders, by all means, don't come. If your project is a way to "eliminate racism" and you feel strongly about the validity of your research, argue your position. (Gobert 1998)

Although representatives of NIH's Environmental Genome Project as well as other program directors at the NIH involved in human genetic variation research did attend, no representative of the Diversity Project took part in the Montana meeting. Project organizers, as in the past, refused to recognize that their project might be linked to the histories of both racism *and* anti-racism.

Subjectivity and Subjection

Experiences at these three meetings made it clear to members of populations recommended for sampling that researchers and bioethicists had in many ways already set the terms of their participation in the dialogue. Rather than grant them novel forms of entré into the formation of scientific research, it appeared that efforts to encourage their "participation" would only continue to act to deny them access to the very spaces in which rules are set and frameworks for designing and regulating scientific research become institutionalized. Accordingly, many viewed with great skepticism offers to enter into "partnerships" with Diversity Project organizers. Instead of an honest effort to address community needs, they suspected these efforts would only act to legitimate the sampling initiative in an age shaped by the logic of affirmative action and the need for "institutionalized exhibitions of representation" (Amit-Talai and Knowles 1996:99). Indeed, their fears were not unfounded. Project organizers arranged their first meetings with potential subjects only after the MacArthur Foundation told organizers that it would grant the project funds only if it consulted with Native and indigenous communities (Reardon 2004). In this and other instances it became clear that project organizers would need to establish that they had listened to "the voices" of the groups their research wished to sample.

This institutional demand for representation raised the specter of tokenism. As one participant at the Stanford meeting explained:

> One of the problems I would say is tokenism. It's easy to say that we talked with our Indians and so now we are moving forward with our stuff. The question is what do you mean by talk to and what authority do they have to speak for anyone? (Interview J)

Or as an indigenous rights leader observed, if you talk to them, "by osmosis you help [the project]" (Interview K). The problem of tokenism presented a dilemma for those asked to represent groups: one wanted to find out what was happening with the project, but to enter into dialogue with its organizers was to risk lending the project legitimacy. To handle this problem, one indigenous rights leader reported that on occasion he would accept invitations to lunch with Diversity Project organizers. He tried to be in touch only enough to stay abreast of what was happening.

Members of tribes in the United States and First Nations in Canada were not alone in their worries about Diversity Project organizers' construction of "representative" voices. Like those asked to speak for "Native American" communities, some of those asked to speak for "African Americans" feared that organizers would strategically construct an "African American voice" of support for the project. This concern became public in January 1994 when a consortium of African American social and biological scientists met in Washington, D.C., to draft the *Manifesto on Genomic Studies among African Americans* (Jackson 1998). Among other things, these researchers sought to respond to a variety of negative experiences with the Diversity Project, including an episode that transpired at the 1993 World Council of Indigenous Peoples (WCIP) meeting in Guatemala. At that meeting, the chair of the NAmC ethics subcommittee, Henry Greely, reportedly stated that African Americans supported the Diversity Project. Given African-American scientists' critiques of the project (to the dismay of some of these scientists, the project had originally not included "African-Americans" as "populations" of interest to the project), authors of the *Manifesto* interpreted Greely's statement as nothing more than a political statement designed to generate the support of racial minorities and indigenous people.[27] They concluded that the "misrepresentation of African American interests regarding genomics studies by Hank Greely (Stanford University) . . . at the World Council meetings in Quetzeltenango" as a reason for concern that African Americans would not be fairly represented in the genomics revolution.

In short, potential research subjects did not fault Diversity Project organizers for not attempting to include them in the design and oversight of the Diversity Project. Instead, they questioned the mode of inclusion. The effort began only after many of the scientific questions guiding the project had been set, entailed inclusion only on Diversity Project organizers' terms using resources they controlled (for example, the MacArthur Foundation grant monies), and appeared oriented toward legitimating the project. Consequently, many feared that the production of participatory subjects would only subject disempowered groups to further objectification and marginalization.

Conclusion

Although efforts to include research subjects in the design and regulation of research hold great promise, this chapter has demonstrated that these efforts cannot be viewed as innocent acts or a priori goods. Rather, the ability of these efforts to overcome divides between scientists and laypeople and to promote the goals of social justice, require making procedural choices that are sensitive to the role participatory structures play in constituting subjects and expressing particular relations of power. The Diversity Project case provides valuable lessons about how to proceed.

First, if researchers are to create projects that reflect the interests and goals of research subjects, then these subjects need to be included earlier in the process, before research has been funded. In particular, subjects need to participate not just in the research process but in the policy process that sets research priorities. Other scholars have noted the absence of lay involvement in this domain (Brown et al. forthcoming:37). The Diversity Project highlights the problems created by this lack of research subject participation in the meetings and policy settings where agendas form and gain institutional support.

Second, efforts to garner participation in a manner that reveals researchers' insensitivity to the histories and present realities of institutionalized racism and colonialism will only undermine efforts to build partnerships with subjects whose lives are shaped by these histories and present structures of oppression. Researchers must be willing to learn from and accept as legitimate the expressed concerns and interests of those with whom they seek to partner. In the case of structurally disadvantaged groups, it should not come as a surprise to researchers that discrimination and racism will be issues of central interest that cannot be dismissed by appealing to claims of objectivity. In the case of the Diversity Project debates, organizers refused to participate in the meeting held on the Flathead Reservation in Montana on the grounds that the meeting was biased. This served to only drive a deeper wedge between scientists and subjects.

Third, the capacity to hear and learn from disempowered groups will not emerge from the good intentions of researchers. Rather, it will require institutional support for meetings designed by these groups and policies that encourage organizers of projects such as the Diversity Project to attend. More fundamentally, it will require researchers to broaden their conception of the "knowledge" needed to conduct human genetic diversity research. Rather than conceive of themselves as gatekeepers who grant diverse populations access to the knowledge needed to ask scientific questions, genetic researchers would do well to recognize that potential research subjects play a role in the very formation of knowledge, and thus might also educate scientists. Further, given that initiatives such as the Diversity Project highlight the ways in which knowledge production practices "loop back" to construct the identities of research subjects, such initiatives would benefit from institutional frameworks that bring into view these entanglements of knowledge and subject formation (Hackling 1999).

Finally, researchers cannot assume that they and their research subjects will find questions that interest them both if only researchers do a good enough job of explaining their research. Participation, in this sense, is not an a priori good that will be valued by all. Rather, it will remain what it has long been: a technique of governance that shapes what counts as knowledge, democracy, and a valued human life.

Acknowledgments

This essay is based on a chapter from a much larger project (Reardon 2005), and has benefited from the critical engagement of those too numerous to acknowledge here. I would like to thank in particular Susan Conrad, Elisabeth Clemens, and the editors of this volume, Kelly Moore and Scott Frickel, for their astute suggestions and careful readings of drafts of this essay. The National Science Foundation (NSF) supported research on which this essay is based under grant SBR-9818409. The views, conclusions and recommendations expressed in this article are those of the author, and do not necessarily reflect the views of the NSF. Fellowships and grants provided by the Science and Technology Studies Department at Cornell University, the Program on Science, Technology and Society at the Kennedy School of Government, and the Genome Ethics, Law and Policy Center in the Institute of Genome Sciences and Policy at Duke University also provided support for the work that gave rise to this chapter.

Endnotes

1. Not all organizers agreed that the "diversity" of the human species could be represented by racial and ethnic groups in the United States, and opposed this vision of the Diversity Project. In later sections of this chapter, I address the importance of these disagreements. Also see Reardon 2005.

2. As the political theorist Barbara Cruikshank (1999:18) explains: "Democratic relations are relations of power and as such are continually recreated, which requires that democratic theory never presuppose its subject but persistently inquire in the constitution of that subject."

3. Additionally, Cavalli-Sforza hoped "local workers" would be enrolled to work at regional collection centers established by the Diversity Project. As he stated in an address to a special meeting of UNESCO in September of 1994: "Local participation in all areas of the world will be essential and the success of the project will be entirely dependent on international collaboration and cooperation" (Cavalli-Sforza 1994). Samples needed to be transformed into cell lines within a day or two, and often transportation alone took a day or two from some of the remote places in which project organizers proposed sampling. Thus, Cavalli-Sforza hoped that this work could be done locally.

4. See Reardon 2005 for a detailed account of these charges.

5. The possibility that the issue was not just one of the *appearance* of the project, but of its material structure, was not one organizers acknowledged at this time.

6. The proposal for an International Forum and the regional committees came out of this Sardinia meeting (Cavalli-Sforza 1994:4).

7. I use align here in the sense that Steven Epstein uses it in his discussion of "categorical alignment" in his chapter in this volume.

8. The Wenner-Gren Foundation funds anthropological research.

9. In his account of the project, Blakey uses the term "descendant or culturally-affiliated community" interchangeably with "African-American public."

10. In an effort to respond to their critics, members of the NAmC drafted the Model Ethical Protocol in 1994 and 1995. It was later published in the Houston Law Review (NAmC 1997).

11. This model of enrolling members of identified populations to sample themselves has become a popular strategy for dealing with the ethical problems raised by population-based research. This approach raises many problems that are beyond the scope of this chapter. Most significantly, it shifts the ethical burden from the main researchers and project organizers to the ethnic researchers they have enrolled to do the sampling. The Harvard/Millennium research collection effort in China is one such example (Pomfret and Nelson 2000). A detailed analysis of the problems of such an approach is needed.

12. A second key difference is that in the case of the ABGP, those participating in the design of the research were not themselves research subjects. This made the political and ethical issues at stake in the Diversity Project significantly different from those in the ABGP.

13. This is not to say that Blakey's efforts did not raise their own problems. For example, as I note elsewhere (Reardon 2005:chap. 4), Blakey holds "African-Americans" to be a coherent, already constituted group, a group that could just be included in research. He does not explicitly address questions about the identity of "African-Americans" and complex questions about who (if anyone) can speak for a group called "African-American."

14. The authors do state that the "very point of the HGDP . . . is to collect information about communities that can be useful in improving understanding of the history and health of that community" (NAmC 1997:1472). However, the question of whether this "understanding" is of interest to the community itself is never raised.

15. These meetings were funded by the MacArthur Foundation, the foundation that supported the drafting of the Model Ethical Protocol.

16. For scientists, on the other hand, the Out-of-Africa theory was just the beginning of what genetics could explore. One might even say that the theory that humans originated in Africa was the biologists' big bang; a whole universe of migrating and mixing genes was waiting to be explored. (Note: In physics, the big bang theory is the theory of the origin of the universe.)

17. This resistance is not to say that they believed that genetic research could under no circumstances help Native people. Other human genetic variation projects have also raised grave concerns, but even the harshest Native and indigenous critics hold out the possibility of designing genetic research that might help their communities. See, for example, Benjamin 2000.

18. Later organizers recognized that some populations might not want to participate (Interview A4, Interview R).

19. The Indigenous Peoples Council on Biocolonialism formed in the mid-1990s in response to concerns created by the Human Genome Diversity Project (at that time, the group was known as the Indigenous Peoples Coalition Against Biopiracy). It now has extended its concerns to include "the protection of

[indigenous peoples'] genetic resources, indigenous knowledge, cultural and human rights from the negative effects of biotechnology" (www.ipcb.org [accessed June 12, 2002]).For concerns about the use of a genetic confirmation of the Bering Strait theory, also see Schmidt 2001: A218.

20. Many misinterpreted the legislation as saying that the state would *require* DNA testing to determine tribal identity. In fact, the legislator who proposed the bill intended that DNA tests be voluntarily requested by individuals as a way to build their case that they were "Native American." As Kimberly Tallbear astutely explains, despite the "relatively benign intent" of the legislator, the bill does accept the notion that biology can determine who rightly claims political and cultural authority" and "may be a forewarning of future laws and policies based on assumptions that a person's or a people's political rights and cultural identity are biologically determined" (Tallbear 2003: 85–86).

21. Some Native Americans also worried that pharmaceutical companies were the primary actors interested in disease research. Indeed, one participant in the first MacArthur meeting reportedly argued that the research would be about wealth, not health.

22. One organizer recognized: "To do the full medical things would cost like the Human Genome Project in each case" (Interview B). See also Marks 1998.

23. This language, "the facts," is that of one organizer of the meeting. As she explained: "You've got to get the facts out there first. You can't come in and say at a meeting, the first international meeting on sampling, the first meeting on sampling period, and say sampling is wrong when you don't even know what sampling is" (Interview H).

24. The meeting had originally caught the attention of those at the CBD meeting when it became known that the Diversity Project organizer, Luca Cavalli-Sforza, would speak. As one conference organizer recalled, these members of indigenous and public-interest groups argued that the "Diversity Project coming to Canada" would promote biocolonialism and exploit Native peoples (Interview H).

25. This was the rate I paid. I arrived on the second day and entered the conference with ease. No trace of protest remained.

26. Many faulted Diversity Project organizers for not making more efforts to meet with Native peoples. After all, they argued, this is what the project had received funding from the MacArthur Foundation to do. For their part, Diversity Project organizers argued that they had not met more often because there was no project and organizers were in a holding pattern waiting for the findings a National Research Council (NRC) committee asked to review the Diversity Project. The meeting did follow directly in the wake of the release of the NRC report.

27. As the organizer of the *Manifesto* explained, if organizers could claim that the initiative had the support of African Americans, a group perceived by many outside the United States as occupying an adversarial relationship to the dominant power structures, then it might become more palatable to indigenous groups (Interview with *Manifesto* author, 1999).

Interviews

The research on which this paper is based included interviews with key actors in the proposal for the Diversity Project. Interviewees were promised that their identities would be kept confidential. The interviews are coded as follows:

Interview B, early Diversity Project organizer, Palo Alto, July 2, 1996.
Interview E, early Diversity Project organizer, July 2, 1996.
Interview F, Native attorney, phone interview, November 5, 1998.
Interview G, Native scientist, phone interview, April 3, 2001.
Interview H, lawyer and bioethicist, April 1, 1999.
Interview J, Indigenous rights activist, April 16, 1999.
Interview K, Indigenous rights leader, April 13, 1999.
Interview M, Indigenous rights leader, June 13, 2000.
Interview R, early Diversity Project organizer, April 7, 1999.
Interview A2, early Diversity Project organizer, August 25, 1998
Interview A4, early Diversity Project organizer, April 19, 1999.

References

Akwasasne Notes. 1978. *Basic Call to Consciousness*. Summertown, TN: Book.
Amit-Talai, Vered, and Caroline Knowles, eds. 1996. *Re-Situating Identities: The Politics of Race, Ethnicity and Culture*. Peterborough, ON: Broadview Press.
Benjamin, Craig. 1996. "Check Your Ethics at the Door: Exclusion of Indigenous Peoples' Representatives from DNA Sampling Conference Casts Doubts on Gene Hunters' Sincerity." http://nativenet.uthscsa.edu/archive/nl/9609/0060.html (accessed February 15, 2001).
———. 2000. "Sampling Indigenous Blood for Whose Benefit?" *Native Americans* 17:38–45.
Blakey, Michael. 1998. "Beyond European Enlightenment: Toward a Critical and Humanistic Biology." In *Building a New Biocultural Synthesis: Political-Economic Perspectives on Human Biology*, ed. Alan Goodman and Thomas Leatherman, pp. 379–406. Ann Arbor: University of Michigan Press.
Bowcock, Anne and Luca Cavalli-Sforza. 1991. "The Study of Variation in the Human Genome." *Genomics* 11 (Summer): 491–498.

Brown, Phil, Sabrina McCormick, Brian Mayer, Stephen Zavestoski, Rachel Morello-Frosch, Rebecca Gasior Altman, and Pamela Webster. In press. "Ahab of Our Own: Environmental Causation of Breast Cancer and Challenges to the Dominant Epidemiological Paradigm." *Science, Technology & Human Values.*

Butler, Judith. 1992. "Contingent Foundations: Feminism and the Question of 'Postmodernism.'" In *Feminists Theorize the Political,* ed. Judith Butler and Joan Scott, pp. 3–21. New York: Routledge.

Cavalli-Sforza, Luca. 1994. "The Human Genome Diversity Project." An Address Delivered to a Special Meeting of UNESCO. Paris, France, September 12.

Cavalli-Sforza, Luca, Allan C. Wilson, Charles R. Cantor, Robert M. Cook-Deegan, and Mary-Claire King. 1991. "Call for a Worldwide Survey of Human Genetic Diversity: A Vanishing Opportunity for the Human Genome Project." *Genomics* 11 (Summer): 490–491.

Cruikshank, Barbara. 1999. *The Will to Empower: Democratic Citizens and Other Subjects.* Ithaca, NY: Cornell University Press.

Cultural Survival Canada et al. 1996. *Open Letter to First International Conference on DNA Sampling.* Montreal, Quebec. www.yvwiiusdinvnohii.net/articles/dna-con.htm (accessed May 24, 2005).

Dukepoo, Frank. 1998. "Sensitivities and Concerns in Native American Communities." Hearing of the National Bioethics Advisory Commission, December 2. http://bioethics.gov/transcripts/jul98/native.html (accessed June 12, 2000).

Epstein, Steven. 1991. "Democratic Science? AIDS Activism and the Contested Construction of Knowledge." *Socialist Review* (April–June): 35–64.

———. 1996. *Impure Science: Aids, Activism, and the Politics of Knowledge.* Berkeley: University of California Press.

First International Conference on DNA Sampling. 1996. *Conference Program.* Organized by Centre de recherché en droit public, Université de Montréal and the Health Law Institute, University of Alberta.

Foucault, Michel. 1997. "Technologies of the Self." In *Michel Foucault / Ethics: Subjectivity and Truth,* ed. Paul Rabinow, pp. 223–252. New York: New Press.

Gobert, Judy. 1998. Letter to North American Committee of the Human Genome Diversity Project. October 9.

Hacking, Ian. 1999. *The Social Construction of What?* Cambridge, MA: Harvard University Press.

Haley, Alex. 1976. *Roots.* Garden City, NY: Doubleday.

Human Genome Organization. 1993. The Human Genome Diversity (HGD) Project—Summary Document. Report of the International Planning Workshop held in Porte Conte, Sardinia, Italy, 9–12 September.

Indigenous Peoples Council on Biocolonialism. 2000. "Indigenous Peoples, Genes and Genetics: What Indigenous People Should Know about Biocolonialism." www.ipcb.org/publications/primers/htmls/ipgg.html (accessed June 13, 2005).

Jackson, Fatimah. 1998. "Scientific Limitations and Ethical Ramifications of a Non-representative Human Genome Project: African American Responses." *Science and Engineering Ethics* 4:155–170.

Kitcher, Phillip. 2001. *Science, Truth, and Democracy*. Oxford: Oxford University Press.

Klawiter, Maren. 2000. "From Private Stigma to Global Assembly: Transforming the Terrain of Breast Cancer." In *Global Ethnography: Forces, Connections and Imaginations in a Postmodern World*, ed. Michael Burawoy et al., *pp*. 299–334. Berkeley: University of California Press.

Marks, Jonathan. 1998. "Letter: The Trouble with the HGDP." *Molecular Medicine Today* (June): 243.

North American Regional Committee of the Human Genome Diversity Project (NAmC). 1997. "Proposed Model Ethical Protocol for Collecting DNA Samples." *Houston Law Review* 33:1431–1473.

Pomfret, John, and Deborah Nelson. 2000. "In Rural China, a Genetic Mother Lode." *Washington Post*, December 20, A1.

Reardon, Jenny. 2005. *Race to the Finish: Identity and Governance in an Age of Genomics*. Princeton, NJ: Princeton University Press.

Roberts, Leslie. 1991. "Scientific Split over Sampling Strategy." *Science* 252 (June 21): 1615–1617.

Schmidt, Charles W. 2001. "Spheres of Influence: Indi-gene-ous Conflicts." *Environmental Health Perspectives* 109:A216–A219.

Sclove, Richard E. 1995. *Democracy and Technology*. New York: Guilford Press.

Tallbear, Kimberly. 2003. "DNA, Blood, and Racializing the Tribe." *Wicazo Sa Review* 18(1): 81–107.

Vizenor, Gerald. 1978. *Wordarrows: Indians and Whites in the New Fur Trade*. Minneapolis: University of Minnesota Press.

Weaver, Jace. 1997. *That the People Might Live: Native American Literatures and Native American Community*. Oxford: Oxford University Press.

Weiss, Kenneth M. 1993. "Letter dated April 30, 1993 to Senator Akaka." In Committee on Governmental Affairs, *Human Genome Diversity Project: Hearing before the Committee on Governmental Affairs*, U.S. Senate, 103rd Cong., 1st sess., April 26, 43–44.

Yona, Nokwisa. 2000. "DNA Testing, Vermont H. 809, and the First Nations." *Native American Village. Abolitionist Examiner*, April/May 2001. www.multiracial.com/abolitionist/word/yona-partone.html (accessed May 23, 2005).

14

On Consensus and Voting in Science

From Asilomar to the National Toxicology Program

DAVID H. GUSTON

"You can't vote to repeal the law of gravity!" is a trump card that defenders of the scientific establishment play when battling critics who would open science up to more democratic impulses. What the former imply by this claim is that the existence of any particular scientific truth is independent of the collective will of even democratic politics. Coupled with the common histories of holdouts against political or ecclesiastical authority, like Galileo, or persistent, solitary voices that finally triumphed over entrenched scientific opinion, like Alfred Wegener (for continental drift) or Stanley Prusiner (for prions), proclaiming the inefficacy of voting speaks to why science and politics inevitably clash. Dismissing voting as appropriate for democracy but not for science buttresses the belief that the transience of popular sovereignty must yield before the inevitable solidity of cumulative knowledge.

While specifically true, however, the claim neglects the great variety and even necessity of voting in science. As a procedure for determining what science gets done, who gets to do it, and what gets to count as knowledge, voting is ubiquitous. It occurs in arenas traditionally defined as political, such as the U.S. Congress, where votes determine the allocation of resources, which in turn helps determine what research gets done and, therefore, what knowledge is produced. When voting to kill the superconducting supercollider (SSC), Congress voted to remain ignorant of a possible route to a Grand Unified Theory. Although often contested as penny-wise but pound-foolish or even as anti-science, Congress's legitimate right to vote on appropriations for science is rarely questioned.

Many arenas traditionally defined as scientific also require the aggregation of individual views—just as a legislature does. Reviewers of grant proposals or manuscripts, for example, vote on whether or not scarce resources (such as budgets or journal pages) should be expended on a project. To be sure, this process is no plebiscite. The aggregated votes of the peer reviewers are not formally binding on most sponsors or editors, and the reviews carry more information than "aye," "nay," or "abstain." But, as Arie Rip (2003:391) recognizes, peer review functions as an "aggregation machine" by compiling the perspectives of separate individuals into the recognized voice of a community. The criteria and procedures employed—the escapement of this aggregation machine—influence the content of science by helping to arbitrate who holds scientific authority, who is privileged to conduct which research, and how those who attempt to speak for science assemble their views. Voting is thus among a small set of procedural elements that should attract our attention if we are interested in decision making within the scientific community and, as will be seen, between politics and science as well.

This chapter is a preliminary examination of the role of voting in science and the relationship between voting and the more familiar concept of consensus. My central argument is that an uncomplicated view of scientific consensus is inadequate for many roles that people desire science to play, especially informing policy. Rather, voting and other procedural elements are critical if science is to be a source of enlightenment and legitimation for policy. Arguing in favor of this role for voting may be controversial because it implies a greater integration of politics and science. In a sense, it argues that scientific consensus is too important to be left to the scientists.

In pursuit of this argument, the first section provides a brief argument justifying discussion of voting, or what will be introduced as "social choice," in science. The second section provides a selective overview of the literature on consensus in science, pointing toward why consensus is an inadequate concept and thus why a discussion of voting is necessary. The third section of the chapter describes in more detail the possibility of voting as a mechanism of social choice in science. It leads to two case studies of voting in science, the Asilomar conference on the hazards of recombinant DNA and the National Toxicology Program's *Report on Carcinogens.* Not only do these examples highlight the critical role voting has played in decision making, but they also suggest that more formal consideration of voting might lead to more democratic outcomes. The conclusion discusses why it might be worthwhile to abandon any concept of scientific consensus that is not firmly rooted in procedural rules like voting.

Social Choice in Science

In political and social theory, the concept of "social choice" means the aggregation of individual preferences into a group or "social" expression, or "choice." Voting is one example of social choice in which citizens of a body politic indicate preferences, e.g., for which person should govern the polity. The outcome, based on previously articulated rules and procedures for aggregating these preferences, is the authoritative designation of who governs. A second example of social choice is the market, in which consumers indicate preferences in the form of demand for goods and services. The outcome, in a market that functions by ideally articulated rules, is equilibrium between demand and supply and an efficient distribution of goods and services. Economist Kenneth Arrow (1963) argues that voting and markets are the two primary forms of social choice in modern capitalist (liberal) democracies. They have supplanted other forms of social choice, such as dictatorship and convention (or tradition), for most public purposes, although dictatorial and conventional decision making may still exist in certain nondemocratic or illiberal subsystems even within liberal democracies.

Science is a crucial subsystem within, and among, most liberal democracies, as well as other nations. Does social choice happen within science, and if so, what form does it take?

One perspective is that the concept of "choice" does not make sense in science, because scientists' views are fundamentally different from citizens' political opinions or consumers' preferences. A full analysis of the similarities and differences among scientific views, political opinions, and consumer preferences is beyond the scope of this chapter, but the following points are at least suggestive. If science were different, perhaps it would be because scientists are compelled by nature to hold views in quite a different way than citizens are compelled to hold opinions or consumers to hold preferences. This perspective is misguided because, as the great body of research in the social studies of science has shown (Jasanoff et al. 1995), scientific views are constructed from a great deal more (e.g., material, social, psychological, ideological) than immediate compulsion from nature. Scientific views are thus compelled by many of the same elements as are political opinions and commercial preferences.

Moreover, scientists cannot be said to hold their views with any more conviction than citizens hold political opinions or consumers hold preferences. Certainly conflicts over political opinions, and even consumer preferences, lead to acts of aggression and violence by people unwilling to compromise. Scientists, too, would likely "rather fight than switch," but perhaps luckily for them they are rarely called on to fight other than metaphorically for their scientific views (except in the most harsh regimes). Neither could it be said definitively that individual scientists reach their own views through processes that are so distinctive from those used by citizens or consumers as to be immune from the operation of social choice on them. Consumer preferences may be constructed by advertising, but also by experience, testing, and evaluation. Political opinions may be formed by ideology, but also by analysis, debate, and deliberation.

Finally, scientists' views are not identical—as one might expect if they were immediately compelled by nature. As Robert Dahl (1989) crucially argues, even a cohort of well-trained technocrats will disagree on some issues, even within their realm of expertise. They would thus require procedures for social choice in those circumstances.

It is thus reasonable to conclude that there is nothing essential in scientists' views that so distinguishes them from political opinions or consumer preferences to exclude them from some process of social choice. If, then, we are justified in thinking in terms of social choice in science, which types of social choice do we find?

Among the most prominent characterizations of the scientific community, Michael Polanyi's famous article "The Republic of Science: Its Political and Economic Theory" (2002) speaks of mutual adjustment, à la the free economic market, as the mechanism of social choice among scientists. Polanyi (467) writes: "[T]he co-ordinating functions of the market are but a special case of co-ordination by mutual adjustment. In the case of science, adjustment takes place by taking note of the published results of other scientists; while in the case of the market, mutual adjustment is mediated by a system of prices broadcasting current exchange relations, which make supply meet demand." It follows that scientific consensus—analogized by Polanyi as an assembled jigsaw puzzle—is the equilibrium outcome of this mutual adjustment among individual scientists.

Polanyi's model of mutual adjustment in science is deficient, however, in ways that make his overall analogy between science and economics seem remarkably tight. His argument, as abstracted above, presumes the existence of academic journals that publish scientific findings. In science again as in the market, firm-like institutions are required to provide some requisites of mutual adjustment. The social studies of science has addressed the dynamics of these firm-like institutions (with reference to the economic metaphor) in the "credibility cycles" of Latour and Woolgar (1976), which enlist academic journals and research patrons in these market interactions, and the research councils of Rip (1994).

Neither journals nor research councils operate by mutual adjustment, but rather, as suggested above, these "aggregation machines" operate by voting and the like. A short catalogue of similar activities requiring aggregation, and thus voting, includes the following: informal decisions among a research group (e.g., whether to attempt another experimental run); formal personnel decisions ranging from passing doctoral exams to gaining tenure in academic departments; the distribution of prestige through awards, prizes, and memberships in honorific societies; and the operations of some scientific and technical advisory committees.

Of course, it is true that these activities may have limited influence on Polanyi's jigsaw puzzle. But as the Asilomar case reveals, they can be influential even without consideration, and as the National Toxicology Program case demonstrates, they can be as influential as they are allowed to be. While this section has established only the sense of considering social choice and voting in science, it has merely hinted at its necessity—which the next section considers more fully.

Consensus and Consensus Formation

THE VALUE OF CONSENSUS

Scientific consensus is an elusive subject, both in scholarship that relies on science as an object of study and in politics that relies on science as an instrumental and ideological resource. To avoid lengthy definitional discussion, I have tried above to limit my use of scientific consensus only to Polanyi's rather intuitive jigsaw puzzle endpoint. In this section, I explore a variety of ideas about consensus in science in an admittedly episodic fashion. Each episode, however, shows the necessity of including with consensus discussions of procedures like voting.

There are both "internal" and "external" reasons why an idea of scientific consensus, as completed jigsaw puzzle, is valuable. Its internal value is, according to Thomas Kuhn (1977:231–232), that consensus is a developmental turning point within a discipline that primes it for more rapid advance. By allowing scientists who have settled on a worldview to focus their researches on jointly uncovering the unknown, consensus prevents them from unproductively disputing the known. Such institutions as journals, granting agencies, and university departments enforce the consensus by denying resources to those who—even for good reason—reject it. Journals, patrons, and departments often use voting mechanisms to make such decisions.

The external value of scientific consensus is apparent in the array of legislative, bureaucratic, and judicial decisions that attempt to predicate themselves upon it. The Food Additives Amendments of 1958 established the category of "generally recognized as safe" (GRAS) for food additives in an attempt to codify the consensus of their history and use (see Jasanoff 1990:218–222). The U.S. Supreme Court declared in its 1993 *Daubert* opinion that "general acceptance within the relevant scientific community" of a scientific claim was an important desideratum, among several, for judges to apply to assess whether expert testimony is reliable enough to admit before a jury.[1]

Science policy studies also identify consensus as a predicate for determining the responsibilities, memberships, or other boundary demarcations of the scientific community. The political scientist Yaron Ezrahi (1980) argues that the type of advice that scientists provide for politicians should vary based on the presence or absence of consensus within the scientific and the political communities. Scientists should engage in a kind of technology assessment of different options when there is scientific consensus but no political consensus; when there is political consensus but no scientific consensus, they should instead specify the extent of agreement and sources of disagreement.[2]

All these instances of the proposed instrumentality of scientific consensus, however, presume an uncomplicated articulation of how to demarcate the relevant scientific community or observe the level of agreement around, or the general acceptance of, a tenet within it. Their performance accords with the argument of Richard Barke (2003), who, when considering the political aspect of Polanyi's metaphor, charges the "republic of science" with failing to specify the nature of citizenship and, more importantly for the concerns of my argument, to respect majority rule. For consensus to be an important resource on which both scientists and policymakers reasonably draw, some process for demarcating and identifying it is necessary. Voting, of course, does not create these demarcations, but it makes clear the necessity and facilitates the practical conditions for their consideration. Further, it identifies the level of agreement.

OUTCOME AND PROCESS

In common parlance, "consensus" has two primary meanings: "The consensus of the group is against the proposal" contrasts with "The group operates by consensus." I have until now been relying on Polanyi's jigsaw puzzle as a stand-in for the former: "The consensus of the scientific community is that the jigsaw puzzle depicts N." I will call this "consensus-as-outcome." The latter meaning refers to an operating procedure—how a group achieves an outcome rather than the outcome itself. This "consensus-as-process" connotes that some deliberation occurs. What consensus precisely denotes either as outcome or process, however, is largely a function of the size and nature of the group. As sociologist James Coleman (1990:857) defines it, for a small group with dense social relations among the members, consensus denotes that all members come to agreement without an explicit voting procedure, and the agreement is recognized when no member of the group continues to voice dissent. Consensus thus implies a rule of decision or closure as well: no continued dissent, rather than, for example, unanimous assent—which in turn would require an explicit voting procedure to determine that assent.

For decisions in a larger group, consensus loses its kinship with unanimity and tends to denote a sufficiently large majority. There are no set rules, however, for how large a group needs to be before consensus decays into supermajority, or how large that supermajority needs to be in order to constitute a consensus rather than, simply, a big majority. Common practice in public opinion polling, for example, views a supermajority of approximately two-thirds, or perhaps 70 percent, as a consensus (Kay 1998).

Science, however, is not a small group with dense social relations, but rather an internationally dispersed network with social relations of greatly varying densities. To seek a consensus among such a population, one would presumably resort to opinion polling. Scientists frequently convene and operate as Coleman's small groups with dense relations, as in peer review panels or advisory boards, and might produce a consensus there. But such a consensus would be only among a nonrepresentative (albeit elite) sample of the larger community. It would no more be a consensus of the scientific community than a focus group is the consensus of a political party or consumer population. Moreover, the outcome of such a group does not necessarily depend on the fact of consensus within the larger group, or the congruence between the smaller and larger group, but merely in the ability of well-positioned actors in the smaller group to act as if their view represents the consensus of the larger group.

These details are important because the internal and external value of scientific consensus relies not only on consensus-as-outcome but also on consensus-as-process. One might suspect that knowing what the relevant scientific community thinks, rather than the process by which its members got there, would be most critical, particularly for the internal value of consensus. Similarly, external users of scientific consensus presumably attribute to scientists the expertise of their methods, and it is only the conclusions—once judged sound by the community—that are of interest to policymakers.

But in practice this all-too-linear deployment of the results of a scientific method does not and cannot satisfy those who seek to use consensus. Minimally, scientists who are part of the consensus have a modest stake in identifying the processes of consensus formation in order to take part in them. As in the peer review panel example above, elite scientists may use the authority of their position to extend and consolidate a consensus that is incomplete. More importantly, scientists who do not partake of the consensus have a major stake in identifying processes in order to disrupt, dispute, or construct alternatives to them. Scientists are also not apolitical entities, and they may have ideas about fairness and justice against which they evaluate even scientific institutions. For the internal value of consensus, process may be as critical as outcome.

Policymakers may have a similar pair of incentives, as supporters of the content of consensus-as-outcome are not apt to dispute the process that formed it, while interests at odds with any alleged scientific consensus are likely to challenge the process of its creation in an attempt to undermine its political authority.[3] But, further, they find it helpful and convenient to rely on consensus-as-outcome precisely because of the authority of consensus-as-process that it is taken to represent. As Ezrahi (1990) argues, the role of science as a model of progressive, nonviolent consensus formation is as important for its influence in democratic politics as is its instrumentality. The *Daubert* court held, for example, the judge's sole interest in the general acceptance desideratum is the methodology of a study and not its conclusions. Even though this holding does not address how the consensus around the methodology may have been formed, it declares that decision makers are open to more than outcomes. How consensus-as-outcome manifests itself through process is, therefore, of interest to scientists and politicians alike.

PROCEDURALLY ENFORCED CONSENSUS

Philosopher Steve Fuller (1988:chap. 9) elaborates a typology of consensus that helps make sense of this need to address process as well as outcome. Most importantly, Fuller describes the "procedurally-enforced consensus," in which the relevant group's deliberations are highly constrained by rules, language, or other procedures that cordon off areas of potential conflict and thus facilitate agreement. The hypothesis implicit here maintains that, at least in any scientific area salient enough for attention in a policy dispute, any consensus to appear will be procedurally enforced. Voting is, of course, one of these procedural mechanisms. Among others are rules of the admissibility of evidence, the burden of proof, and criteria for the application of judgment.

Two additional types of consensus that Fuller defines are the "accidental consensus," in which a group comes to agreement by each individual's coming to the same conclusion independently (as in a public opinion poll), and the "essential consensus," in which the group comes to a collective decision by deliberation as a group, as in committee deliberations. Fuller argues that an accidental consensus is less stable than an essential consensus because any of the contributing individuals in the former may be free at some future point to change their potentially idiosyncratic interpretation of the evidence; whereas, individuals in the latter must produce publicly acceptable rationales for their beliefs, which then cannot so easily be changed. The essential consensus may also be normatively preferable because it better represents ideals of deliberative decision making.

There are clues to the importance of the procedurally enforced consensus elsewhere in the literature. Extending Fuller's work empirically, his student Kyung-Man Kim (1994) writes of the consensus that developed around Mendelian genetics in the first half of the twentieth century. Kim (1994:23) defines scientific consensus as "the resolution of an issue of fundamental epistemological importance manifested in the scientific transformations of the structure of an evolving network of scientific allies and enemies within a specified period of time." Critical nuances are, of course, defining "resolution" and "fundamental epistemological importance"—not to mention the extent of the network (which corresponds to Barke's citizenship concern) and the "period of time," which could be "specified" by either a contender or an observer. The precise areas that go unspoken in Kim's definition of consensus importantly include those involved in procedural enforcement.

Karin Knorr-Cetina (1995:120) suggests that scholars seek variation and talk "not in terms of *one* model and vocabulary of consensus formation but in terms of many" (emphasis in the original). She argues that consensus in high-energy physics (HEP) is an emergent property of the characteristics of a HEP collaboration as a "superorganism," which perpetuates itself by "sum[ming] up states of the equipment, display[ing] processes of measurements and simulation and articulat[ing] the methods and technologies other experiments use" (Knorr-Cetina 1995:122). In this genealogical effort, HEP displaces the temporal crux of consensus from the end of the experiment to its beginning, making controversies about the results of experiments rare but challenges in establishing oneself primary. But then one must ask what processes are important in establishing an experiment. By considering, for example, the selection of projects and sites through Rip's "aggregation machines," even Knorr-Cetina's organic view of consensus formation implies attention to formal procedures for the aggregation of scientific views. Harry Collins (1998) takes a slightly more explicit step toward procedurally enforced consensus by examining how consensus occurs within and, at times, between laboratories engaged in the search for gravitational waves. His consensus is the consequence of an "evidential culture" composed of three dimensions: 1) evidential collectivism versus individualism, which is the dominant attitude in a laboratory about the responsibility of the larger scientific community to assess research results from an early stage versus the responsibility of individual researchers or laboratories to validate and interpret the meaning of results as fully as possible before publicizing them; 2) high versus low evidential significance, which is the dominant preference in the laboratory for either a long chain of inference from data, leading to high interpretative risk but potentially great significance, or a shorter chain of inference and consequently lower risk and significance; and 3) high versus low evidential thresholds, which is the statistical (as opposed to interpretative) risk that a laboratory is willing to assume (e.g., by being willing to talk publicly or attempt to publish at a level of certainty of two or three standard deviations or waiting until results are certain to seven or eight standard deviations.) Collins's case shows that laboratories with different evidential cultures may collaborate, but also that one laboratory may attempt to force its culture on collaborators as a condition of sharing data—thus attempting to force a consensus around a particular experimental approach. These evidential cultures are akin to the rules by which agencies that apply science to regulatory and other policy problems operate.

It is, then, not only appropriate to talk in terms of voting and social choice in science, but in many circumstances it may be necessary because consensus, as emergent or epiphenomenal, is unsatisfying both conceptually and practically. But even if discussion of voting is necessary, how satisfying could voting in science be?

Voting in Science

BENEFITS OF VOTING

As attractive as deliberative consensus may be, there are sound reasons to prefer voting and majoritarianism, as political theorist Robert Dahl (1989) lays out. First, majority rule maximizes the self-determination of the participants, because any hurdle greater than a simple majority offers the possibility of a large group of people held hostage to obstructionism from a minority. Although potentially attractive in a situation with the leisure for the truth to will out, it is much less so where, for example, the law imposes deadlines or where environmental or public health crises loom.

Second, majority rule maximizes utility over any other decision rule for a given set of participants with equivalent stakes. Even the narrowest majority would return a positive overall utility as the utility and disutility of voters on either side of the issue would cancel each other out, but the tie-breaking voter would have utility in the outcome. Any other rule under the same circumstances would yield a negative utility. Even if utility is less important in scientific contexts than political ones, scientists obviously have interests that can be at issue in scientific advisory committees and other decision contexts (e.g., journal and grant review).

Third, as Dahl argues by drawing on the work of mathematician Kenneth May, majority rule uniquely satisfies a set of important criteria for decision rules. Only majority rule is decisive among possible options; independent of which persons prefer which options; neutral with respect to outcomes, including the status quo; and positively responsive to marginal preferences. May's conclusion, however, is particularly dependent upon limiting the number of voting options to two. Otherwise, as Arrow showed, voting cycles can prevent majority rule from being decisive. Findings of scientific advisory boards, however, need not be decisive, because their work is only advisory, and politically responsible actors make the legally decisive judgments. One could also write science policy rules that insist on pair-wise voting to limit cycles. Moreover, scientific information is constantly changing, and nondecisive rules may interact with new information in a productive fashion by increasing the flexibility and novelty of a decision system.

Finally, Dahl argues—from Condorcet—that majority rule is the rule most likely to be substantively correct in a world of similarly fallible human beings. Condorcet calculated that when individuals are sometimes and similarly fallible, majority rule is more likely to be right than minority rule, and the probability of the majority's being right increases with its size. But if one were to insist on a supermajority, then the smaller minority, which has a smaller likelihood of being right, could prevail.

Dahl emphasizes the assumptions necessary for these arguments in favor of majoritarianism to hold: commitment to political equality; well-defined boundaries of the decision-making group; the need for collective decisions; and the requirements that the decision rule be decisive, feasible, and acceptable. These assumptions are attractive for decision making in science, as in politics. The commitment to political equality, potentially the most controversial, presumably plays out in the following way: assuming a rigorous gatekeeping process, any member of a decision-making group—regardless of demographic, disciplinary, or ideological background—should be given equal treatment before the rules of the group. That is, once appointed to a scientific advisory committee, ethicists or lawyers should have equal say as do scientists.

That some of the assumptions embedded in this argument fail in politics, too, does not mean we reject voting. Citizens do not have equal stakes in outcomes, for example, thus threatening the utilitarian purity of majority rule. Citizens are not identically fallible, thus threatening the substantive correctness of majority rule. Two general types of solutions to these problems exist: representation and supermajorities. Representation, particularly by organized interests or stakeholders, helps account for the intensity of preferences and unequal stakes that citizens have in outcomes. The requirement of supermajorities, rather than simple majorities, to alter closely held rights works in a similar fashion. Representation helps leverage knowledge, giving representatives a greater stake in gaining substantive knowledge (and, perhaps, being more correct) than citizens at large. Supermajorities raise the hurdle for the presumptive correctness of decisions, as certain decisions might be so potentially challenging (e.g., altering settled constitutional claims or overriding an executive veto) that a greater level of certainty represented by the supermajority is required.

There are, of course, reasons for preferring supermajorities—usually the protection of other values like stability, comity, or the human and political rights of minorities. A committee that seeks more than a majority will be more apt to discuss and refine its language to keep all its members on board. Such a committee will also tend to permit members with outlying preferences to extract compromises from the majority. Moreover, any move to clarify decision making by introducing explicit voting rules must confront the traditional image in which scientists are passive observers of nature and are compelled by the evidence to aggregate into a consensus. Nevertheless, procedures for the aggregation of preferences must be part of any decision-making system with more than one actor, and their appeal in science is not appreciably less than their appeal in politics. The following case studies of Asilomar and the National Toxicology Program illustrate this necessity and appeal, respectively.

ASILOMAR

The 1975 International Conference on Recombinant DNA Molecules, known as "Asilomar" for the California conference center in which it took place, is often hailed as an episode of "unprecedented process of scientific self-regulation" (Rogers 1977:4).[4] Asilomar is seen as unprecedented because participants and observers alike contrast the relatively public deliberations about the hazards of recombinant DNA with the more secretive behavior of atomic scientists during the first fission experiments. Asilomar is seen as scientific self-regulation because the researchers who were involved in the recombinant DNA techniques themselves raised the issue of safety, gathered other scientists together, deliberated, and published draft plans for regulating the conduct of their own experiments. Asilomar is also an example of scientists adopting voting rules in order to achieve outcomes.

The Asilomar story generally begins at the Gordon Research Conference on Nucleic Acids, held in June 1973, where Herbert Boyer reported the first experiments with hybrid DNA molecules. According to Rogers (1977:42), Maxine Singer—who had heard someone say in reaction to Boyer, "well, now we can put together any DNAs we want to"—brought up for discussion the question of safety. Rogers (1977:42) relates, "In an unprecedented move, seventy-eight of the ninety-odd participants voted to send a letter of concern to the prestigious National Academy of Sciences. And then a far closer, and ultimately more critical, vote of 48–42 called for wider publication of that same letter." The letter, authored by the Gordon Conference cochairs Singer and Dieter Soll, was sent to the NAS and later, under the authority of the slim majority of Gordon conferees, published in *Science*.

Paul Berg, who with Singer had been concerned about biohazard issues for some time, convened a small group of scientists at MIT. This organizing committee planned the meeting at Asilomar and drafted a second letter, requesting a voluntary moratorium on certain experiments, which was published in *Science, Nature*, and *Proceedings of the National Academy of Sciences* and after many drafts as an official report of the NAS. During this time, consequential decisions were being made in an ad hoc way—for example, Stanford's refusal to provide reagents to some scientists and warnings accompanying commercial reagents to abide by the moratorium. Relations with the press were also strained, and as the scientists planned the Asilomar meeting, they debated among themselves and with the press the number of reporters who would be allowed to attend and the rules by which they were to operate. Indeed, one of the first votes taken at Asilomar was whether to allow the sixteen reporters to keep their tape recorders. There were many abstentions, but "a majority of those who voted were in favor of permitting members of the press to keep their recording equipment" (Rogers 1977:54).

An international collection of experts in pathogens, infectious disease, and the emerging recombinant techniques gathered at Asilomar. Boyer relates that "there were probably 150 scientists, and there were probably 150 different opinions. And all of them could be equally authoritative or full of nonsense."[5] Prior to the meeting, several working groups had convened to draft reports that the larger group would then review during the meeting. After one group reported with a recommendation to divide experiments into six classifications for regulation, including one class to be forbidden altogether, Berg moved that the group adopt the report as written. Quoted in Rogers (1977:60), Berg "thought at the time, that a roomful of the leading minds on the leading edge of science can't agree on how to run a meeting. But the proceedings that bright morning began to resemble some obscure primitive tribe, eons ago, accidentally stumbling by trial and error onto the secrets of parliamentary procedure."

After more discussion, including provocative presentations by lawyers that were at times critical of the scientists' conceptions of academic freedom, workplace safety, and the like, the organizing committee began drafting a final, five-page Statement of the Conference Proceedings. David Baltimore, for one, fretted "that it would not be possible to generate enough of a consensus, that the meeting would not have a resolution."[6] Berg shared this fear: "I was worried largely [because] I could not hear a common thread. I could not see the consensus. People, I think, were being very self-serving."[7] Presenting the draft statement on the final morning of the meeting, Berg announced a thirty-minute reading period and set a noon deadline to reach a consensus by means other than voting. "This statement represents [the organizing committee's] assessment of the consensus as it exists," Berg said (Rogers 1977:83). An unidentified questioner from the audience persisted: "Do you have any idea as to the mechanism for determining this representative consensus? Are you talking about a vote on each paragraph?" Berg replied, "Nope. . . . We're not talking about a vote. I think we'll get an idea about consensus in some way without a vote" (85). According to Sydney Brenner, David Baltimore "answered by saying that a consensus will be achieved by the way the consensus will be achieved."[8]

According to Baltimore, Berg wanted to downplay the idea of consensus around any guidelines coming out of the meeting: he wanted "not to attempt to develop guidelines in a meeting. . . . And I think that Paul was very leery of the possibility of developing consensus in a meeting at all. I had a little more background in—kind of, political conventions and stuff, and I knew you could do it. But I also knew it was a difficult job."[9] Singer was even more explicit: "[W]e did not expect [our draft] to have a very good reception. . . . [W]e had this big discussion as to whether we would have a vote, and we decided we wouldn't allow a vote . . . because we thought we would be voted down."[10] There was also discussion over the relationship between any statement issued by the organizers and the input from the group, and to what extent the former would be taken to represent the latter. The organizers decided that any statement would represent their own opinions and not be a formal statement from the group, and thus any "failure to achieve a consensus position would not immobilize the committee from issuing its recommendations to the NAS" (Krimsky 1982:143).

Berg informed the group of this plan, but he was challenged on how any determination of consensus could be achieved without voting, and in the end, he was unable to avoid a show of hands. David Botstein, supporting a vote, offered that "[s]omehow—maybe we can tell the press to go away—we should get some feeling of whether eighty percent of the people in this place like, or hate it, or *something*. . . . I think maybe an overwhelming majority of the people here are willing to commit themselves to something of this sort" (Rogers 1977:85; emphasis in the original). Sydney Brenner rose and called for a show of hands on the principle expressed in the first paragraph of the statement that research should continue, but with significant controls. There was a consensus in accepting the statement, as Berg called for nay votes and found none. As Brenner recalls, "Then we just started voting."[11] Emboldened, Berg called additional votes on additional sections of the statement, but on one issue—whether to recommend that some class of experiments not be conducted with current methods, rather than recommend that they should only occur in the highest containment facilities, Berg concluded without a vote that a majority favored the former position. It became apparent, as abstentions increased, issues became more controversial, and the meeting neared its appointed conclusion, that confusion over the options for any vote was widespread. Nevertheless, Berg took a hasty vote on the overall document after the second lunch bell rang, with perhaps four hands rising in opposition, and declared "substantial agreement" on the statement (Rogers 1977:100). As participant Ephraim

Anderson observed, "in the end we voted positively in order to meet the deadline and get out, for a document the composition of which we don't precisely know yet."[12]

The final statement did, in fact, differ from the version discussed at the conference. According to Robert Sinsheimer, "It seemed to me that they had incorrectly summarized the results of the conference and that the final report was different in several significant respects from the actual draft which everyone asked and voted on the last day."[13] But "'[e]ven if they voted our statement down,' Maxine Singer said later, 'we'd agreed to send it in ourselves'" (Rogers 1977:83).

NATIONAL TOXICOLOGY PROGRAM

Scientific advisory committees are accessible loci to study procedures like voting for the aggregation of scientific opinion. The bulk of the scholarship on scientific advisory committees, at least in the United States, has been historical and institutional in nature (e.g., Smith 1992), although some of it has sought to articulate more closely what makes them successes or failures (Jasanoff 1990). This work points to the importance of science policy decisions, including rules of evidence and burdens of proof for decision making, but it does not discuss in detail the aggregation of views among committee members.

Some of these advisory committees seek consensus; others vote. Among the latter is the National Toxicology Program's Board of Scientific Counselors' Report on Carcinogens Subcommittee. The National Toxicology Program (NTP) is part of the National Institute of Environmental Health Sciences, one of the twenty-odd institutes of the National Institutes of Health within the U.S. Department of Health and Human Services. Congress created NTP in 1979 to implement a law (PL 95–622) mandating, inter alia, the regular publication of a list of substances either known or reasonably anticipated to be human carcinogens. NTP publishes this list, the *Report on Carcinogens*, biennially after conducting an elaborate deliberative process. This deliberation includes voting by three advisory committees, including the Report on Carcinogens Subcommittee, which advise the director of NTP, who then determines the final content of the report.

Scrutiny begins with nominations by the public of substances as either known human carcinogens or reasonably anticipated to be human carcinogens (substances may also be nominated for de-listing from these categories).[14] After the collection and review of extensive information on the substance from only the peer-reviewed literature, two committees composed solely of federal employees—one of NTP employees, the other of representatives from such interested agencies as the Food and Drug Administration and the National Cancer Institute—vote on the nominations. Although the overall tally from these votes is public, it is released only later in the process, and public records do not show how individual members voted. However, the Report on Carcinogens Subcommittee has members who are not federal employees, and therefore it falls under the jurisdiction of the Federal Advisory Committee Act (FACA, PL 92–463), which requires open meetings and records. The votes from the Report on Carcinogens Subcommittee may thus be analyzed in some detail.

Members of the advisory committees are also given relatively strict and concise criteria to guide their scientific judgment. The advisors are instructed to find a "known human carcinogen" if and only if "[t]here is sufficient evidence of carcinogenicity from studies in humans that indicates a causal relationship between exposure to the agent, substance or mixture and human cancer" (DHHS 1996). They may find a "reasonably anticipated to be a human carcinogen" if and only if the following conditions are met:

> There is limited evidence of carcinogenicity from studies in humans, which indicates that causal interpretation is credible, but that alternative explanations, such as chance, bias or confounding factors, could not adequately be excluded,
>
> or
>
> There is sufficient evidence of carcinogenicity from studies in experimental animals which indicates there is an increased incidence of malignant and/or a combination of malignant and benign tumors: (1) in multiple species or at multiple tissue sites, or (2) by multiple routes of exposure, or (3) to an unusual degree with regard to incidence, site or type of tumor, or age at onset;
>
> or
>
> There is less than sufficient evidence of carcinogenicity in humans or laboratory animals, however; the agent, substance or mixture belongs to a well defined, structurally-related class of substances whose members are listed in a previous Report on Carcinogens as either a known to be human carcinogen or reasonably anticipated to be human carcinogen,

> or there is convincing relevant information that the agent acts through mechanisms indicating it would likely cause cancer in humans.

NTP also instructs the advisory committees that

> Conclusions regarding carcinogenicity in humans or experimental animals are based on scientific judgment, with consideration given to all relevant information. Relevant information includes, but is not limited to dose response, route of exposure, chemical structure, metabolism, pharmacokinetics, sensitive sub populations, genetic effects, or other data relating to mechanism of action or factors that may be unique to a given substance. For example, there may be substances for which there is evidence of carcinogenicity in laboratory animals but there are compelling data indicating that the agent acts through mechanisms which do not operate in humans and would therefore not reasonably be anticipated to cause cancer in humans.

These criteria are, in part, responsible for the great deal of agreement among the members of the advisory groups. Guided by these criteria, more than 80 percent of the votes in the *Ninth Report on Carcinogens* (NTP 2000)—which considered such important and controversial substances as saccharin, dioxin, and tamoxifen—were cast the same way (Guston 1999). Without such clear science policy rules to guide scientific judgment, advisors can lapse into unproductive conflict (Jasanoff 1990).

To further see the influence of these science policy criteria on outcomes, consider the case of saccharin. Prior to 1995, the listing criteria for "reasonably anticipated to be a human carcinogen" were more concise than above and did not specifically include reference to mechanistic data. But after NTP altered the criteria to include the consideration of "mechanisms which do not operate in humans and would therefore not reasonably be anticipated to cause cancer in humans," the committees—in very close votes—recommended de-listing saccharin based on a finding in experimental rats that the carcinogenic mechanism would not be reasonably expected to occur in humans (Guston 2004).

Analyzing the votes shows that voting rules may allow advisory committees to reach decisions that may not have been reached by consensus rules. Even though NTP has no rules for interpreting the three advisory committees' votes, one can classify their voting on each substance by hypothetical rules: *unanimity*, meaning that all present voted for the same outcome; *strict consensus,* meaning that no one present voted against the outcome preferred by the rest (but may have abstained); *supermajority,* meaning at least two-thirds but not all of those present voted for the same outcome; or *simple majority,* meaning that more than one-half but fewer than two-thirds voted for the same outcome. For the *Ninth Report on Carcinogens,* the three advisory committees voted a total of seventy-three times (there was one repeat vote, so the total is not divisible by three). Thirty-five committee votes (48 percent) achieved unanimity, six (8 percent) achieved strict consensus, twenty-two (30 percent) achieved supermajority, and ten (14 percent) achieved simple majority. Thus, if the committees had a formal consensus rule, they might have failed to make a recommendation in about 40 percent of the cases, and 14 percent of the cases might still have been unresolved even under a supermajority rule (Guston 1999).

NTP underscores how a procedurally enforced consensus, including voting, can operate relatively well. NTP formally implements what Asilomar struggled ignorantly to obtain. Yet NTP's formality also makes it easier to determine how outcomes, even of a process guided by the peer-reviewed literature and conducted by a small group of scientists, can be influenced by the procedures used to elicit and aggregate their views.

SCIENTIFIC CONSENSUS THROUGH VOTING

The cases of Asilomar and NTP reveal distinct ways in which scientists do vote on matters of scientific and societal consequence. The Asilomar case is perhaps as close to a native version of Dahl's critique of guardianship or technocracy as possible. The best-trained scientists raised the issue of biohazards, assembled the best-trained minds to talk about it, and deliberated to come to some consensus on action. Yet even these guardians needed voting—a mechanism to aggregate the views of the group—in order to agree. Other considerations (e.g., "citizenship" in the group, agenda-setting issues, amendments) were entirely ad hoc.

One might argue that had the Asilomar organizing committee—the best of the best-trained—simply written a report, they would not have needed voting rules. But one could counter quite powerfully that a report from the organizing committee alone would have neither been as legitimate nor as technically proficient as the report from the larger assembly. If the organizing committee alone had been involved, who could have said that the resulting document was achieved through true deliberation, that is, represented an essential consensus in Fuller's sense or was merely the accidental consensus of a few like-minded gadflies?

One might also object that what was being decided at Asilomar was not science but rather regulation or science policy, and that it is therefore not surprising to see traditionally political modes of decision making appear. But one cannot so neatly demarcate science from science policy (Gieryn 1999; Jasanoff 1990). And even if such a boundary could be clearly drawn, then the scientists arrogated to themselves regulatory or policy decisions, and they naively conducted their politics and their voting like a "primitive tribe," reinventing parliamentary procedures from the start and on the fly.

Asilomar was deeply flawed in its organization for consensus. Krimsky (1982:151) writes, "Asilomar was severely limited in the following ways: selection of participants, clarity of the decision-making process; boundaries of discourse; public participation; and control of dissent." More specifically, he concludes that "it was unclear how a consensus would be reached [and p]articipating scientists questioned the meaning of a vote," and that "[t]he final report issued by the Organizing Committee did not reflect the level of dissent that existed at the meeting [and] it may have provided a false impression about the degree to which scientists at Asilomar, especially members of the working panels, had supported the documents" (Krimsky 1982:152). Clear voting rules and accompanying discussions about who gets to vote how and on what questions would have helped on many of these counts.

The experience with NTP provides a welcome contrast in which clear rules and procedures do not demean the scientific nature of the activities, while they also improve the prospects for both consensus and fairness. The question "is it science?" has less currency at NTP than it might with Asilomar because there is little dispute that its task, defined as identifying carcinogenic hazards, is more science than risk assessment or management. Yet this scientific task provides a transparent example of Fuller's procedurally enforced consensus. There are rules and criteria, without which outcomes would have been different, as the saccharin example shows—if achieved at all. The perception of agreement among the NTP advisory committees varies depending on which channel—consensus, unanimity, supermajority, or majority—the view is set.

But all is not perfect at NTP. Although the criteria to guide scientific judgment in identifying carcinogens give credence to the idea that an

Table 1. Votes of the Members of the Report on Carcinogens Subcommittee on All Substances, in Comparison to the Majority

Name	More	As	Less
Bailer	1	21	1
Belinsky	0	18	5
Bingham	2	9	1
Frederick	1	18	2
Friedman-Jimenez	0	17	0
Henry	0	7	3
Hooper	3	19	1
Mirer	3	18	0
Hecht	0	12	0
Kelsey	1	14	0
Medinsky	0	8	2
Russo	1	7	2
Zahm	0	13	0

essential or deliberative consensus emerges through the NTP process (even if voting is the mechanism to manifest it), additional evidence suggests that accidental consensus has not entirely been dispelled. Table 1 shows how the members of the Report on Carcinogens Subcommittee voted on all the substances considered in the *Ninth Report.*[15] Aggregating the votes by the sectoral or disciplinary affiliation of the subcommittee member suggests that these accidental qualities influence outcomes.

In Table 1, "more" represents how many votes the individual cast that were "more protective" than the majority of subcommittee members cast for any substance, "as" means how many votes were "as protective" as the majority, and "less" means how many votes were "less protective" than the majority. A more protective vote is voting to list a substance as "reasonably anticipated to be a human carcinogen" when the majority voted not to list the substance, or voting to list it as a "known human carcinogen" when the majority voted to list it as "reasonably anticipated." A "less protective" vote is the other way around.[16]

Table 2 sums the votes by sectoral and disciplinary affiliation. The industry members are less protective overall than other members, and the single labor member is more protective. In the disciplinary analysis, subcommittee members affiliated with laboratory disciplines (e.g., toxicology) are less protective, those affiliated with populations and statistics (e.g., biostatistics, epidemiology) are in the middle, and those affiliated

Table 2. Votes of the Members of the Report on Carcinogens Subcommittee on All Substances, Aggregated by Sector and Disciplinary Group

	More	As	Less
Academic	5	80	4
Government	3	50	6
Industry	1	33	7
Labor	3	18	0
Stats/Pop	1	34	1
Organismal	7	65	3
Laboratory	4	82	13

with organismal studies (zoology, medicine) are more protective. At least for some of the more controversial substances considered, the consensus at NTP has an undeniably accidental quality.

Conclusion

As this chapter has shown, the concept of social choice in science makes sense, and the incomplete nature of scholarship on scientific consensus makes inquiries into voting in science necessary. The case studies provide evidence that not only does voting in science occur, but that how it occurs is scientifically and politically important.

An Asilomar participant exclaimed to a historian, *"Sogar ein blindes Huhn findet auch mal ein Korn!"* "Sometimes even a blind chicken can find a seed!"[17] But which seed? "A lot of other things could have come out," David Baltimore recalled, "that would have made many people just as satisfied as what did come out."[18] A lot of things could have come out that would have made many people much less satisfied, too. The point is that any decision-making process that fails to account for how the process might influence the array of potential outcomes is seriously flawed. As Sheila Jasanoff (2003:160) writes, "Expertise, like other forms of democratically delegated power, is entitled to respect only when it conforms to norms of transparency and deliberative adequacy." Consensus without further procedural specification does not satisfy these norms.

The NTP case shows that voting can facilitate outcomes in a scientific context but does not guarantee consensus. There are many other contexts to study the procedural enforcement of consensus, including domestic scientific advisory committees in the United States as well as international bodies such as the International Agency for Research on Cancer (IARC) and the Intergovernmental Panel on Climate Change (IPCC). Further research into these organizations, as well as into social choice in more informal scientific settings, would significantly enhance our understanding of scientific consensus-as-process, which—as has been argued—is critical for scientists, scholars, and policymakers to understand.

Scientists vote—sometimes in well-orchestrated elections and sometimes in chaotic, ad hoc ones. A scientific subsystem that does not partake in such a politics of transparent social choice—one that hides both its substantive disagreements and its disciplinary and sectoral interests beneath the cloak of consensus—is not a fully democratic one. The claim about voting to repeal natural laws is a red herring. Scientists need to vote in well-orchestrated ways if they want substantive influence on public laws.

Acknowledgments

The author wishes to thank Mark Brown, Arie Rip, Ed Hackett, Elisabeth Clemens, and the editors for their helpful comments. Some of this material is based upon work supported by the National Science Foundation under Grant # SBR 98-10390. Any opinions, findings, and conclusions or recommendations expressed in this material are those of the author and do not necessarily reflect the views of the National Science Foundation.

Endnotes

1. In *Frye v. the United States* (1923), the Supreme Court established the *Frye* rule, which declared general acceptance within the relevant scientific community to be the sole criterion for admitting expert testimony. The Court ruled in *Daubert* that the Federal Rules of Evidence, passed by Congress in 1975, instead govern such testimony. For the construction of expertise in the legal context, see Jasanoff (1995:chap. 3).

2. Ezrahi avoids the complication that scientific consensus and political consensus can influence each other as, for example, Collingridge and Reeve (1980) address.

3. See Jasanoff (1992) for this behavior in the courtroom and Bimber and Guston (1995) for the same behavior in the U.S. Congress.

4. Sources of this account include Rogers (1977), Krimsky (1982), and oral

histories and the "Historical Note" from the Recombinant DNA History Collection, MC 100, pp. 3–10, Institute Archives and Special Collections, MIT Libraries, Cambridge, MA (hereafter RDHC). Oral histories were conducted in late 1975 by MIT historian Charles Weiner and colleagues.

5. Herbert Boyer, RDHC, p. 31.

6. David Baltimore, RDHC, p. 95.

7. Paul Berg, RDHC, p. 93.

8. Sydney Brenner, RDHC, pp. 52–53.

9. David Baltimore, RDHC, p. 83.

10. Maxine Singer, RDHC, p. 71.

11. Brenner, RDHC, p. 53.

12. Ephraim S. Anderson, RDHC, p. 52.

13. Robert Sinsheimer, RDHC, p. 15.

14. Not only "substances" but also exposure circumstances are included in the process.

15. Tables 1 and 2 are taken from Guston (2004).

16. Individuals do not have the same number of votes because some may have joined the committee at different times in its deliberations, some may have missed meetings, and some may have abstained or declared conflicts of interest. All votes, however, are on substances considered for the *Ninth Report.*

17. Anderson, RDHC, p. 58; my translation.

18. Baltimore, RDHC, pp. 112–113.

References

Arrow, Kenneth J. 1963 [1951]. *Social Choice and Individual Values.* 2nd ed. New Haven, CT: Yale University Press.

Barke, Richard P. 2003. "Politics and Interests in the Republic of Science." *Minerva* 41:305–325.

Bimber, Bruce, and David H. Guston. 1995. "Politics by the Same Means: Government and Science in the Unites States." In *Handbook of Science and Technology Studies,* Sheila Jasanoff, Gerald E. Markle, James C. Petersen, and Trevor Pinch, pp. 554–571. Thousand Oaks, CA: Sage.

Coleman, James. 1990. *Foundations of Social Theory.* Cambridge, MA: Harvard University Press.

Collingridge, David, and Colin Reeve. 1980. *Science Speaks to Power: The Role of Experts in Policy Making.* London: Pinter.

Collins, H. M. 1998. "The Meaning of Data: Open and Closed Evidential Cultures in the Search for Gravitational Waves." *American Journal of Sociology* 104:293–338.

Dahl, Robert. 1989. *Democracy and Its Critics.* New Haven, CT: Yale University Press.

Department of Health and Human Services. 1996. "Notice: Revised Criteria and Process for Listing Substances in the Biennial Report on Carcinogens." *Federal Register* 61(188, 26 September): 50499–50500.

Ezrahi, Yaron. 1980. "Utopian and Pragmatic Rationalism: The Political Context of Science Advice." *Minerva* 18:111–131.

———. 1990. *The Descent of Icarus: Science and the Transformation of Contemporary Democracy*. Cambridge, MA: Harvard University Press.

Fuller, Steve. 1988. *Social Epistemology*. Bloomington: Indiana University Press.

Gieryn, Thomas F. 1999. *Cultural Boundaries of Science: Credibility on the Line*. Chicago: University of Chicago Press.

Guston, David H. 1999. "Constructing Consensus in Food Safety Science: The National Toxicology Program's Report on Carcinogens." Paper presented at the annual meeting of the American Association for the Advancement of Science, Anaheim, CA, 22 January.

———. 2004. "Institutional Design for Robust Knowledge: The National Toxicology Program's *Report on Carcinogens*." In *Yearbook in the Sociology of Science, 2004*, ed. P. Weingart et al.

Jasanoff, Sheila. 1990. *The Fifth Branch: Science Advisors as Policymakers*. Cambridge, MA: Harvard University Press.

———. 1992. "What Judges Should Know about the Sociology of Knowledge." *Jurimetrics Journal* 43:345–359.

———. 1995. *Science at the Bar: Law, Science, and Technology in America*. Cambridge, MA: Harvard University Press.

———. 2003. "(No?) Accounting for Expertise?" *Science and Public Policy* 30: 157–162.

Jasanoff, Sheila, Gerald E. Markle, James C. Petersen, and Trevor Pinch. 1995. *Handbook of Science and Technology Studies*. Thousand Oaks, CA: Sage.

Kay, Alan F. 1998. *Locating Consensus for Democracy: A Ten-Year U.S. Experiment*. St. Augustine, FL: Americans Talk Issues.

Kim, Kyung-Man. 1994. *Explaining Scientific Consensus: The Case of Mendelian Genetics*. New York: Guilford Press.

Knorr-Cetina, Karen. 1995. "How Superorganisms Change: Consensus Formation and the Social Ontology of High-Energy Physics Experiments." *Social Studies of Science* 25:119–147.

Krimsky, Sheldon. 1982. *Genetic Alchemy: The Social History of the Recombinant DNA Controversy*. Cambridge, MA: MIT Press.

Kuhn, Thomas. 1977. *The Essential Tension: Selected Studies in Scientific Tradition and Change*. Chicago: University of Chicago Press.

Latour, Bruno, and Steve Woolgar. 1979. *Laboratory Life: The Social Construction of Scientific Facts*. Beverly Hills, CA: Sage.

National Toxicology Program (NTP). 2000. *Ninth Report on Carcinogens*. Research Triangle Park, NC: National Institute of Environmental Health Sciences.

Polanyi, Michael. 2002 [1962]. "The Republic of Science: Its Political and Economic Theory." *Minerva* 1:54–73. Reprinted in *Science, Bought and Sold: Essays in the Economics of Science,* ed. Philip Mirowski and Esther-Mirjam Sent. Chicago: University of Chicago Press.

Rip, Arie. 1994. "The Republic of Science in the 1990s." *Higher Education* 28: 3–23.

———. 2003. "Aggregation Machines: A Political Science of Science Approach to the Future of the Peer Review System." In *Knowledge, Power, and Participation in Environmental Policy Analysis,* ed. William Dunn, Matthias Hisschemoller, Jerome R. Ravetz, and Robert Hoppe. Policy Studies Review Annual, vol. 12. New Brunswick, NJ: Transaction Press.

Rogers, Michael. 1977. *Biohazard.* New York: Alfred A. Knopf.

Smith, Bruce L. R. 1992. *The Advisers: Scientists in the Policy Process.* Washington, DC: Brookings.

15

Learning to Reflect or Deflect?

U.S. Policies and Graduate Programs' Ethics Training for Life Scientists

LAUREL SMITH-DOERR

During a family vacation to Washington, D.C., in the summer of 2003, I visited a new Smithsonian exhibit entitled "Genome: The Secret of How Life Works." To enter the exhibit, prominently sponsored by the drug company Pfizer, the museumgoer is directed into a narrow passageway that tells "the story of you." At the first turn, a large-sized picture of a blastocyst is labeled, "this WAS you." On the wall of the next turn, a mirror greets the patron along with the statement, "this IS you." Finally, the narrow, winding passage ends with a picture of a double helix that claims, "this is the SECRET of you." As the museum patron enters the exhibit proper, she walks by a twenty-five-foot-long strand of DNA. Meanwhile, a nearby video screen shows a vaguely familiar-looking white man who effusively exclaims: "Hello! My name is Eric Lander. What you see before you is the DNA double helix. It's the secret of you, the secret of me, and actually the secret of all life on this planet. The DNA double helix is made up of genes which are sort of a recipe for who you are. The only problem is, that recipe is written in a secret code. It took a very long time to figure out how to read that code—but now we have it!!" (The exhibit is also published online at http://genome.pfizer.com/menulong.cfm, last accessed November 15, 2004). The remainder of the exhibit takes a similar tone, cheerfully describing the wonders of the human genome and its probable future applications to human health.

One could spend an entire chapter deconstructing this exhibit, tracing the ways in which genetic determinism and commercial interests in genetics are intertwined. But for my purposes, the exhibit raises a related question: what was left out? The ethical implications of human genome research were barely mentioned. At the last station of the thirty-two in the exhibit, called "Reality Check Theater," Eric Lander remarks: "[M]ost scientists agree that trying to make a human baby by cloning is a bad idea. The chance of birth defects is just much too high to take the risk and there are serious ethical issue [sic] to think about." He concludes: "The promise is enormous but not everything about genetics is clear cut so it's important to separate fantasy from reality, entertainment from real issues." This token attention to ethical concerns in genomic research nearly denies that they exist (except as "fantasy") and raises the question of how and where biological scientists learn to think about issues such as how their work is communicated to the world and whether it has commercial value.

Since October 2000, the main federal funding agency in the United States for life scientists, the National Institutes of Health (NIH), has required grantees to complete certified training in research ethics, primarily in the treatment of human subjects. In so doing, the NIH asserted that scientists and scientists-in-training, not just clinicians, need to consider the social implications of their work. The logic of the NIH's decision was based on an expectation of rational action on the part of graduate programs. This model of policymaking assumes that if funding is affected, scientists will have incentives to reflect seriously about the social and ethical context of their work. This chapter focuses on the degree to which U.S. doctoral programs in molecular biology have complied with the NIH's new requirements.[1] Is there evidence that these educational policies are enacted and taken seriously by these programs, as the NIH's incentive system suggests that they should be? Or do university scientists deflect the impact of such requirements by appearing to follow the federal guidelines while in fact resisting any real change?

To examine these questions, I use archival and interview data. I collected information on a random sample of fifty molecular biology and biochemistry PhD programs in 2001 and 2003 and conducted pilot interviews with life scientists about the implementation of the NIH guidelines. I explore when the ethical and social aspects of science might be discussed informally rather than being a part of the formal, visible curriculum, and why some scientists express an adverse reaction against government funding agencies legislating a top-down approach to research ethics. The results of this study have implications for understanding political and organizational institutions of science and suggest reasons why noncompliance with science policies might arise.

We might expect that PhD students would at least receive some training in research ethics in response to the NIH policy of 2000. Because so many predoctoral students are funded through NIH resources, graduate courses in research ethics should also be in evidence. In 1997, 44.9 percent of all biomedical graduate students received federal funding, and about 75 percent of those received funds from the NIH (NRC 2000:24). Indeed, NIH discussion of the policy reminds investigators that graduate students funded by training grants are "required to comply with a program on the responsible conduct in research which includes protection of human subjects as a topic" (NIH, Extramural Research 2002). Because most of the PhD programs either have training grants or plan to apply for them, in effect the NIH policy for investigators serves as another incentive to educate graduate students in research ethics. Accordingly, NIH mandates such as this one for ethics training certification should have an effect on graduate programs.

The monetary-based incentives from NIH signal that politicians and policymakers felt that scientists needed to change their behavior pertaining to ethical issues in biomedical science yet were unlikely to do so voluntarily. As David Guston (2000) argues, the establishment of the Office of Research Integrity and other moves by Congress to regulate science show that the social contract—the trust in science to monitor itself—had broken down by the 1980s. If the trust in science is broken, the political solution seems to be to try to return it through top-down guidelines. Indeed, the history of the NIH policy is one of top-down direction from president to cabinet secretary to policymakers. According to testimony on May 25, 2000, before the U.S. Senate by William Raub, deputy assistant secretary of science policy, the motivation for the education requirement came in part from President Clinton's request that the Department of Health and Human Services assuage public concerns about the safety of clinical trials. In direct response to an initiative from Secretary of Health and Human Services Donna Shalala, the NIH developed the policy that principal investigators had to be certified as passing an educational course in the treatment of human subjects (NIH, Director 2000). The NIH requirement indicates that enforcement is needed to get researchers to pay attention to the ethical implications of science. Does change in the appearance of behavior, however, mean that scientists are receiving the sort of training that was intended?

COMBINING SCHOOLS OF THOUGHT TO UNDERSTAND HOW SCHOOLS TEACH THINKING

Two schools of thought—the new institutionalism in organizational analysis and science and technology studies (STS)—provide insights into why graduate programs seem to resist complying with the mandate to educate life scientists in the ethical implications of their research. While each of these is a powerful perspective in its own right, the combination is particularly useful because the new institutionalism provides analysis at a more macro field level, while STS tends toward smaller-scale analysis.

Consider the complementary perspectives of neo-institutionalism and STS on national policymaking. Both offer culturally sensitive theoretical explanations that take a nuanced perspective on the effects of policy rather than assuming that monetary incentives straightforwardly guide behavior. New institutionalists show how policies are interpreted and implemented through organizational cultures at the field level. Inertia plays a key role in the theory; new laws govern organizational fields that already have a cultural history and network of interests in place (Edelman 1992). The implementation of maternity-leave policies in American firms, for example, depends on earlier organizational interpretations of public policy as much as on broad market forces like women entering the paid labor force (Kelly and Dobbin 1999). And yet while neo-institutionalism can tell us about the rise of science ministries and formal science policies worldwide in the past few decades (Drori, Meyer, Ramirez, and Schofer 2003), it fails to provide analysis of scientific actors who strategically define their domain of expertise (Gieryn 2004).

Science and technology studies, on the other hand, does examine the localized cultures created by science policymakers (Jasonoff 1995; Hilgartner 2000). For example, Kerr's (2003) study of the content of bioethics policies in Europe describes how policy networks come to frame the issues uniformly in terms of individual choice and scientific progress.

Although valuable, the localized STS perspective misses the important effects of organizational-level dynamics and power structures (Klein and Kleinman 2002). Thus, their foci on different levels of analysis mean that neo-institutionalism and STS highlight different aspects of social reality. I employ both schools of thought in order to take advantage of new institutionalism's insights into life science graduate programs as an organizational field and STS's insights into the actions scientists take to deflect unpopular policies.

PERSPECTIVES ON SCIENTIFIC ETHICS

Traditionally, ethics is defined as a rather static set of standards for conduct based on a system of moral values. Casper (1998:138) offers a more practical definition of ethics as "a set of concrete social practices that can be captured [by] examining the social processes and judgments underlying what comes to count as an acceptable practice." My study, rather than focusing on the *content* of ethical codes or practices, investigates whether the visible practice of teaching scientists to consider the ethical implications of their work occurs in university training. Thus the emphasis is on ethics training rather than on ethics, per se. What I mean by ethics training is formal education through a specific university course (e.g., "Research Ethics," "Bioethics," "Responsible Research Conduct"). A critic might argue that research ethics are best "learned by doing," such as by observing a mentor's behavior; this is also true of scientific research. Yet, of course, there are still curricular requirements for students on scientific topics. The visibility of ethics training is measured by course offerings and requirements publicly acknowledged on graduate program Web sites.

An important precedent for the awareness of ethical issues in life science organizations comes from the Human Genome Project. This is also a story of mandates by directors. Early in the history of the Human Genome Project at the Department of Energy (DOE), Charles DeLisi conceived of an ethical component (Cook-Deegan 1994). Later, James Watson mandated that a small percent of all NIH Human Genome Initiative funds go to analyzing ethical and legal issues surrounding the project (Kevles and Hood 1992; Watson 1990). As a result, the Ethical, Legal and Social Issues (ELSI) office of the Human Genome Project was established as a joint NIH/DOE effort. Whether the founding of ELSI was motivated by genuine concern about the moral implications of genetics or by savvy strategy to ensure political support of the National Center for Human Genome Research (Cook-Deegan 1994), the initiative does seem to show that leading natural scientists are willing to support studies by social scientists, legal scholars, and bioethicists on the history and potential moral consequences of the Human Genome Project. Yet the implications of the ELSI initiative are unclear. Does a specialized segment of funding mean that natural scientists are absolved from educating themselves in ethical, legal and social issues; in effect, allowing them to "outsource" concern for the ethical issues raised by their research? The separation of pure science from messy conflicts of interest or other social and ethical issues harks back to a longstanding objectivity ethos in American state-supported science (e.g., Vannevar Bush's report to President Roosevelt; Kleinman 1995).

A basic idea in science and technology studies—that technology (and science) is (are) not neutral—is in direct conflict with the objectivity assumption of science. As Robert Merton (1973) observed, one fundamental ethos of scientists is disinterestedness—ideally, that "pure research" is untainted by subjectivity and interests other than the advancement of knowledge. This objectivity assumption seems to imply that a well-intentioned, careful specialist is not responsible for the outcomes of his science. "Objective science" is an idea that resides at a deep cultural level, as apparent in Nelkin and Lindee's (1995) description of the road map metaphor used by scientists. The Human Genome Project uses mapping imagery, which "suggests that once a gene is located, its interpretation will be objective" (8). The STS perspective, what Langdon Winner (1986) calls "technological politics," in contrast, conceives of every technology as having an ideology embedded within it. Ironically, the objective, disinterested ethos of science actually is technological politics—an ideology that affects the science. For example, Nelkin and Lindee's (1995) argument illustrates that the perspective of the Human Genome Project as a neutral map is a kind of political stance; one that probably benefits the status quo if it means that questions are not raised about genetic discrimination and other potential social/ethical dilemmas.

As an example of how political interests shape science, Paul Rabinow (1999) presents an ethnographic account of the aborted collaboration between Millennium Pharmaceuticals and the French government. The government ultimately balked at "giving away French DNA" to an American company. In discussing the circumstances of the impasse, Rabinow found that France had an early concern with ethical issues surrounding genetics and had established a national ethics board. Attention was paid to ethical issues, but only by bioethicist "experts." The existence of a bioethics specialization, however, raises the question of whether scientists are responsible for thinking through humanistic (or nationalistic, in France's case) issues stemming from their research. Rabinow (1999:110) argues that such a division of knowledge could mean that the use of ethics is only reactive while "the work on exploring, constructing, discovering, and inventing 'the genomic' is left to scientists and physicians (whose striving for advancing knowledge and health always carries the potential for excess) as well as to venture capitalists and the large multinational pharmaceutical firms. Consequently, the distinct risk is present that the knowledge and truths about living beings that are emerging in molecular biology and genomics will be formulated by, and in the interest of, barbaric and/or decadent forces." Whether pharmaceutical corporations are "barbaric" is arguable, but they certainly have economic interests that might affect the work of "objective" scientists who are imperceptive of ethical issues.[2] This STS argument that science is not neutral although it claims to be provides one explanation for why courses on ethics might be unpopular: they contradict a deeply held belief about the nature of the scientific enterprise.

Another way to think about scientists' reluctance to embrace required ethics education is to consider the organizational context in which technological politics emerge. Organizations are difficult to change, and when change occurs it is often only on the surface. Once an organization starts moving in one direction, it continues down the same path and can be changed only with great difficulty (Stinchcombe 1965). For example, once a restaurant is founded as a fast food place, it would be extremely difficult to switch into a fine dining establishment. A graduate program founded on the objectivity ethos is unlikely to be willing or able to quickly embrace an ethics requirement into its core curriculum.

Organizational inertia is compounded by the fact that organizations are not isolated entities. The broader institutional environment, the field in which organizations are situated, exerts influences as well. Once a particular way of setting up a graduate program is legitimated, other programs will probably look quite similar. This isomorphism may be due to copying other prestigious programs or to normative pressures to look professional (DiMaggio and Powell 1983). Similarities between organizations worldwide arise through the isomorphic pressures inherent in institutionalized ways of doing things. The threat of withholding resources produces coercive isomorphism in new institutionalist terms, which means that organizations come to look similar because they are coerced into doing so (DiMaggio and Powell 1983). At the same time, a new commonly used structure or organizational behavior may be a "rational myth" or the means that an organization uses to communicate its legitimacy as an "efficient" or "effective" collective but that does not represent real change (Meyer and Rowan 1977). For example, Lauren Edelman (1992) illustrates how corporations instituted affirmative action offices in response to Equal Employment Opportunity legislation but did little to actually change their hiring and promotion practices.

Although changing the design of a scientific organization is a conscious decision, the social processes guiding the decision may not be explicit or rational (i.e., who will use the technology and how they will use it is not always clearly considered). The result is usually unanticipated consequences. Decision makers are not context-less rational calculators, able to mold organizations and technologies optimally regardless of past and ongoing social processes and structures. For example, rather than profit-maximization, unethical business practices can lead to technical disasters like the Challenger explosion. Vaughan (1996:68; emphasis mine) shows that this tragedy was a result of organizational inertia: "Amorally calculating managers intentionally violating rules to achieve organization goals does *not* explain the Challenger disaster. . . . [P]roduction pressures became institutionalized . . . influencing decision making by managers and engineers *without* requiring any conscious calculus." Observing organizational context reveals the complexity of ethical issues in reality. Decision makers design both organizations and technology based on institutionalized ways of doing things.

Clearly, though, people working in organizations are not puppets. New institutionalist studies of U.S. policies underline how organizational actors negotiate and interpret the social forces of legitimacy, resources, and legal constraints, instead of taking a knee-jerk response to these incentives. Dobbin, Sutton, Meyer and Scott (1993) provide a convincing institutional argument for the development of internal labor markets (ILMs). In contrast to rationalist theories that argue ILMs arose simply because they were more efficient employment structures, especially for larger, technology intensive organizations, Dobbin et al. (1993) show how ILMs emerged as an organizational response to Equal Employment Opportunity Commission (EEOC) policies. Further, the interpretation of EEOC laws in the courts shaped organizational behavior toward bureaucratic hiring and promotion through ILMs rather than the more contested use of hiring quotas or standardized testing. Personnel department professionals created normative isomorphism in firms by instituting ILM solutions, at the same time the state fostered coercive isomorphism through the policies. Edelman, Fuller and Mara-Drita (2001) show further how managers institutionalize policy in reinterpreting civil rights laws. Federal EEOC policies designed to promote workplace diversity by gender and race have been "managerialized" to mean diversity in the broader sense of people with differing viewpoints or regional origins.

As Clemens (1997) illustrates, implementing policy depends on having actively supportive constituencies, especially in a pluralist system like the United States. In her study of the Progressive Era, the combined effects of institutional environment and stakeholder groups figure prominently in the outcomes for new policies. In Washington State, populists entered office and had their agenda pass through the legislature, but there was little follow-through—as in the lack of enforcement of new worker's compensation laws. But in Wisconsin, corporatist coalitions of farmers and organized labor in the Socialist Party forged a voting block that supported state agricultural funding for small farms rather than University of Wisconsin and the larger agribusiness interests.

From empirical studies like those described above, neo-institutionalism's underlying assumption of the duality of structuration (Giddens 1984) is clear. Institutions in any kind of organization, whether firm or university, are a point of both constraint and agency.[3] Taken-for-granted assumptions, such as the institution that objectivity provides a mechanism for scientists' self-governance, shape the cognitions of organizational actors. Yet organizational actors also sustain and re-create institutions by enacting them.

With regard to training life science PhD students about the social and ethical consequences of their research, these insights from STS and new institutional theory suggest that graduate programs would be unlikely to implement full-fledged ethics training into their courses of study. A more likely effect of the federal grant agency incentives is the appearance of change in graduate programs. A related question concerns which universities have ethics education, by the prestige of the school. Perhaps elite schools are more likely to feel external pressures to include ethical training for graduate students, since they are more visible to funding agencies, policymakers, and the media. On the other hand, the more elite programs may be those that have the power to resist outside pressures to shape their educational policies, and thus they can omit any "non-scientific" training. If there is evidence that courses in research ethics or social issues in biological sciences are offered in response to policies, the second question is whether the curriculum is taken seriously or provides a rational myth. This paper answers the first question and explores the second.

One way to measure inertia in the ethics education of life science PhDs is by similarities in the approach of different universities. In particular, I would expect that schools within the same level of prestige would look like each other in their training of scientists in ethical issues. Elite schools might be more likely than others to include ethics because of their greater visibility and number of resources. Alternatively, the field of life science programs may be uniform throughout in the treatment of ethics education because an elite model has diffused across programs. Mainly, some pattern of similarity based on institutional and organizational pressures (e.g., other schools may mimic elites' treatment of ethics) is expected.

FEDERAL POLICIES, UNIVERSITIES, AND THE INSTITUTIONALIZATION OF SCIENCE

To assess how the 2000 ethics certification policy has affected universities, similar cases can be considered. A related U.S. policy change occurred when federal granting agencies required universities to design a systematic way to deal with research misconduct (e.g., publishing fraudulent results) during the 1980s. After the NIH announced in 1987 that grantee universities would have to describe their procedures for investigating misconduct, congressional hearings were held in 1988–89. Chubin and Hackett (1990:134–135) describe the proposals from these hearings as ranging from the "drastic" measure of establishing an independent agency to evaluate research quality to the "ineffectual" requirement of establishing research ethics courses in graduate education. Presumably, graduate curriculum changes would be ineffective because these ethics courses are usually not taken seriously, and lessons learned seem to have little effect on students' later careers.

The outcome of congressional concern about research misconduct was the establishment of the Office of Research Integrity (ORI) within the Office of the Secretary of Health and Human Services by 1995. ORI is not an independent agency per se but is what Guston (2000) calls a "boundary organization," straddling the divide between organized politics and institutionalized science. ORI has given universities a mandated definition of misconduct and accountability for scientists. Ethics training policy is more diffuse and decentralized, and seems to be less salient to Congress, perhaps because the issues (e.g., stem cell cloning, genetic discrimination) present more difficulty in labeling an individual "bad guy" than high-profile misconduct cases of the 1980s did. As such, the ethics certification policy may be more comparable to the research integrity issue prior to its centralization in ORI. In this earlier period before the research integrity policy was well defined, scientists continued to take the long-established objectivity ethos for granted.

Despite the establishment of ORI, the popular press still tends to portray science as self-governing, including in ethics education. A *New York Times* article (Kolata 2003) noted that the government does not provide data on the number of research ethics courses offered to biological scientists. Yet based on observation of one in-depth, required, full-credit course in research ethics, the *Times* journalist opined that "more and more, universities are instituting *real* courses . . . and requiring students to take them" (emphasis mine). Is this a fair summary of graduate science curriculum, or is it overly optimistic?

Institutions have inertia not just within the boundaries of relevant organizations, but also within the expectations of broader publics. The objectivity ethos of science is not limited to scientists but becomes the rhetoric for justifying an elite specialist rather than a democratic role for science in society. Chubin and Hackett (1990:4) question this basis of legitimacy for science: "[W]e usually delegate to experts the authority for making decisions in areas we do not understand or have not been trained to know. We trust the experts to bear our best interests in mind. We hope that if our trust is misplaced the expert's own profession will take swift and decisive corrective action on our behalf. But is this an appropriate relationship between science and a democratic society?"

Data and Methods

How does someone become certifiably trained in research ethics? One way is for principal investigators (PIs) to visit the Web site accessible through the NIH homepage (www.nih.gov). This method reaches leaders of research teams, but not the graduate students who are working in the lab supported by the grant. Closer investigation of the certification Web site reveals that passing the course is accomplished by clicking through dozens of Web pages on codes of conduct and case studies and by answering some brief multiple choice questions along the way.

Indeed, the most common mode for teaching research ethics is the use of case-based lessons, including the curriculum PIs encounter on the Web. The following is one example from the National Cancer Institute site (http://cme.cancer.gov/c01/g05_01.htm, last accessed November 11, 2004): "Investigators wish to study an association between a particular gene and shyness. They have designed a study as follows: Research is to be performed on a cohort of second-grade children at a large public school. The children will be observed in the classroom by the research team and samples of their saliva will be collected for genetic analysis." The student receiving this Internet education is prompted to respond to a question about the risks of this case. The following four issues are provided as answers: "The children's privacy may be violated when their classroom behavior is observed; children who are identified as 'shy' may be stigmatized; [there may be] state laws governing genetic testing or use or disclosure of genetic information; and, uncertainty about what will be done with genetic samples at the conclusion of the study." Thus, the ethical and social problems with looking for a gene for shyness are narrowly identified as issues of maintaining confidentiality, following legal guidelines, and avoiding individual harm. The site does not, however, tell the reader exactly how to protect subjects from harm. Also, the broader problem of recognizing how shyness is socially constructed in given situations and cultures (i.e., behavior considered "shy" in one context might just be good manners in another), and how this study would thus be a misdirection of scientific resources is not raised. In fact it *cannot* be raised by the Internet student in this online course.

Overall, the test-based Internet certification seems a rather superficial way to ensure the diffusion of knowledge about protecting human subjects and research ethics more generally. Scientists seem to agree: typically, when asked about this training requirement they roll their eyes and shake their heads to signal disdain, as if it is not even worth speaking about. When prompted by further questions, the verbal responses I heard also ridiculed it (e.g., "silly," "laughable," "a joke").

Although the Internet is not the best place to conduct training on complex topics like research ethics, it is a good place to search for information on the training available in graduate programs. To assess the extent of ethics training in life science graduate programs, I conducted a content analysis of course information available on the Web sites of U.S. universities in 2001 and replicated the analysis with the same programs in 2003. While the content analysis was designed to be a qualitative look at a manageable number of graduate programs and their requirements as a whole, part of the sample was selected randomly to provide a more representative picture. The sampling frame was provided by the list of all American universities with life science PhD programs ranked by the National Research Council (1995). I selected a stratified random sample of about 25 percent of the schools (N=50). The sample was stratified by the prestige of the university's ranking in biochemistry and molecular biology. I coded all of the top ten programs because these are the most visible. If mimetic isomorphism is in progress, these are the schools that would be the source of copying. Also, the elite ten confer nearly a third of life science PhDs (see Smith-Doerr 2004 for further discussion of the graduate program categories). I randomly selected twenty schools in each of the other two categories: schools ranked 11–50 and schools ranked 51+. Life science is an evolving area of knowledge, and the lack of standard departmental/program names reflects the interdisciplinary nature of the enterprise. Based on conversations with knowledgeable interlocutors, I chose to focus on molecular biology as the most central to life science. When searching a university selected from the NRC list, I searched for the department most closely representing the study of molecular biology.

One caveat of studying Web site content is reliance on a particular type of information. Perhaps some schools have more resources for Web design than others. In 2001 and 2003, every school sampled had a professional-looking Web site for a biochemistry-molecular biology–related program. There were some differences in the extent of information provided about the programs on the Web, and this was taken into account in the analysis. I used a rather conservative measure of whether a program had ethics training. For example, Harvard was coded as having an ethics elective because one course, "Stem Cells and Cloning," was described as: "An advanced course in developmental biology. Embryonic and adult stem cells in different organisms will be examined in terms of their molecular, cellular and potential therapeutic properties. . . . Current findings will be considered in a historical context; *ethical and political considerations will not be ignored*" (www.mcb.harvard.edu/Education/Graduate/courses.html, accessed September 30, 2003; emphasis mine).

The real question is whether a Web site does an adequate job of reflecting the priorities of an educational organization. Certainly online data are not available on the informal organization of graduate programs;[4] such data can only be gathered through qualitative observation. What a Web site does provide is information on the public face of an organization. It may be the case that ethical issues are taught to PhD students informally although they are not mentioned on a program's Web site. Nonetheless, a program's Web site provides an interesting message about "what's important to us." The publicity given to ethics training for scientists is information about organizational goals and the framing of proper roles for life scientists. Qualitative content analysis also reveals whether including some ethics training is mostly "window dressing" to appear legitimate to funding agencies, university administration, or other audiences.

Discussion of Results

One year after the 2000 NIH policy, relatively few courses appeared in the ethical and social implications of research in biochemistry and molecular biology graduate programs. Two-thirds of the fifty universities sampled in 2001 did not list any required or elective ethics courses in their graduate program information online. By 2003, there was some improvement. The programs not offering any visible ethics training decreased to two-fifths of the total.

Table 1 summarizes the data on how the prestige of programs is related to the requirement of formal ethics training, regardless of how extensive the credit is for courses. In 2001, programs at research universities ranked in the top fifty were more likely to have requirements, while lower-ranked schools had proportionately more elective courses (see Figure 1). By 2003, more of the lower-ranked schools had required courses, and more of the top-ranked schools had elective courses that included some aspect of social or ethical aspects of research (see Figure 2). Nearly all schools listed courses on their Web sites. In 2001, lower-ranked

Table 1. Number (and Percentage) of Life Science Ph.D. Programs with Required Ethics Training by NRC Ranking, 2001 and 2003

NRC Ranking	N	Ethics Required		Ethics Electives Offered		List Courses on Web Site	
		2001	2003	2001	2003	2001	2003
1–10	10	3 (30%)	5 (50%)	0	4 (40%)	9 (90%)	10 (100%)
11–50	20	7 (35%)	9 (45%)	2 (10%)	3 (15%)	20 (100%)	19 (95%)
51–200	20	1 (5%)	6 (30%)	4 (20%)	4 (20%)	14 (70%)	19 (95%)
Total	**50**	**11 (22%)**	**20 (40%)**	**6 (12%)**	**11 (22%)**	**43 (86%)**	**48 (96%)**

Source: data collected and analyzed by the author.

schools were less likely to have their programs fully articulated online but had caught up to other programs by 2003.

One possible explanation of these findings is that in the earlier observation, there was simply not enough information about graduate programs available on Web sites, especially for the less highly ranked departments. From the last column of Table 1, we see that 88 percent of the programs listed their courses on the web in 2001, but some did not. In particular, schools like Kansas State University (KSU) and the Medical College of Georgia (MCG) are not highly ranked by the NRC and also did not provide specific information on course numbers, titles, and descriptions. Both did describe course requirements generally, though. At KSU, students take a "core (two-semester) biochemistry course." The core courses at MCG include topics such as "molecular cloning, gene isolation and analysis, gene therapy, transgenic animals, cell culture and transfection, patch clamp technology." Yet in these specific descriptions of the programs, research ethics issues are not mentioned. Thus, although these two Web sites are less explicit about numbered course requirements, the assumption that ethics is not part of the curriculum is based on a holistic analysis of all available information. In 2003, both of these schools did list courses and requirements clearly (and at that time, MCG had a required ethics course, and KSU had an elective one).

Even if some programs have ethics requirements that do not appear on their Web sites, the fact that they are not showcasing their ethics training is illustrative. Keep in mind that the primary audience for Web sites describing graduate programs is prospective students. Are schools

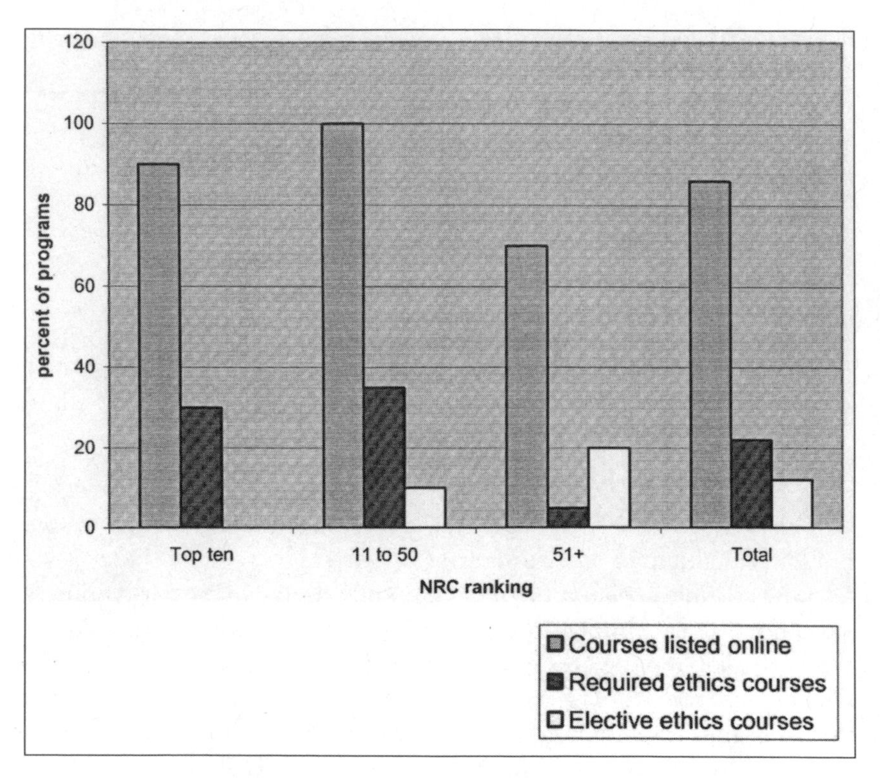

Figure 1. Percentage of Ph.D. programs with ethics courses, by NRC ranking, 2001. Data collected and analyzed by author.

assuming that biologists with a bachelor's degree would recoil from discussion of ethics in graduate training? If so, this preference probably comes from their socialization as life science majors in college (perhaps with little ethics training at that level as well).

Recent trends that arose in the 2003 content analysis indicate that the newly offered courses may be pro forma attempts to appease the NIH rather than change the professional socialization of graduate students. Four programs now explicitly state that their courses meet the NIH requirement, making clear that meeting the requirement, rather than learning about ethics, is the key goal. For example, the University of Southern California gives a 1 credit course in "Ethics and Accountability in Biomedical Research," which is described as follows: "This course is designed as an option for meeting current federal regulations which require that all predoctoral and postdoctoral fellows paid from

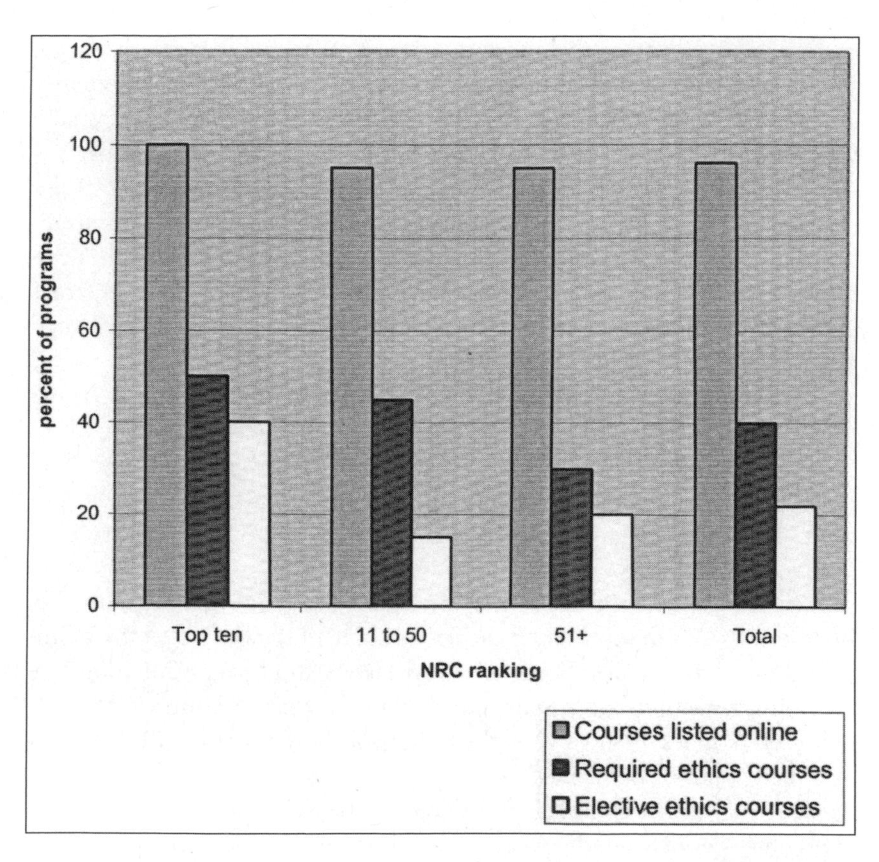

Figure 2. Percentage of Ph.D. programs with ethics courses, by NRC ranking, 2003. Data collected and analyzed by author.

federal contracts and grants have a component of ethical training." These legalistic course descriptions signal that the programs are taking care to be seen as rule followers, familiar with the letter of the law. Another trend new to the 2003 data is that some programs are following the NIH's lead and offering these courses online, rather than spending regular class time and faculty hours covering ethical and social issues. Three universities offer online options. The University of Massachusetts at Lowell, for example, offers "Bioethics" as a one-credit course. Students can take the regular course, which features guest lecturers from industry, or take the online version. Either option seems problematic.

Another way to gauge how seriously programs take the mandate to educate students in the larger social issues surrounding their research is how much credit they give for courses that meet the NIH standard. Of the twenty schools requiring ethics training in 2003, only three clearly give full credit for these courses (this includes Loma Linda University, which requires a course in "Religion" in its Seventh Day Adventist tradition). Two schools do not specify the credits for the research ethics course. The remainder give no credit or a fraction of regular credit (mostly one credit). In other words, rather than the *New York Times'* estimate of "more and more" universities requiring "real" research ethics courses (Kolata 2003), a mere three out of fifty visibly demonstrate such a commitment.

The content analysis data show some initial evidence supporting the idea that life science graduate programs construct a rational myth of ethics training rather than responding to the funding incentives of NIH policy with real change. Institutional isomorphism also seems to be taking place, as there is relative uniformity in the kind of ethics training across organizations.

In interviews with academic life scientists, there was variation in how seriously individuals took the importance of the broader ethical implications of their work. Some are very serious and thoughtful about the issues but view their interest as something to be pursued outside of NIH-enforced policies. Others view the policies as bothersome and the issues as outside of their purview.

A thoughtful, self-described biologist who was a postdoctoral fellow at an elite school argued: "Even in the absence (or ridiculed presence) of ethics training per se, the conversations are still going on. I must admit that the prevalent attitude is that any new [scientific] knowledge is good knowledge. Even so, I have noticed that more and more, ethical issues are being discussed in science journals, over coffee, in classrooms."

Her point was that ethics training was not the place where the real thinking about the broader issues goes on, so whether or not it is offered is of little relevance in the day-to-day life of the average student. The educational goal of the NIH policy seems to miss the intended target, even among those students who do keep the larger issues in mind. One graduate director at a research university saw the policies as bothersome and opined that most graduate students also viewed attempts at ethics training as a nuisance: "Most of them see it as a joke, frankly." When asked if he thought that attitude reflected faculty sentiments as well, the director smiled sheepishly and admitted, "Yeah, faculty are no better." The way scientists treat the ethics requirement does not look like the way other organizational actors interpret policies to fit within existing institutions, using new policies for their own organizational or political agendas (Edelman et al. 2001; Clemens 1997). Instead, life scientists seem to focus on how to get by the rules from the state—either in order to take ethics more seriously or to avoid "extra-scientific" demands.

The attitudes toward graduate ethics training that I encountered were similar to the responses Chubin and Hackett (1990) found in their interviews with scientists about the problems of peer review: denial that there was any problem with the status quo; belief that only scientists can govern science; and when acknowledging that scientists may not always play by the rules, still remaining confident that scientists recognize the best work and thus require no outside input. They quote an academic administrator on the requirement of establishing misconduct hearing procedures: "Like so many [universities], we have been tardy in setting up official mechanisms for handling this problem because of what might be called a form of psychic numbing. It never occurred to us that it could happen to us" (133). When the taken-for-granted legitimacy of science is called into question, the institutional features and inertia became more visible.

Another graduate director I interviewed who was also the dean of his college discussed his experience with inertia in setting up a brand new graduate program in the life sciences. He personally felt it was important to include formalized ethics training from the inception of the program, but he faced conflict from faculty opposed to any "non-science" course. In recalling how his program initiated its course in research ethics, he said: "In fact, all the faculty here did not immediately buy into it [the ethics course]. I was strongly in favor of requiring a course in ethics. Some faculty were saying, 'Why should we sustain something that's not about the methods or content of science?' But we're trying to have our students become leaders here and go out and do important things, and they need to be aware of the implications of what they're going to do." Once he had convinced his colleagues of the necessity of the course, whether it was worth the same as a "real" course was the next issue: "Some faculty only wanted to make it a two credit course. You know, not quite up to the regular course—only two credits." He used his influence as the founder of the program (and dean of his college at the time) to make sure the ethics requirement counted as a "real" course: "So it's there as a full fledged course, the same amount of credits as anything else." This graduate director used moral suasion and his prestige as a nationally known scholar to convince his colleagues to take training students in the ethical and social implications of science seriously. This departmental conflict occurred before the 2000 NIH policy was in place, or the policy might also have been used as an argument in his favor. This may be a major benefit of the federal grant policies—they can be used as tools for those who would bargain in their departments for a broader graduate education.

I also spoke with an "activist scientist" (on this role see Moore 1996; Frickel 2004) who had successfully implemented an interdisciplinary seminar to explore the social and ethical aspects of science at his university. He indicated that this initiative was "outside of and despite" federal policies. To accomplish bringing together this lively campus group, he used his experience in social movement organizing to identify key faculty and students around campus who had a strong interest in the social and ethical implications of recent developments in the biological sciences. This kind of local-level, grassroots organizing of scientists, together with interested others, may be a more effective way to spark the interest (and education) of neophyte life scientists in the complexities of the social and political issues surrounding their work. Whether this kind of movement could be sustained at the national level would likely depend on the broader political opportunity structure (McAdam 1982).

Why is the attitude of life scientists toward the ethics education policy fundamentally one of minimal compliance? The unique combination of new institutional and STS perspectives taken in this study can help answer this question. A new institutional perspective on the content analyses is that graduate programs interpret the federal policies such that while ethics courses are formally offered, departments have isomorphically adopted a model of giving no or low credit (and legitimacy) to them. In graduate programs we see a decoupling of the formal policy from the partial implementation of ethics courses on the margins of curricula. Conversations with scientists show deflection of the federal requirements, in some cases in order to enact more local solutions for reflecting on the place of life science in society. An STS perspective on the interviews reveals that scientists, rather than uniformly treating the social aspects of science as unimportant, can create grassroots strategies to provide legitimate discourse about ethical issues.

Conclusions

Through the way ethics training is being integrated into universities, life science graduate students may be learning an unintended lesson: how to deflect policies that come from the state in a top-down fashion. The adoption of research ethics courses may be increasing, yet there is huge variation in how scientists learn to reflect on the social-ethical issues in their work. The active, continual contestation of what is within the purview of science also defines what is *not* science (Gieryn 1999). Both the institutional arrangements of graduate programs and the discourse of many individual scientists define ethical issues as outside of science. And there is no universal way to change this professional culture. A narrowly rational assumption in national-level policy does not work. In this case, monetary incentives have only led to partial, incomplete, and widely varying adoption of ethics training.

This study may have implications for other research that reveals unexpected contradictions in governmental science policy. Consider the puzzle raised by two chapters in this volume that seem to point in conflicting directions. On the one hand, Steven Epstein describes how social movement activists are able to help shape "biopolitical paradigms," such as how governmental science agencies define minority groups. On the other hand, Jenny Reardon illustrates the inability of groups of indigenous peoples to enter dialogue with scientists in federal agencies during the process of defining racial categories. This seeming contradiction in the U.S. science policy to involve citizen groups in definition processes could perhaps be further illuminated by an institutional view of the decoupling between policy formation and implementation. Large, bureaucratic, loosely coupled systems like the NIH allow ample room for differential application and interpretation of procedures. Activists may encounter this organizational space differently; middle-class professionals may find it a resource for maneuvering while members of indigenous rights or organizations find it a place where informal discrimination is hidden behind a formal facade. Rationalized policy systems are interpreted, used, and sometimes rejected in unexpected ways. Policymakers may have a machine model in mind, where cost-benefit ratios are welded to desired behavioral outcomes. In reality, scientists (and other relevant publics whom institutions permit to play) treat policy more like Lego bricks to take apart, stack, or cast aside.[5]

In this study, pilot interviews with scientists provide suggestive data rather than definitive answers for how informal, local organization of discourse on science in society issues might occur. Further research is needed on how scientific actors use governmental and academic organizational contexts to deflect state policies. Recent scholarship has focused on increasing democratic participation in science policymaking

(Kleinman 2000; Lengwiler 2004; Moore this volume). The institutionalization of activities such as the credentialing of scientists' ethics training, however, may be one way that democratization of science is deflected. Once a credentialing system and the attendant organizational routines are in place, public discourse is unlikely to enter into science ethics education, just as governance of equal opportunity (Edelman et al. 2001; Dobbin et al. 1993) and science misconduct (Chubin and Hackett 1990) seem to be sealed off from public discourse. Again, combining STS analysis of scientists' boundary-defining negotiations along with neo-institutional examination of constraints and opportunities presented by organizational and state contexts could be fruitful.[6]

A substantive factor that may add to scientists' ability to deflect educational requirements, and for university programs to decouple the requirements from their actual activities, is that the policies concern bioethics. The policy context of bioethics in the United States, as John Evans (2002) charts, is one in which an earlier, deeper debate thinned into a rationalized formula. Part of this thinning can be attributed to the role played by Congress in calling for simple ethical principles that could be enforced by federal funding agencies. In any case, bioethics is now an institutionalized niche and has been criticized for its narrow and shallow treatment of a set of standardized ethics issues such as informed consent (Evans 2002; Corrigan 2003).

The problem with a critical perspective on science policy, however, is that it can leave a vacuum in terms of recommending what should be done. I believe there are at least some hints in this study toward productive directions for educating life scientists about the context of their work. In the United States, it seems that less top-down (i.e., more local) methods for discussing science in society are better received. Practically, this means that universities would do well to provide the resources to allow for creative, grassroots solutions to the question of how best to educate graduate students. The NIH policy is actually broad enough to permit this creativity; it has just become institutionalized practice to employ narrow bioethics case studies on the Internet for certification. (A factor contributing to this symbolic response to the policy may be that the enforcement is a letter signed by the principal investigator promising that all key personnel on the project have been properly trained in the ethical treatment of subjects.) As one possibility of a more creative approach to teaching the social and ethical issues in science, consider an example. Instead of distributing a curriculum of standard cases with the "right" ethical solutions, imagine a seminar in which PhD students design a course for undergraduate science majors. Collectively, graduate students could collate a reading list and discuss how to select pedagogical themes that emerge from the literature on science, ethics, and society. Through teaching, they would learn. Unfortunately, as this chapter has shown, there are no easy, straightforward, universal policy solutions.

Encouraging scientists to engage in reflection is a goal that will require creative local efforts but will be eminently worthwhile if the result is enlightened scientific agendas and practices. A differently trained life science community might produce a different message in a Smithsonian exhibit on genetic discoveries, for example. Instead of giving the message "you are your genes" sponsored by Pfizer, an exhibit might ask up front, "what do you think?" Such exhibits already work well in asking museum attendees what they think about theories of dinosaur living conditions and morphology. Asking museum patrons what should be done with genetic information could lead to provocative discussions about genetic discrimination, public funding, and commercial outcomes of science. Surely, for whomever engages in thinking about the thorny ethical issues in the life sciences, reflection will arise from asking hard questions. It is unlikely, however, to originate in highly rationalized institutional systems.

Acknowledgments

A version of this chapter was presented at the 2003 annual meetings of the Society for Social Study of Science, Atlanta, GA. I would like to thank the editors, Kelly Moore and Scott Frickel, for their encouragement to pursue this project and for insightful comments on the paper. Kelly Moore went above and beyond the call of duty in providing detailed suggestions for revisions. I am also grateful for the helpful comments I received from Maren Klawiter, Dave Guston, Daniel Kleinman, and Ed Hackett and Elisabeth Clemens, and for research assistance from John Underwood.

Endnotes

1. The molecular biology/biochemistry area of study was selected because of its breadth and centrality in the life sciences (Smith-Doerr 2004) and its relevance for human therapeutic applications and the biotechnology industry, where an understanding of the social/ethical context of research is particularly important.

2. To be sure, Merton (1973) also argued that science conducted in for-profit organizations could not fit the ideal ethos. In particular, the norm of "communism"—shared knowledge—is violated by proprietary, patented science. But as many have noted before, Merton's formulation is more a theoretical discussion of what science should be than an empirical assessment of disinterestedness and other norms in action.

3. Perhaps, then, the juxtaposition of incentives and rational myths is also a false dichotomy. One might say that federal policies provide the institutional material for both incentives and rational myths. However, it is still important to recognize the distinction between a simplistic cost/benefit reaction to an incentive and the construction of a myth of rational rule following while behavior remains inert.

4. But interestingly enough, Harvard publishes entire interviews with students and postdocs about the program and life in the lab. These interviews include discussion of informal norms such as whether students hang out with people from their labs, or if faculty believe jobs in industry are as good as academic jobs.

5. Indeed, similar processes have been called "bricolage" (Levi-Strauss 1966).

6. For example, comparing the U.S. case to nations in the European Union (in terms of how scientists are educated to think about the social and ethical contexts of their work) might reveal some interesting differences in both the institutionalization of science policy and local meanings of science curriculum.

References

Casper, Monica J. 1998. *The Making of the Unborn Patient: A Social Anatomy of Fetal Surgery*. New Brunswick, NJ: Rutgers University Press.

Chubin, Daryl E., and Edward J. Hackett. 1990. *Peerless Science: Peer Review and U.S. Science Policy*. Albany: State University of New York Press.

Clemens, Elisabeth S. 1997. *The People's Lobby: Organizational Innovation and the Rise of Interest Group Politics in the United States, 1890–1925*. Chicago: University of Chicago Press.

Cook-Deegan, Robert. 1994. *The Gene Wars: Science, Politics, and the Human Genome*. New York: W. W. Norton.

Corrigan, Oonagh. 2003. "Empty Ethics: The Problem with Informed Consent." *Sociology of Health & Illness* 25:768–792.

DiMaggio, Paul J., and Walter W. Powell. 1983. "The Iron Cage Revisited: Institutional Isomorphism and Collective Rationality in Organizational Fields." *American Sociological Review* 48:147–160.

Dobbin, Frank, John R. Sutton, John W. Meyer, and W. Richard Scott. 1993.

"Equal Opportunity Law and the Construction of Internal Labor Markets." *American Journal of Sociology* 99:396–427.

Drori, Gili S., John W. Meyer, Francisco O. Ramirez, and Evan Schofer. 2003. *Science in the Modern World Polity: Institutionalization and Globalization.* Stanford, CA: Stanford University Press.

Edelman, Lauren. 1992. "Legal Ambiguity and Symbolic Structures: Organizational Mediation of Civil Rights Law." *American Journal of Sociology* 97: 1531–1576.

Edelman, Lauren, B., Sally Riggs Fuller, and Iona Mara-Drita. 2001. "Diversity Rhetoric and the Managerialization of Law." *American Journal of Sociology* 106:1589–1641.

Evans, John H. 2002. *Playing God? Human Genetic Engineering and the Rationalization of the Public Bioethical Debate.* Chicago: University of Chicago Press.

Frickel, Scott. 2004. *Chemical Consequences: Environmental Mutagens, Scientist Activism, and the Rise of Genetic Toxicology.* Rutgers, NJ: Rutgers University Press.

Giddens, Anthony. 1984. The Constitution of Society: Outline of the Theory of Structuration. Cambridge: Polity Press.

Gieryn, Thomas F. 1999. *Cultural Boundaries of Science: Credibility on the Line.* Chicago: University of Chicago Press.

———. 2004. "Tom Gieryn's Rant and Rave on *Science in the Modern World Polity.*" Handout summarizing presentation during "Authors Meet Critics" session at the American Sociological Association annual meetings, San Francisco, August.

Guston, David H. 2000. *Between Politics and Science: Assuring the Integrity and Productivity of Research.* Cambridge: Cambridge University Press.

Hilgartner, Stephen. 2000. *Science on Stage: Expert Advice as Public Drama.* Stanford, CA: Stanford University Press.

Jasonoff, Sheila. 1995. *The Fifth Branch: Science Advisors as Policymakers.* Cambridge, MA: Harvard University Press.

Kelly, Erin, and Frank Dobbin. 1999. "Civil Rights Law at Work: Sex Discrimination and the Rise of Maternity Leave Policies." *American Journal of Sociology* 105:455–492.

Kerr, Anne. 2003. "Governing Genetics: Reifying Choice and Progress." *New Genetics and Society* 22:141–156.

Kevles, Daniel J., and Leroy Hood, eds. 1992. *The Code of Codes: Scientific and Social Issues in the Human Genome Project.* Cambridge, MA: Harvard University Press.

Klein, Hans, and Daniel Lee Kleinman. 2002. "The Social Construction of Technology: Structural Considerations." *Science, Technology, & Human Values* 27:28–52.

Kleinman, Daniel Lee. 1995. *Politics on the Endless Frontier: Postwar Research Policy in the United States.* Durham, NC: Duke University Press.

——, ed. 2000. *Science, Technology, and Democracy*. Albany: State University of New York.

Kolata, Gina. 2003. "Ethics 101: A Course about the Pitfalls." *New York Times*, October 21.

Lengwiler, Martin. 2004. "Shifting Boundaries between Science and Politics: New Insights into the Participatory Question in Science Studies." *Techno-science* 20(3): 2–5.

Levi-Strauss, Claude. 1966. *The Savage Mind*. Chicago: University of Chicago Press.

McAdam, Doug. 1982. *Political Process and the Development of Black Insurgency, 1930–1970*. Chicago: University of Chicago Press.

Merton, Robert K. 1973. "The Normative Structure of Science." In *The Sociology of Science: Theoretical and Empirical Investigations*, ed. N. W. Storer, pp. 267–278. Chicago: University of Chicago Press.

Meyer, John W. and Brian Rowan. 1977. "Institutionalized Organizations: Formal Structure as Myth and Ceremony." *American Journal of Sociology* 83: 340–363.

Moore, Kelly. 1996. "Organizing Integrity: American Science and the Creation of Public Interest Organizations, 1955–1975." *American Journal of Sociology* 101:1592–1627.

National Institutes of Health (NIH). Office of the Director. 2000. "Required Education in the Protection of Human Research Participants." OD-00-39, June 5.

——. Office of Extramural Research. 2002. "Frequently Asked Questions for the Requirement for Education on the Protection of Human Subjects." http://grants1.nih.gov/grants/policy/hs_educ_faq.htm (accessed September 24, 2004).

National Research Council (NRC). 1995. *Research-Doctorate Programs in the United States: Continuity and Change*. Washington DC: National Academy Press.

——. 2000. *Addressing the Nation's Changing Needs for Biomedical and Behavioral Scientists*. Washington, DC: National Academy Press.

Nelkin, Dorothy, and M. Susan Lindee. 1995. *The DNA Mystique: The Gene as a Cultural Icon*. New York: W. H. Freeman.

Rabinow, Paul. 1999. *French DNA: Trouble in Purgatory*. Chicago: University of Chicago Press.

Smith-Doerr, Laurel. 2004. *Women's Work: Gender Equality vs. Hierarchy in the Life Sciences*. Boulder, CO: Lynne Rienner.

Stinchcombe, Arthur L. 1965. "Social Structure and Organizations." In *Handbook of Organizations*, ed. J. G. March, pp. 142–193. New York: Rand McNally.

Vaughan, Diane. 1996. *The Challenger Launch Decision: Risky Technology, Culture, and Deviance at NASA*. Chicago: University of Chicago Press.

Watson, James D. 1990. "The Human Genome Project: Past, Present, and Future." *Science* 248:44–49.
Winner, Langdon. 1986. *The Whale and the Reactor: A Search for Limits in an Age of High Technology*. Chicago: University of Chicago Press.

16

Regulatory Shifts, Pharmaceutical Scripts, and the New Consumption Junction

Configuring High-Risk Women in an Era of Chemoprevention

MAREN KLAWITER

On April 6, 1998, the directors of the National Institutes of Health, the National Cancer Institute, and the National Surgical Adjuvant Breast Cancer and Bowel Project (NSABP) called a press conference to announce that they were terminating the Breast Cancer Prevention Trial (NSABP) fourteen months ahead of schedule because interim results showed a dramatic 45 percent reduction in the incidence of invasive breast cancer among the group of healthy, high-risk women who had been given tamoxifen compared to the women on placebo.[1] The BCPT was a randomized, controlled, double-blind, Phase III clinical trial designed to assess the safety and effectiveness of tamoxifen in reducing the incidence of invasive breast cancer among healthy women at increased risk of developing invasive breast cancer. Launched in 1992, the BCPT was funded with $68 million from the National Institutes of Health and enrolled over 13,000 women at 270 centers in the United States and Canada. According to Richard Klausner, the director of the National Cancer Institute, the results exceeded their expectations by so far that the data monitoring committee felt ethically obligated to stop the trial and unblind it so that the women on placebo could have access to tamoxifen.[2] With the analysis of additional data, the results were soon changed to a 49 percent reduction in the incidence of invasive breast cancer (see Fisher et al. 1998). Although tamoxifen had been used since the 1970s to treat women with breast cancer and was currently the best-selling breast cancer treatment drug in the world, this was the first time in the United States that an anti-cancer drug had been tested on healthy individuals. Thus, the results were not only exciting; they were historically unprecedented.

The press conference garnered extensive media attention. The following morning, the front-page headline of the *New York Times* read: "Researchers Find the First Drug Known to Prevent Breast Cancer" (Altman 1998). In the Midwest, the *Detroit News* quoted "scientific experts" who claimed: "The news that a simple pill can prevent breast cancer . . . may be the most dramatic victory against cancer since America declared war on the disease a quarter-century ago" (Poe 1999:658). On the West Coast, the *San Francisco Examiner* scooped the press conference and published an Associated Press article the day before under the headline "Breast Cancer Drug Reportedly Can Prevent Disease." The article read: "For the first time, a drug has been shown to prevent breast cancer among women at high risk for the disease, jubilant Federal health officials said here today" (Associated Press 1998). Unfortunately, tamoxifen also doubled the risk of several rare but serious side effects, including uterine cancer, blood clots, and stroke. In fact, there was no statistically significant difference in the mortality rate of women taking tamoxifen compared to the women on the placebo.

Despite these unfortunate side effects, the results of the BCPT were promoted with unbridled enthusiasm by past and present NCI officials, executive officers of the American Cancer Society, and medical researchers involved in the clinical trial. Bernard Fisher, the principal investigator of the trial and a giant in the world of breast cancer research, called the announcement "probably the most emotional of my career. . . . This is the first time in history that we have evidence that breast cancer can not only be treated but can be prevented" (quoted in Smigel 1998:648). Richard Klausner, director of the NCI, asserted: "The results are remarkable. They tell us that breast cancer can be prevented. This study is not an end. It is rather a propitious beginning" (Klausner 1998). Harmon Eyre, M.D., executive vice president of research and medical affairs for the American Cancer Society, asserted: "The magnitude of these results was greater than anyone dreamed of"

(American Cancer Society 1998:3). Several months later, reflecting on the impact of the announcement, Bernardine Healy, former director of the National Institutes of Health and editor-in-chief of the *Journal of the National Cancer Institute*, wrote: "The results were lauded around the world. . . . After so much heat, almost any light shining from this study would have meaning. But this is not just any light; the abundant blessing of this trial is the brilliant new knowledge that *breast cancer can be prevented*" (Healy 1998:280). As part of its consumer and media-friendly approach to the historic event, the NCI posted the results of the Breast Cancer Prevention Trial to a Web site dedicated for this purpose along with accessible summaries, explanations, and visual aids (charts, graphs, pictorials), which could be downloaded and used for presentations.

Three weeks after the trial's conclusion Zeneca Pharmaceuticals, the manufacturers of tamoxifen under the brand name of Nolvadex, filed a Supplemental New Drug Application (SNDA) with the FDA and was granted a priority review. On September 2 the Oncologic Drugs Advisory Committee (ODAC) met to consider Zeneca's application.[3] During the intervening months, interim results from two randomized, controlled, double-blind placebo trials conducted in Italy and the United Kingdom had been published in the *Lancet*. Both trials failed to confirm the findings of the BCPT: no tamoxifen-induced reduction in the incidence of invasive breast cancer among healthy women was demonstrated in either study (Powles et al. 1998; Veronesi, Maisonneuve, et al. 1998). In the FDA's presentation to ODAC, Susan Honig, who presented the FDA's case, argued that "Although the European trials reported negative results, we believe that there were design differences that resulted in lower risk populations being entered into those studies. Overall, the size, the statistical power and the internal consistency of P-1 [the Breast Cancer Prevention Trial], we think, make its results robust and believable, and we also believe that the results are consistent with all of the other published reports of the ability of tamoxifen in the realm of prevention, most notably preventing contralateral breast cancer. So, we would feel that the weight of the evidence favors what was observed in P-1" (Oncologic Drugs Advisory Committee 1998).

After listening to the testimony of several breast cancer activists and advocacy organizations, considering the presentations of the drug's sponsor (now AstraZeneca), the FDA, and the NSABP, and debating the evidence, the ODAC recommended to the FDA that tamoxifen be approved "for the short-term reduction of the risk of breast cancer among women at increased risk." After negotiating with AstraZeneca, who wanted Nolvadex to be approved for the "prevention" of breast cancer, not "the reduction of risk among women at increased risk," the FDA struck a compromise and, on October 28, 1998, a short six months after Zeneca filed its SNDA, the FDA approved tamoxifen "to reduce the incidence of breast cancer in women at high risk for breast cancer." According to NCI calculations, approximately 29 million women in the United States were "high risk" as defined by the BCPT and the FDA.

The conditions for the development of a new market of healthy women seemed auspicious. Surveys consistently showed that women regularly overestimated their risk of being diagnosed with breast cancer and feared this disease above all others (Phillips, Glendon, and Knight 1999). In addition, women in the U.S. had been instructed for almost three decades in the importance of taking precautions—in the form of breast self-exams, clinical exams, and mammographic screening—to minimize the seriousness of this disease and arrest its growth before it became a threat (Aronowitz 2001; Reagan 1997; Lerner 2000; Leopold 1999; Gardner 1999). Furthermore, to a great extent, breast cancer had already been reconceptualized as a disease continuum (Klawiter 2002). Third, tamoxifen was already a well-known and widely prescribed drug with a twenty-year history of use in the United States. And finally, this new technology was launched in the wake of regulatory changes that relaxed prohibitions against advertising prescription drugs directly to consumers (see especially Palumbo 2002, Wilkes, Bell and Kravitz 2000).

In 1999 I began the research on which this chapter is based. I expected it to be a study of the rapid development of a new market for tamoxifen comprised of healthy, high-risk women. But the new market never materialized, at least not in the way that I expected. Prescriptions for this new indication were certainly written, but the anticipated flood of women streaming into doctors' offices never materialized. In 1997 Zeneca's worldwide Nolvadex sales were $502 million (Zeneca 1997). In 1998, they rose slightly to $526 million (a growth of 6 percent); in 1999 Nolvadex sales grew to $573 million (a growth of 7 percent); and in 2000 worldwide sales of Nolvadex were $576 million (a growth of 1 percent) (AstraZeneca 1998, 1999, 2000). Exact figures for the number of new prescriptions written for healthy, high-risk women are not available, but they probably constitute only a small percentage of the overall increase in sales (which had been steadily growing since the 1970s).

In 2001 worldwide sales of Nolvadex jumped to $630 million—a 12 percent increase overall, and an 18 percent increase in U.S. sales (AstraZeneca 2001). But this increase in sales was mostly likely attributable not to new prescriptions written for healthy, high-risk women, but rather, to the fact that in July of 2000 the FDA approved Nolvadex to reduce the risk of invasive breast cancer in women with ductal carcinoma in situ—a non-invasive form of breast cancer that accounted for nearly 20 percent of all newly diagnosed breast cancer cases (see Fisher et al. 1999). In 2002 sales of Nolvadex decreased by 21 percent, to $480 million, and in 2003 sales declined to $178 million (AstraZeneca 2002, 2003). This dramatic decline in sales reflected the expiration of a distribution agreement with Barr Laboratories in 2002 (freeing Barr to sell a generic version of tamoxifen) and the expiration, in February 2003, of AstraZeneca's U.S. patent for Nolvadex. During the more than four years that transpired between the FDA's approval of Nolvadex for healthy, high-risk women and the expiration of the U.S. patent, however, AstraZeneca was unsuccessful at generating a significant demand for breast cancer chemoprevention among women who qualified as high risk, according to the FDA-accepted definition. In retrospect, the big splash turned out to be a minor belly flop. But why? Why did tamoxifen fail as a technology of primary prevention?

In this chapter I examine how the process of rescripting tamoxifen and configuring new users was shaped by important changes in the drug regulatory regime and the emergence of the breast cancer movement that reorganized the pharmaceutical field of consumption. First, I argue that the liberalization of rules constraining direct-to-consumer advertising of prescription drugs and the streamlining and acceleration of the drug approval process did indeed (as many have argued) enhance the power of the pharmaceutical industry and its access to consumers. At the same time, these regulatory changes also strengthened the voices of patient and consumer organizations and social movements by multiplying the opportunities for them to intervene in the process of market formation. Second, the breast cancer movement emerged during the same period of time that the BCPT was being launched. Second, during the 1990s breast cancer activists and organizations joined the National Women's Health Network in monitoring, testifying against, and publicly-criticizing the design and conduct of the Breast Cancer Prevention Trial. Thus, long before tamoxifen was approved by the FDA for reducing the risk of breast cancer among healthy, high-risk women, it had been subjected to recurring public criticism across multiple arenas. In effect, while Zeneca, and later AstraZeneca, sought to rescript tamoxifen and configure end-users for this new indication, breast cancer and women's health activists sought to "de-script" tamoxifen as a technology of primary prevention and disrupt the configuration of healthy women as high-risk end users.

Where pharmaceutical forms of breast cancer prevention and risk reduction were concerned, women's health and consumer rights organizations actively intervened in and contested the production of knowledge and markets. At every step in the process, the BCPT was debated and translated and shaped and reshaped across multiple sites and arenas, including professional meetings, medical journals, scientific publications, daily newspapers, popular magazines, radio shows, television broadcasts and Web sites. Health activists developed analyses that challenged and complicated the official narrative, and disseminated their analyses through newsletters, press releases, the internet, and other media. They published articles in medical journals (see, for example, Fugh-Berman and Epstein 1992a, 1992b), they testified at three different congressional hearings, and before the ODAC, and they organized an anti-DTC advertising campaign.

They were unsuccessful at stopping the trial and they failed to prevent FDA approval, but they increased the noise and negative attention, they fomented skepticism and dissent, they urged caution, and they gummed up the works. The argument sketched above is not a complete answer to the "why" question, but it is one part of the story that engages themes and issues relevant to the concerns of this volume (see also Hogel 2001; Kalwiter 2002; Wooddell 2004; Fosket 2004). In doing so, however, it necessarily brackets other dimensions of the story. I do not, for example, address the role of physicians (see Klawiter 2001). or discuss the experiences of healthy women who were and were not reconfigured as high-risk end users of tamoxifen (see Tchou 2004, 2005). Other dimensions of the story are presented as faits accomplis, with no sense of process or history. The breast cancer movement, for example, whose complexities I have written about elsewhere (Klawiter 1999, 2000, 2003, 2004), appears here as an already constituted entity devoid of internal conflict and diversity. And although I identify key developments in the recent history of the FDA, I do not discuss the role of AIDS activists and reform coalitions in spurring those changes.

The analysis in this chapter draws on the following data: transcripts of congressional hearings and meetings of the Oncologic Drugs Advisory Committee, NIH and NCI reports and press releases, direct-to-consumer advertisements, warning letters issued by the FDA's Division of Drug Marketing, Advertising, and Communication (DDMAC), and materials produced by activists and advocacy organizations in the women's health, consumer, and breast cancer movements. The remainder of this chapter is organized in the following manner. I begin with a discussion of research on the pharmaceutical industry and frameworks in science and technology studies that have informed my analysis. The next section provides an overview of important changes in the regulatory regime that reconfigured the field of consumption for prescription drugs. Following that, I turn to the case study of tamoxifen. I begin with an analysis of the configuration of high-risk women within the Breast Cancer Prevention Trial. In the next three sections I describe some of the ways in which women's health and breast cancer activists intervened in the rescripting of tamoxifen and the configuration of high-risk women. I conclude with some comments on further implications of this research.

Scripting Pharmaceutical Technologies and Configuring Users

The pharmaceutical industry is an important source of scientific knowledge, technological innovation, and medical therapeutics; a focal point of government oversight and regulation; a global industry of vast proportions; a generator of economic growth; a major source of health care costs; the wealthiest and most powerful political lobby in the United States; and a lightening rod for controversy. Over the course of the last decade a growing number of scholars of science, medicine, and technology have turned their attention to pharmaceutical drugs and the pharmaceutical industry. These investigations have ranged from studies that focus on laboratory research to those that focus on clinical trials, regulatory practices, and the configuration of users (see, for example, Abraham and Sheppard 1999; Clarke and Montini 1993; Corrigan 2002; Davis 1996; Epstein 1996, 1997; Fishman 2004; Geest, Whyte, and Hardon 1996; Goodman and Walsh 2001; Greene 2004; Kawachi and Conrad 1996; Lakoff 2004; Mamo and Fishman 2001; Marks 1997; McCrea and Markle 1984; Metzl 2003; Vuckovic and Nichter 1997).

Scholars working within the interdisciplinary field of science and technology studies have developed a language for talking about the process of fitting technologies to end users and end users to technologies as scripting the technologies and configuring the users (Akrich 1992; Bijker, Hughes, and Pinch 1987; Cowan 1987; Oudshoorn 2003; Oudshoorn and Pinch 2003; van Kammen 2003). Technologies are not only designed for a particular purpose but are also "scripted" for particular users. These scripts—the images, ideas, and guiding assumptions about the potential users, both conscious and unconscious—are incorporated into the technology at different stages and with different degrees of flexibility. Likewise, the construction of markets for new technologies involves a process—often many different processes—of constructing, or configuring, the users.

All told, this body of scholarship deepens and broadens our understanding of the production and consumption of pharmaceutical drugs. In short, we know that the production of pharmaceutical drugs, like that of other techno-scientific products, is a thoroughly social and historical process. It is affected by funding structures and priorities. It is dependent upon the legacy of doable problems, the state of scientific knowledge, and the versatility of scientific instruments. It is embedded within social worlds that are structured as working groups, networks, disciplines, professions, organizations, agencies, universities, and industries. It is stabilized and destabilized through the translational activities of actor networks. And it is deeply shaped by regulatory regimes, consumer markets, and social movements.

Regulatory regimes are not, however, just one more actor or variable that needs to be added to the explanatory mix. As Scott Frickel and Kelly Moore point out in their introduction to this volume, changes in regulatory regimes can have profound effects on the practices of knowledge production and the distribution of power. Pharmaceutical drugs, just like their chemical cousins in industrial production, are subject to state regulation. But the regulation of pharmaceutical drugs takes place through a different set of mechanisms. Unlike the chemicals used and produced in industrial processes, the chemicals used to treat patients are licensed by the FDA before they can enter the market, which requires a lengthy and expensive process of clinical trials testing safety, dosage, and effectiveness. Finally, access to prescription drugs is regulated by physicians, who are licensed to write prescriptions, and pharmacists, who are licensed to dispense them.

Because of the way in which access to prescription drugs is regulated by the FDA and mediated by physicians, the scripting of pharmaceutical technologies and the configuration of users is an extremely complex process. Writing almost twenty years ago, Ruth Schwartz Cowan famously coined the term "consumption junction" to describe "the place and time at which the consumer makes choices between competing technologies" (1987:263). This proved to be a particularly powerful way of opening the black box of consumption in science and technology studies, and many scholars have continued to build upon and broaden this conceptual framework (see especially Oudshoorn and Pinch 2003). My analysis is heavily indebted to this body of scholarship, but I have found it useful to reconceptualize the *consumption junction* as a *field of consumption.*

The concept of fields, which derives from the work of Pierre Bourdieu (1977), has been productively taken up and developed by social movement scholars (Ray 1998, 1999; Crossley 2003, 2005; Goldstone 2004) and scholars in the "new institutionalism" tradition (see especially Scott et al. 2000; Kleinman and Vallas, this volume; Epstein, this volume). Drawing on Bourdieu's understanding of fields as configurations of forces and sites of struggle, Raka Ray defines fields as the "structured, unequal, and socially constructed environment within which organizations are embedded and to which organizations and activists constantly respond. Organizations are not autonomous or free agents, but rather they inherit a field and its accompanying social relations, and when they act, they act in response to it" (Ray 1999:6). The concept of fields, as Ray notes, makes it particularly useful for the study of the relationship between states, social movements, and the formation of pharmaceutical markets.

Regulatory Changes Reorganized the Field of Consumption

Important changes in the regulatory practices of the Food and Drug Administration have occurred over the course of the last fifteen to twenty years. These changes are twofold. The first set of changes concerns the drug approval process. The second involves drug promotion activities. First, the new drug approval process and the process for approving supplemental new drug applications was streamlined and speeded up as a result of the implementation of the Prescription Drug User Fee Act

(PDUFA) of 1992 and the Food and Drug Modernization Act (FDAMA) of 1997 (FDA n.d.; Lasagna 1989). Drug sponsors, especially prior to the implementation of the FDAMA, were often reluctant to submit SNDAs—applications to license already approved drugs for new indications or in new treatment populations—because the process was seen as difficult, costly, and time-consuming. Drugs approved for one purpose or category of patients can be prescribed for any purpose in any patient population. This is known as "off-label" prescribing. Some drugs, in fact, are prescribed more frequently for their "off-label" uses than they are for their labeled, or indicated, use (see Beck 1998). Pharmaceutical firms are prohibited, however, from promoting off-label uses in drug advertisement. FDA approval of new indications was more consequential when pharmaceutical firms were allowed to advertise these new indications directly to consumers.

Second, long-standing FDA policies prohibiting the promotion of prescription drugs directly to consumers were formally reinterpreted by the FDA in the late 1980s to allow the use of print advertising in newspapers and magazines and were thoroughly rewritten in 1997 so that pharmaceutical companies could use broadcast media to advertise prescription drugs directly to consumers, thus opening the floodgates for television advertising. These changes shifted the field of consumption, the dynamics of decision making surrounding the consumption of drugs, and the relationship of the consumer/patient/end user to prescribing physicians and pharmaceutical technologies. It also broadened the scope of the FDA's regulatory reach, heightened public awareness and scrutiny of the drug industry, and created a new arena of activism for consumer, health, and disease-based activism.

The roots of today's direct-to-consumer (DTC) advertising lie in changes proposed to the FDA by the pharmaceutical industry in 1981, shortly after DTC ads first started to appear (Mirken 1996; Palumbo and Mullins 2002). In response to the pharmaceutical industry's proposal to use advertising to "educate" the public about prescription drugs, the FDA asked the industry for a voluntary moratorium so that it could study the proposal. The voluntary moratorium stayed in place until 1985, when the FDA published a notice in the *Federal Register* stating that the current regulations were "sufficient to safeguard the consumer from false or misleading promotional material" (Wilkes, Bell, and Kravitz 2000:113) The FDA's 1985 ruling constituted a reinterpretation rather than a formal revision of its guidelines, and it took some time for the regulatory environment to thaw and for the pharmaceutical industry to take advantage of the changed environment. The amount of spending on DTC prescription drug advertising rose from $12 million in 1989 to $313 million in 1995 (Palumbo and Mullins 2002).

Although these reinterpreted guidelines made print advertising a feasible option, the guidelines continued to inhibit the pharmaceutical industry from advertising its products over television or radio because the summary of warnings and contraindications that they were required to include was incompatible with the shortness of radio and television advertisements. The pharmaceutical industry continued pushing for new guidelines on broadcast advertising, and in 1997 the FDA drafted new guidelines for DTC advertising of prescription drugs using broadcast media. It was the 1997 revisions that opened the floodgates for television advertising. In 1998, a year after the regulations were revised, the amount of money spent on DTC advertising grew to $1.17 billion; in 1999 DTC advertising grew to $1.58 billion; and in 2001, the pharmaceutical industry spent $2.38 billion on DTC advertising (Palumbo and Mullins 2002). There have been a number of public hearings concerning direct-to-consumer advertising of prescription drugs, and dozens of studies have examined the effects of DTC on physician attitudes, physician-patient relations, consumer awareness, consumer behavior, and overall prescribing patterns (for recent syntheses see Aiken 2003, FDA 2003, Rados 2004). The evidence points in different directions, no doubt at least in part because practices and attitudes are still changing. At the very least, this much is clear: DTC has reorganized the prescription drug field of consumption by reorganizing relationships between physicians, pharmaceutical companies, patients, and consumers.

These changes in the regulatory regime broadened the scope of the FDA's regulatory reach; created a heightened public awareness and scrutiny of the drug industry; intensified the reach of the pharmaceutical industry and its impact on popular culture, the mass media, and consumers; and increased the number of opportunities for health and disease-based social movements to intervene, thus enhancing their influence. We turn now to the case of tamoxifen for healthy women.

The BCPT's Configuration of High-Risk Women

The NCI's chemoprevention program was established in the early 1980s, the result of recommendations put forth by several working groups that met between 1978 and 1981 for the purpose of investigating "the feasibility of a research effort in clinical and experimental chemoprevention" (Greenwald and Sondik 1986). The concept of a breast cancer prevention trial had been under discussion at the NCI since 1984, when the NCI's chemoprevention program was just getting started. The rationale for testing the chemopreventive properties of tamoxifen on healthy women emerged from the results of previous studies of tamoxifen on breast cancer patients that had been ongoing since the early 1970s. These trials convincingly showed that tamoxifen significantly increased recurrence-free survival, reduced mortality from breast cancer, reduced the incidence of new cancers in the contralateral (opposite) breast, and was well tolerated by breast cancer patients.[4] In June of 1990, the NCI issued a call for proposals, and in February of the following year, the NCI and the National Cancer Advisory Board approved the trial protocol submitted by the National Surgical Adjuvant Breast and Bowel Project (NSABP), the highly respected clinical trials group headed by Bernard Fisher that had designed and conducted many of the most important clinical trials of breast cancer treatments. Recruitment of subjects began shortly thereafter.

Eligibility for the BCPT was determined by level of risk. Women 60 years of age or older were eligible on the basis of age alone. Women between the ages of 35 and 59 were eligible if their risk was equal to or greater than the five-year risk of an *average* sixty-year-old woman. The average risk of sixty-year-old women turned out to be 1.66 percent over a five-year period. Thus, sixty-year-old women of average, or normal, risk set the bar for entry and became the material referent for the working definition of "high-risk" women.

Risk scores were calculated using the Gail model, an algorithm for estimating breast cancer risk that was developed by a group of statisticians at the NCI in the 1980s and named after the lead biostatistician, Mitchell Gail. The Gail model was chosen for use in the BCPT because it was one of the only models that could handle combinations of risk factors and express absolute risk in terms of probability over time. The Gail model used the following risk factors to calculate risk: age, number of first-degree relatives with breast cancer, previous breast biopsies, presence or absence of atypical hyperplasia (a noncancerous breast condition), reproductive history (number of births and age at first birth), and age at menarche (the onset of menstruation).

Originally designed to include 16,000 women with elevated risks of breast cancer, the trial stopped recruitment at 13,388 because the risk scores of the women enrolled were higher than anticipated, which meant that a smaller number of subjects were needed to power the study. The average risk score of the women who actually participated in the BCPT (as opposed to the average risk score anticipated in the study design) was well above the 1.66 percent minimum. For women participants between the ages of 35 and 49, the average five-year risk score was 3.22 percent; for those aged 50–59, the average risk score was 3.75 percent; and for participants 60 years of age and older, the average five-year risk of breast cancer was 3.92 percent. Despite the risk profile of the study's participants, "high risk" continued to be defined by the parameters of the study's design.

The Gail model is an algorithm, not a classification system, so how did 1.66 become the definitional parameter of the category? Why that precise risk? Or to phrase it differently, why did a 98.44 percent likelihood of *not* being diagnosed with breast cancer in the next five years qualify as high risk? Why choose, as the benchmark for entry into the trial, the five-year risk of an average sixty-year-old woman? The answer to this question is far from obvious. Articles, editorials, and commentaries published in medical journals reproduce and disseminate the "high-risk" category according to the BCPT's definition, but rarely problematize or historicize the category.

A five-year risk of 1.66 or greater was chosen for reasons that have to do with time, expense, and the anticipated supply of eligible, available, experimental subjects, not with any sort of medical consensus regarding the meaning or parameters of high risk. Biostatisticians working with the NSABP calculated that a study designed to follow participants for five years would need about 16,000 women with an average five-year risk of 1.66 in order to produce significant results, given the size of the effects of tamoxifen that they expected to observe. The Breast Cancer Prevention Trial did not reflect a pre-existing consensus or a pre-existing classification system. The figure of 1.66 was embedded in the design of the BCPT, and as a result, high risk came to be understood in these terms.

There was nothing arbitrary about the choice of 1.66, pragmatically speaking, yet there was nothing scientifically necessary about it either. The British-based Royal Marsden trial, for example, recruited women with a strong family history of breast cancer rather than women with a specific risk score (Powles et al. 1998). Thus, high risk was understood in terms of family history rather than absolute risks predicted by a mathematical algorithm. The Italian trial, on the other hand, recruited women from the general population. Thus, the breast cancer risk level of participating subjects was much lower in the Italian Tamoxifen Prevention Study (Veronesi, Maisonneuve, et al. 1998) than it was in the BCPT. The Italian Tamoxifen Prevention Study also, however, exclusively recruited "hysterectomised women." This was done to minimize the risk of endometrial cancer (cancer of the lining of the uterus), which was known to be a serious but rare side effect of tamoxifen. Finally, recruitment mechanisms for the Italian trial did not create a new risk category or classification system.

Prior to the design of the BCPT, "high risk" was a loose term with an unspecified meaning. But as a result of the BCPT, "high risk" became a category with a specific meaning. In this new classification system, 1.66 was the signifier and "high risk" the signified. Together, they constructed a new category: high-risk women. The definition of high-risk women embraced by the NCI and embedded within the BCPT, however, was not accepted unproblematically in other arenas.

Breast Cancer Activists Testify at Public Hearings

Prompted by the concerns of activists and feminist politicians, congressional hearings on the BCPT were held in 1992, 1994, and 1998. Prompted by activists' concerns, Representative Donald Payne called a special hearing of the House of Representatives Human Resources and Intergovernmental Relations Subcommittee on October 22, 1992, to hear testimony about the safety and soundness of the study. Tellingly, an article about the hearings was published in the *Journal of the Cancer Institute* entitled "Breast Cancer Prevention Trial under Scrutiny (Again)." In it, the author commented, "Although the trial is well under way with more than 3,300 women enrolled at over 270 centers . . . dissenting voices once again criticized the $70 million federal study to determine whether or not tamoxifen can prevent breast cancer in women at increased risk of the disease" (Smigel 1992:1692). Already by 1992, "again" is being used to refer to the barrage of criticism coming from activist quarters. The article quotes the National Women's Health Network as opposing the trial on the grounds that "there is little evidence to support the protective claims . . . women eligible for enrollment are not truly at high risk for breast cancer; and . . . tamoxifen is too toxic for use in healthy women" (1692). Helen Rodrigues-Trias, president elect of the American Public Health Association, "also questioned the potential dangers of the drug versus its benefits, as well as problems that might arise from the excessive publicity about the study" (1693). Despite these criticisms, the trial continued.

In 1994, Senators Dianne Feinstein and Connie Mack of the Senate Cancer Coalition convened another public hearing "to explore the ethical and procedural issues surrounding the $68 million, two-year old study." The impetus for the public hearing was the release of data indicating that tamoxifen was not as safe as had been assumed. Problems with the informed consent procedure were also discussed. Once again, the public hearing was widely covered by the mainstream media. This time Amy Langer of the National Alliance of Breast Cancer Organizations (NABCO) was quoted in the *Washington Post* saying that stopping the trial could mean "women would conduct their own private trials in kitchens and offices" by taking the drug "on their own" (Sawyer 1994)—a statement clearly meant to resonate with the symbolism of "back alley abortions." Carolyn Aldige, of the Cancer Research Foundation of America, is quoted as saying that the trial "must proceed" because one cancer expert told her that "there are trials in progress that involve drugs 1,000 times more dangerous than tamoxifen." Both Langer and Aldige were identified by the *Washington Post* as "independent advocates" who expressed "strong support for the study." Both NABCO and the Cancer Research Foundation of American were, however, heavily financed by the pharmaceutical industry.

The 1994 hearings called by Senators Feinstein and Mack were sandwiched between hearings organized on April 13 and June 15 by the House of Representatives Subcommittee on Oversight and Investigations to investigate "a number of important issues associated with serious falsification and fabrication of data in some of the Nation's most important clinical trials on the treatment and prevention of breast cancer" conducted by the NSABP. Dr. Poisson, who admitted to falsifying data, was debarred for life by the FDA from performing research involving investigational drugs, and recruitment for the BCPT was temporarily suspended. The informed consent procedure was revised, a stronger warning about risks and side effects was included, and the trial continued, but with an increasingly tarnished reputation.

On April 22, 1998, the Senate's Subcommittee on Labor, Health and Human Services of the Committee of Appropriations held a hearing to review the decision to terminate the BCPT early. On April 30, 1998—just before Zeneca filed its SNDA—the Congressional Caucus for Women's Issues held a public forum to discuss the process that the FDA would use to review Zeneca's application. Thus, when the ODAC convened on September 2, its members were well aware of the heightened level of interest in, and public scrutiny of, their proceedings. Testimony from patient, consumer, and various advocacy organizations, many of whom recommended against approval for prevention, were submitted and read into the record. Following the public testimony, ODAC began its deliberations, grappling with two central questions: (1) Was there evidence of prevention or only of risk reduction? (2) For whom was this drug indicated? Elsewhere, I have written in greater detail about ODAC's deliberations concerning these questions (Klawiter 2002; see also Wooddell 2004). Here, however, I want to focus on the testimony of the women's health activists and spokespeople for advocacy organizations who attended the meeting.

A letter submitted by Barbara Brenner, executive director of Breast Cancer Action was read into the record. The letter began: "Dear Committee Members, based on the data currently available, both from the NCI and from the recently released European results, Breast Cancer Action opposes approval of the proposed indication for Nolvadex. Women are entitled to expect that any drug approved for the prevention of breast cancer will both actually prevent the disease and carry benefits that outweigh the risks of taking it. As far as we know now, neither is true for Nolvadex. We urge you to 'just say no' to this application. . . . Breast Cancer Action long ago summarized the trials as 'bad research, bad drug, bad news for women'" (ODAC 1998). Cynthia Pearson, the executive director of the National Women's Health Network testified as follows: "The Network urges the FDA and the committee that has been asked by the FDA to give it advice not to approve tamoxifen for prevention or even for risk reduction at this time" (ODAC 1998).

Eight activists representing eight women's health and breast cancer organizations urged ODAC to advise the FDA not to approve tamoxifen for use on healthy women. Each representative described her organization's mission, the size of its membership, and its independence from the pharmaceutical industry. The testimony they gave included an exhaustive and meticulous discussion of the BCPT data and its significance. Only one representative urged ODAC to recommend approval, and she acknowledged that the organization she represented was funded by the pharmaceutical industry.

Breast Cancer Activists Contest DTC Advertising

Like the rest of the pharmaceutical industry, AstraZeneca took advantage of the new regulatory environment for direct-to-consumer advertising. No longer forced to rely exclusively on the attitudes and attentions of physicians, AstraZeneca designed a DTC advertising campaign to promote the newly approved use of tamoxifen directly to potential consumers. AstraZeneca's first foray into DTC advertising was a "health-seeking ad" on network television that introduced the recurring refrain that "there is something you can do" and provided a toll-free number where additional information and a free video could be obtained. Unlike television advertising campaigns for "blockbuster drugs," however, AstraZeneca focused most of its DTC advertising campaign on print media.

Elsewhere I have written about direct-to-physician and direct-to-consumer advertisements placed by AstraZeneca that resulted in the issuance of warning letters by the FDA's Division of Drug Marketing, Advertising and Communications (DDMAC) (Klawiter 2002). Here, however, I highlight some of the actions taken by women's health and breast cancer activists to "descript" (Akrich 1992) tamoxifen as a technology of primary prevention and disrupt the configuration of healthy women as high-risk end users. Recognizing that direct-to-consumer advertising is one of the key sites in which the rescripting of tamoxifen and the configuration of high-risk women would take place, women's health and breast cancer activists carefully monitored AstraZeneca's advertisements, and when they identified advertisements that appeared to violate FDA guidelines, they filed complaints with DDMAC. Recognizing that DDMAC was understaffed and unable to effectively enforce its guidelines, activists also fought these battles in the cultural arena.

The National Women's Health Network (NWHN) has been at the forefront of efforts to monitor and contest the discourses and practices of risk reduction as they developed with regard to tamoxifen. In January of 2000, for example, the NWHN filed a complaint with the DDMAC regarding an advertisement for tamoxifen that appeared in *Newsweek* magazine. In a letter dated January 12, 2000, Cindy Pearson, executive director of the National Women's Health Network, asked DDMAC to require Zeneca to pull the ad, asserting that despite previous warning letters issued by the FDA, Zeneca continued demonstrating a "pattern of misrepresentation" in its promotion of Nolvadex. In the January 12 letter, the NWHN argued that the ad placed by Zeneca was misleading "because of the shifting use of absolute and relative risk numbers." The advertisement indicated, for example, that women on tamoxifen experienced a 49 percent reduction in the number of breast cancers (relative to the women on placebo—a relative risk). But when it came to the representation of risks, Zeneca presented the data on absolute risk, stating that health-threatening side effects "occurred in less than 1% of women." Pearson argued that: "If the ad used relative risk consistently, it would say that women who took tamoxifen had 49 percent fewer breast cancers and 253 percent more endometrial cancers. Or alternatively, if Zeneca wants to use the absolute numbers to assert that women have less than 1 percent chance of being harmed by tamoxifen, the ad should also explain that in absolute terms women have only a 1–2 percent chance of benefiting form the drug depending on their underlying risk of getting breast cancer in the first place." In early February of 2001 the FDA issued a letter stating that it found merit in the NWHN's complaint.

In addition to monitoring AstraZeneca's DTC campaigns and filing complaints with the FDA, the NWHN created a counter-campaign in response to an advertisement that the NWHN found particularly problematic. In April of 1999, Zeneca placed a two-page ad promoting tamoxifen to healthy women in *People* magazine, a popular mass-circulation publication. The first page consisted of a photograph of a young woman sitting on the side of her bed (with rumpled sheets), with her back to the camera. She is wearing a black lace bra and black panties. The text at the top of the page reads: "If you care about breast cancer . . ." and then continues along the top of the facing page, "care more about being a 1.7 than a 36B." Directly beneath it: "Know your breast cancer risk assessment number. Know that Nolvadex® (tamoxifen citrate) could reduce your chances of getting breast cancer if you are at high risk."

This ad was particularly offensive to breast cancer and women's health activists because it sought to appeal to women's desires to be seen as sexually attractive (the black lace bra), managed to do so in a way that was superficial and stereotyping (referring to women's presumed concern with their cup size), and then established a relationship between attractiveness, finding out one's score on the "new risk assessment test," and taking Nolvadex if scoring a 1.7 or higher. The NWHN responded to this ad by recirculating it to its membership in changed form. Across the top of the ad, the NWHN wrote in bold caps: "WHAT'S WRONG WITH THIS AD?" On the reverse side the question was posed again, and beneath it a series of answers, including the analysis that AstraZeneca "is trying to scare healthy women into taking a drug that is known to cause life-threatening health problems, including endometrial cancer and blood clots. Their ads won't tell you that healthy women who took Nolvadex died of pulmonary embolism during the clinical trial." The NWHN's response was widely circulated within the breast cancer movement.

Around this same period of time, Breast Cancer Action and its executive director, Barbara Brenner, took the lead in bringing together a group of organizations that, in various ways and through various means, were developing critical analyses of DTC advertising and shared an interest in promoting a view of public health that stressed cancer prevention through the elimination of toxic substances from the environment rather than "risk reduction through pharmaceutical interventions," which, they argued, were "individual precautions not available to everyone." This collection of organizations became a new coalition, Prevention First. Prevention First applied for and received a two-year, $300,000 grant from the San Francisco–based Richard and Rhoda Goldman Fund, a foundation with a history of funding breast cancer and environmental projects.

In October 2001, Prevention First produced a flyer, which each member organization circulated via their communication networks (inserting it in newsletters, posting it on their websites, etc.). The flyer constituted the public launching of their project against the pharmaceutical industry, and specifically against DTC advertising of tamoxifen to healthy women. The flyer began with a question: "Will Breast Cancer Prevention ever come in a pill?" The answer supplied was this:

> Billions of dollars are spent annually on cancer research—but the vast majority of that research is on drug development, not on prevention. Pills touted to "prevent" cancer can at best only lower the risk of developing the disease—and they often increase the risk of other health problems. True cancer prevention requires understanding and eliminating environmental causes of the disease.

The flyer continued, this time naming names:

> Americans are bombarded with ads for pills to "prevent" diseases like breast cancer. . . . Eli Lilly has promoted Evista® (raloxifene) for breast cancer benefits though the drug has been approved only for osteoporosis. And AstraZeneca, which manufactures the breast cancer drug Nolvadex® (tamoxifen), has marketed the drug to healthy women, even though it's more likely to hurt than help most of them.

The flyer asserted that drug ad campaigns such as those conducted by Eli Lilly and AstraZeneca hurt everyone by generating "misinformation" and driving up the cost of drugs. The authors of the flyer identified themselves as health organizations that maintain their independence from drug companies by refusing to accept funding from them. A 2½-inch-wide column running down the left side of the flyer, labeled "THE FACTS," made three bulleted claims: drug companies spend twice as much on marketing and administration as they do on research and development; the FDA forced AstraZeneca to withdraw several ads; and "by focusing on pills for breast cancer 'prevention,' the drug ads divert attention from the *causes* of disease."

On May 23, 2001, the FDA held a "Consumer Roundtable," and Prevention First presented testimony regarding DTC advertising and standards of approval for new oncologic drugs. Prevention First advocated several changes related to both the structure and functioning of the FDA and DTC advertising. Beginning with the strongest measure, Prevention First argued that "since it appears to be the case that the FDA has neither the resources or the authority to properly protect the public against the dangers of DTCA . . . [Prevention First] urge[s] the outright ban of direct-to-consumer advertising." Second, in lieu of an outright ban, Prevention First advocated a "three strikes you're out" rule whereby any company receiving three "cease and desist" letters from the FDA within a seven-year period, regarding any drug, would be prohibited from advertising that product directly to consumers. Third, Prevention First urged (again in lieu of a ban) the FDA to require companies that run ads that violate the agency's guidelines "to immediately run corrective advertisements in the same manner and markets, for the same amount of time, as the misleading ad was in circulation." Finally, Prevention First urged a restructuring of resources so that the FDA could preview DTC advertisements before they were placed, and so that they could respond more promptly to complaints filed by the public regarding specific ads. These issues continue being debated in Congress and within the FDA.

Conclusions

This chapter focuses on an increasingly important arena of scientific research, technological innovation, regulatory intervention, medical practice, and social movement action: the use of pharmaceutical drugs to treat the condition of being at risk. More specifically, I examine the pharmaceuticalization of risk in the context of cancer chemoprevention, focusing on attempts to rescript tamoxifen as a technology of breast cancer risk reduction and to reconfigure healthy women as high-risk end users. In this chapter I argue that recent changes in the regulatory regime reorganized the *field of consumption* in ways that strengthened the voice of consumers and the influence of social movements at the same time that it enhanced the power, as well as the vulnerability, of the pharmaceutical industry. This paradoxical set of outcomes can help explain why tamoxifen succeeded in gaining rapid FDA approval for this new indication and patient population but failed in the marketplace as a technology of primary prevention.

Organizations like the National Women's Health Network, Breast Cancer Action, and Prevention First did not simply reject or oppose tamoxifen as a technology of primary prevention and to disrupt the configuration of healthy women as high-risk end users. They also produced their own, alternative discourses of breast cancer risk and prevention. Changes in the regulatory regime reorganized the field of consumption, but the women's health and breast cancer movements helped shaped the discourses and practices through which tamoxifen was rescripted as a technology of breast cancer prevention and women were configured as high-risk users.

And they continue doing so. Struggles over tamoxifen and breast cancer risk reduction continue to this day. Through their publications, Web sites, and public presence, organizations in the women's health and breast cancer movements have actively discouraged women from participating in the Study of Tamoxifen and Raloxifene, which was launched shortly after the conclusion of the BCPT, to compare the effectiveness of both drugs in reducing breast cancer incidence among postmenopausal, high-risk, healthy women.

There are a number of factors that make this case study particularly interesting. First, the case of tamoxifen is part of a much broader set of changes connected to the pharmaceuticalization of risk that are transforming the practice of medicine and the production of pharmaceutical knowledge and technologies. As Alexandra Benis noted five years ago in the *Harvard Review of Public Health:* "The story of tamoxifen has become at once a shining symbol of the promise of chemoprevention and a revelation of just how complex and intricate a process drug discovery can be—scientifically, economically, and ethically." That intricacy, as I have tried to demonstrate in this chapter, has been made even more complex by the emergence of the breast cancer movement and changes in the regulatory regime that reorganized the field of consumption. Second, although tamoxifen was the first pharmaceutical drug approved for cancer prevention, it will not be the last. Cancer chemoprevention is an important domain of pharmaceutical research and development, and although the controversy over tamoxifen has died down, the issues that it raises persist in the form of ongoing clinical trials around the world that are testing new pharmaceutical technologies of cancer chemoprevention. Finally, this is an interesting case study because it is a study of market failure rather than success. This serves as a useful corrective to the popular perception that the pharmaceutical industry, through the use of DTC advertising, can dupe the consuming public into popping pills for every possible condition, either real or imagined. While it is true that the pharmaceutical industry has been able to translate its immense power and profits into pharma-friendly policies and regulatory practices, individual firms do not operate as hegemons, and the success of individual drugs is far from guaranteed.

Social movements are not omnipotent players, and I do not mean to suggest otherwise. Time and again, public hearings were held and objections were expressed, but the Breast Cancer Prevention Trial went forward, and six months after Zeneca filed its SNDA, the FDA approved tamoxifen for the treatment of healthy, high-risk women over the protests of the women's health and breast cancer movements. But when it came on the market, breast cancer chemoprevention did not capture the desires of healthy women and AstraZeneca was unable to capitalize on their fears. The women's health and breast cancer movements, contributed to that failure. They were able to do so, in part, because regulatory shifts amplified their influence in a reorganized field of consumption.

Acknowledgments

I thank the Robert Wood Johnson Foundation and the many fine people associated with the RWJ Scholars in Health Policy Research Program for their support of this project, which I had the luxury of pursuing from 1999 to 2001 at the University of Michigan. Special thanks go to Rodney Hayward, Michael Chernew, Renee Anspach, Paula Lantz, Catherine McLaughlin, Dean Smith.

Endnotes

1. Throughout this chapter I use the locution "healthy women" to refer to women who are asymptomatic, appear to be cancer-free, and have never been diagnosed or treated for breast cancer—women who, in theory, might be candidates for breast cancer chemoprevention. This has the unfortunate consequence of implying that women who *are* symptomatic, who show evidence of cancer, or who have been diagnosed or treated for breast cancer are unhealthy women (and vice versa, that cancer-free women are never sick). This is clearly not true. I'm uncomfortable with the locution, but I've been unsuccessful at figuring out an alternative. "Breast Cancer Prevention Trial" (BCPT) is shorthand for "National Surgical Adjuvant Breast Cancer and Bowel Project Breast Cancer Prevention Trial P-1" (NSABP BCPT P-1). The National Surgical Adjuvant Breast Cancer and Bowel Project (NSABP) was the clinical trial cooperative that designed and conducted the BCPT. The NSABP was formed in 1971 to conduct clinical trials in breast and colorectal cancer research. Members of the cooperative group had been involved in collaborative research, however, as early as 1958. With a membership of more than 6,000 medical professionals, and almost 300 medical centers in the United States, Canada, and Australia, and more than forty years of experience conducting clinical trials, the NSABP is the largest clinical trial group working in the area of cancer research. Its primary source of funding is the National Cancer Institute (NSABP n.d.).

2. Shortly after tamoxifen was licensed by the FDA for the treatment of healthy, high-risk women (which occurred on October 30, 1998), the NSABP, in collaboration with AstraZeneca (formerly Zeneca), the manufacturer of Nolvadex, established a program to provide five years of Nolvadex to participants in the BCPT who had been assigned to the placebo arm of the trial (NSABP 1999–2000).

3. The FDA relies on a system of independent advisory committees, made up of professionals from outside the agency, for expert advice and guidance in making sound decisions about drug approval. Each committee meets as needed to weigh available evidence and assess the safety, effectiveness, and appropriate use of products considered for approval. In addition, these committees provide advice about general criteria for evaluation and scientific issues not related to specific products. Each committee is composed of representatives from the fields of research science and medical practice. At least one member on every advisory committee must represent the "consumer perspective." The Oncologic Drugs Advisory Committee (ODAC) is responsible for advising the FDA on cancer-related treatments and preventive agents (NCI 2004).

4. For the rationale behind the BCPT and its development, see Smigel 1992, Fisher et al. 1998, Bush and Helzlsouer 1993, Fosket 2004, and Wooddell 2004.

References

Abraham, John, and Julie Sheppard. 1999. "Complacent and Conflicting Scientific Drug Regulation: Clinical Risk Assessment of Triazolam." *Social Studies of Science* 29:803–843.

Aiken, Kathryn J. 2003. "Direct-to-Consumer Advertising of Prescription Drugs: Physician Survey Preliminary Results." Study presented at Direct-to-Consumer Promotion: Public Meeting, September 22–23, Food and Drug Administration, Washington, D.C. www.fda.gov/cder/ddmac/globalsummit2003/index.htm (accessed June 15, 2005).

Akrich, Madeleine. 1992. "The De-Scription of Technical Objects." In *Shaping Technology/Building Society: Studies in Sociotechnical Change,* ed. W. E. Bijker and J. Law, pp. 205–224. Cambridge, MA: MIT Press.

Altman, Lawrence K. 1998. "Researchers Find the First Drug Known to Prevent Breast Cancer." *New York Times,* April 7.

American Cancer Society. 1998. *Update* 4(2): 3.

Aronowitz, Robert. 2001. "Do Not Delay: Breast Cancer and Time, 1900–1970." *Milbank Quarterly* 79:355–386.

Associated Press. 1998. "Breast Cancer Drug Reportedly Can Prevent Disease." *San Francisco Examiner,* April 5.

———. 2002. "Drug Ads Targeting Patients." *Arizona Republic,* February 14.

AstraZeneca. 1999. "Annual Report." www.astrazeneca.com (accessed June 15, 2005).

———. 2000. "Annual Report." www.astrazeneca.com (accessed June 15, 2005).

———. 2001. "Annual Report." www.astrazeneca.com (accessed June 15, 2005).

———. 2002. "Annual Report." www.astrazeneca.com (accessed June 15, 2005).

———. 2003. "Annual Report." www.astrazeneca.com (accessed June 15, 2005).

Benis, Alexandra. 1999. "A Prescription for Prevention?" *Harvard Public Health Review* (summer). www.hsph.harvard.edu/review/summer_tamoxifen.shtml.

Bijker, Wiebe E., Thomas P. Hughes, and Trevor Pinch. 1987. *The Social Construction of Technological Systems.* Cambridge, MA: MIT Press.

Bourdieu, Pierre. 1977. *Outline of a Theory of Practice.* Cambridge: Cambridge University Press.

Breast Cancer Breakthrough. 1998. *New York Times,* April 8.

Bush, Trudy L., and Kathy J. Helzlsouer. 1993. "Tamoxifen for the Primary Prevention of Breast Cancer: A Review and Critique of the Concept and Trial." *Epideimiologic Reviews* 15:233–243.

Clarke, Adele, and Theresa Montini. 1993. "The Many Faces of RU486: Tales of Situated Knowledges and Technological Contestations." *Science, Technology, & Human Values* 18:42–78.

Corrigan, Oonagh P. 2002. "'First in Man': The Politics and Ethics of Women in Clinical Drug Trials." *Feminist Review* 72:40–52.

Cowan, Ruth Schwartz. 1987. "The Consumption Junction: A Proposal for Research Strategies in the Sociology of Technology." In *The Social Construction of Technological Systems,* ed. W. Bijker, pp. 261–280. Cambridge, MA: MIT Press.

Crossley, Nick. 2003. "From Reproduction to Transformation." *Theory, Culture and Society* 20:43–68.

———. 2005. "How Social Movements Move: From First to Second Wave Developments in the UK Field of Psychiatric Contention." *Social Movement Studies* 4:21–48.

Davis, Peter. 1996. *Contested Ground: Public Purpose and Private Interest in the Regulation of Prescription Drugs.* New York: Oxford University Press.

Epstein, Steven. 1996. *Impure Science: AIDS, Activism, and the Politics of Knowledge.* Berkeley: University of California Press.

———. 1997. "Activism, Drug Regulation, and the Politics of Therapeutic Evaluation in the AIDS Era: A Case Study of ddC and the 'Surrogate Markers' Debate." *Social Studies of Science* 27:691–727.

Fisher, B., J. P. Costantino, D. L. Wickerham, et al. 1998. "Tamoxifen for Prevention of Breast Cancer: Report of the National Surgical Adjuvant Breast and Bowel Project P-1 Study." *Journal of the National Cancer Institute* 90:1371–1388.

Fisher, Bernard, J. Dignam, and N. Wolmark. 1999. "Tamoxifen in Treatment of Intraductal Breast Cancer: National Surgical Adjuvant Breast and Bowel Project B-24 Randomized Controlled Trial." *Lancet* 353:1993–2000.

Fishman, Jennifer R. 2004. "Manufacturing Desire: The Commodification of Female Sexual Dysfunction." *Social Studies of Science* 34:187–218.

Food and Drug Administration (FDA). N.d. FDA Modernization Act, www.fda.gov/cder/guidance/105–115.htm (accessed May 26, 2005).

———. 2003. Direct-to-Consumer Promotion: Public Meeting. September 22–23, Washington, D.C., Presentations Including Public Testimony, Results from National Surveys, and Research Examining Advertising Effectiveness, Effects on Physicians' Prescribing Practices, Utilization and Demand, and Internet Advertising. www.fda.gov/cder/ddmac/DTCmeeting2003_presentations.html.

Fosket, Jennifer. 2004. "Constructing 'High-Risk Women': The Development and Standardization of a Breast Cancer Risk Assessment Tool." *Science, Technology & Human Values* 29:291–313.

Fugh-Berman, Adrian, and Samuel S. Epstein. 1992a. "Should Healthy Women Take Tamoxifen?" *New England Journal of Medicine* 327:1596–1597.

———. 1992b. "Tamoxifen: Disease Prevention or Disease Substitution?" *Lancet* 340:1143–1145.

Gardner, Kirsten E. 1999. "'By Women, For Women, With Women': A History of Female Cancer Awareness Efforts in the United States, 1913–1970s." PhD diss., Department of History, University of Cincinnati.

Geest, Sjaak van der, Susan Reynolds Whyte, and Anita Hardon. 1996. "The Anthropology of Pharmaceuticals: A Biographical Approach." *Annual Review of Anthropology* 25:153–178.

Goldstone, Jack A. 2004. "More Social Movements or Fewer? Beyond Political Opportunity Structures to Relational Fields." *Theory and Society* 33:333–365.

Goodman, Jordan, and Vivien Walsh. 2001. *The Story of Taxol: Nature and Politics in the Pursuit of an Anti-Cancer Drug.* New York: Cambridge University Press.

Greene, Dennis. 2001. "The Shift in Promotional Spending Mix." *DTC Perspectives,* vol. 1, 12–14.

Greene, Jeremy A. 2004. "Attention to 'Details': Etiquette and the Pharmaceutical Salesman in Postwar American." *Social Studies of Science* 34:271–292.

Greenwald, Peter, and Edward Sondik. 1986. "Diet and Chemoprevention in NCI's Research Strategy to Achieve National Cancer Control Objectives." *Annual Review of Public Health* 7:267–291.

Healy, Bernardine. 1998. "Tamoxifen and the Breast Cancer Prevention Trial: Women Helping Women." *Journal of Women's Health* 7:279–280.

Hogle, Linda F. 2001. "Chemoprevention for Healthy Women: Harbinger of Things to Come?" *Health* 5:299–320.

Kawachi, Ichiro, and Peter Conrad. 1996. "Medicalization and the Pharmacological Treatment of Blood Pressure." In *Contested Ground: Public Purpose and Private Interest in the Regulation of Prescription Drugs,* ed. P. Davis, pp. 46–71. New York: Oxford University Press.

Klausner, Richard. 1998. Testimony before the Senate Appropriations Subcommittee on Labor, Health, Human Services and Related Agencies. 105th Cong., 2nd sess., April 21.

Klawiter, Maren. 1999. "Racing for the Cure, Walking Women, and Toxic Touring: Mapping Cultures of Action within the Bay Area Terrain of Breast Cancer." *Social Problems* 46:104–126.

———. 2000. "From Private Stigma to Global Assembly: Transforming the Terrain of Breast Cancer." In *Global Ethnography: Forces, Connections, and Imaginations in a Postmodern World,* ed. M. Burawoy, J. Blum, S. George, Z. Gille, T. Gowan, L. Haney, M. Klawiter, S. Lopez, S. O Riain, and M. Thayer, pp. 299–334. Berkeley: University of California Press.

———. 2001. The Rocky Road to Cancer Chemoprevention: Detour, Delay, or Dead-end? Paper presented at the annual meeting of the Robert Wood Johnson Foundation Scholars in Health Policy Research Program, May 30–June 1, Aspen, CO.

———. 2002. "Risk, Prevention and the Breast Cancer Continuum: The NCI,

the FDA, Health Activism and the Pharmaceutical Industry." *History and Technology* 18:309–353.
———. 2003. "Chemicals, Cancer, and Prevention: The Synergy of Synthetic Social Movements." In *Synthetic Planet: Chemical Politics and the Hazards of Modern Life*, ed. M. Casper, pp. 155–193. New York: Routledge.
———. 2004. "Breast Cancer in Two Regimes: The Impact of Social Movements on Illness Experience." *Sociology of Health & Illness* 26:845–874.
Lakoff, Andrew. 2004. "The Anxieties of Globalization: Antidepressant Sales and Economic Crisis in Argentina." *Social Studies of Science* 24:247–269.
Lasagna, L. 1989. "Congress, the FDA, and New Drug Development: Before and After 1962." *Perspectives in Biology and Medicine* 32:322–341.
Leopold, Ellen. 1999. *A Darker Ribbon: Breast Cancer, Women, and Their Doctors in the Twentieth Century*. Boston: Beacon Press.
Lerner, Barron H. 2000. "Inventing a Curable Disease: Historical Perspectives on Breast Cancer." In *Breast Cancer: The Social Construction of Illness*, ed. S. J. Ferguson and A. S. Kasper, pp. 25–49. New York: St. Martin's Press.
Mamo, Laura, and Jennifer Fishman. 2001. "Potency in All the Right Places: Viagra as a Technology of the Gendered Body." *Body & Society*. 7:13–35.
Marks, Harry M. 1997. *The Progress of Experiment: Science and Therapeutic Reform in the United States, 1900–1990*. Cambridge: Cambridge University Press.
McCrea, Frances B., and Gerald E. Markle. 1984. "The Estrogen Replacement Controversy in the USA and UK: Different Answers to the Same Question?" *Social Studies of Science* 15:1–26.
Metzl, Jonathan. 2003. *Prozac on the Couch: Prescribing Gender in the Era of Wonder Drugs*. Durham, NC: Duke University Press.
Mirken, Bruce. 1996. "Ask Your Doctor." *San Francisco Guardian*, October 23, 45–47.
National Cancer Institute. 2004. Understanding the Approval Process for New Cancer Treatments. www.nci.nih.gov/clinicaltrials/learning/approval-process-for-cancer-drugs, last updated January 6, 2004 (accessed August 6, 2004).
National Surgical Adjuvant Breast and Bowel Project (NSABP). 1999–2000. "No Cost Tamoxifen Program Announced for BCPT Women on Placebo." *Coast to Coast* (newsletter published by the NSABP for BCPT Participants), Winter, pp. 1, 4. www.nsabp.pitt.edu/Coast_to_Coast_Newsletter.pdf (accessed May 27, 2005).
———. N.d. "What Is the NSABP?" www.nsabp.pitt.edu (accessed June 13, 2000).
Oncologic Drugs Advisory Committee (ODAC). 1998. 58th Meeting, Bethesda, MD, September 2.
Oudshoorn, Nelly. 2003. "Clinical Trials as a Cultural Niche in Which to Configure the Gender Identities of Users: The Case of Male Contraceptive Development." *How Users Matter: The Co-Construction of Users and Technologies*,

edited by Nelly Oudshoorn and Trevor Pinch, pp. 209–228. Boston: MIT Press.

Oudshoorn, Nelly, and Trevor Pinch, eds. 2003. *How Users Matter: The Co-Construction of Users and Technologies*. Cambridge, MA: MIT Press.

Palumbo, Francis B., and Daniel C. Mullins. 2002. "The Development of Direct-to-Consumer Prescription Drug Advertising Regulation." *Food and Drug Law Journal* 57:423–444.

Phillips, Kelly-Anne, Gordon Glendon, and Julia A. Knight. 1999. "Putting the Risk of Breast Cancer in Perspective." *New England Journal of Medicine* 340: 141–144.

Poe, Amy. 1999. "Cancer Prevention or Drug Promotion? Journalists Mishandle the Tamoxifen Story." *International Journal of Health Services* 29: 657–661.

Powles, T., R. Eeles, S. Ashley, et al. 1998. "Interim Analysis of the Incidence of Breast Cancer in the Royal Marsden Hospital Tamoxifen Randomised Chemoprevention Trial." *Lancet* 352:98–101.

Rados, Carol. 2004. "Truth in Advertising Rx Drug Ads Come of Age." *FDA Consumer Magazine*, July–August. www.fda.gov/fdac (accessed September 20, 2005).

Ray, Raka. 1998. "Women's Movements and Political Fields: A Comparison of Two Indian Cities." *Social Problems* 45:21–36.

———. 1999. *Fields of Protest: Women's Movements in India*. Minneapolis: University of Minnesota Press.

Reagan, Leslie J. 1997. "Engendering the Dread Disease: Women, Men, and Cancer." *American Journal of Public Health* 87:1779–1787.

Sawyer, Kathy. 1994. "Breast Cancer Drug Testing Will Continue: Potential of Tamoxifen Is Said to Outweigh the Risks." *Washington Post*, May 12.

Scott, Richard W., Martin Ruef, Peter J. Mendel, and Carol A. Caronna. 2000. *Institutional Change and Healthcare Organizations: From Professional Dominance to Managed Care*. Chicago: University of Chicago Press.

Smigel, Kara. 1992. "Breast Cancer Prevention Trial under Scrutiny (Again)." *Journal of the National Cancer Institute* 84:1692–1694.

———. 1998. "Breast Cancer Prevention Trial Shows Major Benefit, Some Risk." *Journal of the National Cancer Institute* 90:647–648.

Tchou, J., N. Hou, A. Rademaker, V. C. Jordan, and M. Morrow. 2004. "Acceptance of Tamoxifen Chemoprevention by Physicians and Women at Risk." *Cancer* 100:1800–1806.

———. 2005. "Acceptance of Tamoxifen Chemoprevention by Physicians and Women at Risk." *Cancer* 103:209–210.

van Kammen, Jessica. 2003. "Who Represents the User? Critical Encounters Between Women's Health Advocates and Scientists in Contraceptive R&D." *How Users Matter: The Co-Construction of Users and Technologies*, edited by Nelly Oudshoorn and Trevor Pinch, pp. 151–171. Boston: MIT Press.

Veronesi, U., P. Maisonneuve, et al. 1998. "Prevention of Breast Cancer with Tamoxifen: Preliminary Findings from the Italian Randomised Trial among Hysterectomised Women." *Lancet* 352:93–97.

Vuckovic, Nancy, and Mark Nichter. 1997. "Changing Patterns of Pharmaceutical Practice in the United States." *Social Science and Medicine* 44:1285–1302.

Wilkes, Michael S., Robert A. Bell, and Richard L. Kravitz. 2000. "Direct-to-Consumer Prescription Drug Advertising: Trends, Impact, and Implications." *Health Affairs* 19:110–128.

Wooddell, Margaret J. 2004. "Codes, Identities and Pathologies in the Construction of Tamoxifen as a Chemoprophylactic for Breast Cancer Risk Reduction in Healthy Women at High Risk." PhD diss., Science and Technology Studies, Rensselaer Polytechnic Institute, Troy, NY.

Zeneca. 1997. "Annual Report." www.astrazeneca.com/article/11221.aspx.

———. 1998. "Annual Report." www.astrazeneca.com/article/11221.aspx.

CONTRIBUTORS

REBECCA GASIOR ALTMAN is a doctoral student in the Department of Sociology at Brown University. She serves as a research assistant on a NIEHS-funded community-based participatory research project, "Linking Environmental Justice and Breast Cancer Activism," coordinated by Brown University, Silent Spring Institute (Massachusetts), and for Communities for a Better Environment (California). Her dissertation explores the relationship between community-based participatory research and the environmental justice movement. Other research interests include: environmental health politics, health social movements, microbilization and activists' careers, "green" health care, and the history of the politics of the tobacco industry.

PHIL BROWN is professor of sociology and environmental studies at Brown University. He studies "contested illnesses" such as asthma, breast cancer, and Gulf War–related illnesses, involving public debates over environmental causes and the impact of social movements on those debates. At present he is examining coalitions between environmental organizations and labor unions and other labor organizations. In another project he is studying connections between breast cancer advocacy and environmental justice organizing. He is the author of *No Safe Place: Toxic Waste, Leukemia, and Community Action* (University of California Press, 1997) editor of *Perspectives in Medical Sociology* (Waveland Press, 2000) coeditor of *Illness and the Environment: A Reader in Contested Medicine* (New York University Press, 2000), and coeditor of *Social Movements in Health* (Blackwell, 2005).

STEVEN EPSTEIN is associate professor of sociology at the University of California, San Diego. He also is affiliated with UCSD's interdisciplinary graduate program in science studies. He is the author of the award-winning book *Impure Science: AIDS, Activism, and the Politics of Knowledge* (University of California Press, 1996). His research and teaching interests include biomedical politics; science, social movements, and the state; and the politics of sexuality, gender, and race.

SCOTT FRICKEL is assistant professor of sociology at Tulane University. His research is in science studies, social movements, and environmental sociology. He is the author of *Chemical Consequences: Environmental Mutagens, Scientist Activism, and the Rise of Genetic Toxicology* (Rutgers University Press, 2004). Other research has appeared in such journals as *American Sociological Review, International Sociology, Organization & Environment, Science as Culture,* and *Social Problems.* Two current research projects involve the historical and comparative analysis of scientific and intellectual movements and an organizational analysis of scientist activism in environmental health and justice movements.

DAVID H. GUSTON is professor of political science at Arizona State University and associate director of ASU's Consortium for Science, Policy, and Outcomes. His book *Between Politics and Science: Assuring the Integrity and Productivity of Research* (Cambridge University Press, 2000) received the 2002 Don K. Price Prize by the American Political Science Association for best book in science and technology policy. He is coeditor of the forthcoming *Science, Technology, and Public Policy: The Next Generation of Research* (with D. Sarewitz; University of Wisconsin Press), coauthor of *Informed Legislatures* (with M. Jones and L. M. Branscomb; University Press of America, 1996), and coeditor of *The Fragile Contract* (with K. Keniston; MIT Press 1994). He is North American editor of the peer-reviewed journal *Science and Public Policy* and is a fellow of the American Association for the Advancement of Science.

CHRISTOPHER R. HENKE is assistant professor of sociology at Colgate University, with research and teaching interests in science, work, and the environment. He received his PhD in sociology/science studies from the University of California, San Diego in 2000 after completing a study of science-industry relationships in California's farm industry, the subject of his book manuscript in progress, entitled *Cultivating Science, Harvesting Power: Science and Industrial Agriculture in California.* He is also currently conducting research on the "monarch butterfly controversy," a dispute regarding genetically engineered food and its environmental impacts.

DAVID J. HESS is professor of science and technology studies at Rensselaer Polytechnic Institute. He is the author or editor of eleven books on science, technology, and society. Recently he served as editor of a special issue of *Science as Culture* on "Health, the Environment, and Social Movements." More information and some publications are available at his Web site, www.davidjhess.org.

MAREN KLAWITER is assistant professor of sociology in the School of History, Technology and Society at the Georgia Institute of Technology. She received her Ph.D. in 1999 from the University of California, Berkeley, and from 1999 to

2001 she was a fellow in the Robert Wood Johnson Foundation Scholars in Health Policy Research Program at the University of Michigan. Her contribution to this volume is part of a larger book project on the pharmaceuticalization of risk, tentatively titled *Diagnosing Risk, Prescribing Prevention: Constructing the Breast Cancer Continuum,* which she began at the University of Michigan. A separate book project, *Reshaping the Contours of Breast Cancer: Disease Regimes, Cultures of Action, and the Bay Area Field of Contention,* is nearing completion.

DANIEL LEE KLEINMAN is professor in the Department of Rural Sociology at the University of Wisconsin–Madison, where he is also affiliated with the Holtz Center for Science and Technology Studies. One line of ongoing research is reflected in his contribution to this volume and in his book *Impure Cultures: University Biology and the World of Commerce* (University of Wisconsin Press, 2003). Another is an exploration of the conditions under which genetically modified crops are regulated according to social criteria. This work, published with Abby J. Kinchy, has appeared in *Science as Culture* and *Sociological Quarterly*. In an effort to speak to audiences beyond academia, he has written *Science and Technology in Society: From Biotechnology to the Internet* (Blackwell, 2005) and coedited, with Jo Handelsman and Abby Kinchy, *Controversies in Science and Technology;* Volume 1: *From Maize to Menopause* (University of Wisconsin Press, 2004).

BRIAN MARTIN is associate professor of science, technology, and society at the University of Wollongong, Australia. He worked as an applied mathematician before moving to social science. He is the author of ten books and hundreds of articles dealing with scientific controversies, dissent, nonviolence, democracy, information politics, and other topics. Further information is available on his Web site: www.uow.edu.au/arts/sts/bmartin/.

BRIAN MAYER is a doctoral candidate in the Department of Sociology at Brown University. His dissertation examines the strategies and conditions leading to successful attempts at building social movement coalitions between labor and environmental groups. He is particularly interested in exploring the role that shared concerns about the hazardous effects of toxic substances can play in creating common identities between labor and environmental activists. As part of his work, he is involved in several collaborative partnerships with environmental health organizations and has assisted in a number of organizing projects.

SABRINA MCCORMICK is assistant professor in the Department of Sociology and the Environmental Science and Policy Program at Michigan State University. She completed her Ph.D. in sociology at Brown University in 2005. Her dissertation compared the anti-dam movement in Brazil and the environmental breast cancer movement in the United States to examine how both movements attempt to contest and shape expert knowledge. She began this work as a member of a research team funded by the Robert Wood Johnson Foundation and continued it as a Henry Luce Foundation Environmental Fellow at the Watson Institute of International Studies. During that time, she studied health social movements in the United States. She is currently writing a book about the relationship between breast cancer and the environment and is directing a documentary film on the same topic.

KELLY MOORE is assistant professor of sociology at the University of Cincinnati. Her past work has examined how social movements have shaped the organization of relationships between scientists and the public. Her work has appeared in *American Journal of Sociology, How Movements Matter,* and *Research in the Sociology of Organizations.* Her book *Disrupting Science: American Scientists and Anti-Militarism, 1945–1975* is forthcoming with Princeton University Press. Her current research compares restoration ecology projects in San Francisco and New York City to understand how political institutions and social movements shape urban landscapes.

RACHEL MORELLO-FROSCH is assistant professor at the Center for Environmental Studies and the Department of Community Health, School of Medicine, at Brown University. Her research examines race and class determinants of the distribution of health risks associated with air pollution among diverse communities in the United States. Her current research and publications focus on comparative risk assessment and environmental justice, segregation and environmental health, models for community-based environmental health research, the role of science and precaution in environmental health policymaking, children's environmental health, and the intersection between economic restructuring and environmental health.

JASON OWEN-SMITH is assistant professor of sociology and organizational studies at the University of Michigan in Ann Arbor. He is interested in institutional change, scientific collaboration, organizational learning, the dynamics of complex networks, and the commercialization of academic research. His current project focuses on the intersection of science, technology, and commerce in the academy with particular emphasis on the ramifications of patenting, licensing, and university-industry collaborations. Research related to these topics has appeared in publications such as the *American Sociological Review,* the *American Journal of Sociology, Social Studies of Science, Research Policy,* the *Journal of Higher Education,* and *Organization Science.*

JENNY REARDON is assistant professor of sociology at the University of California–Santa Cruz. In 2004–2005 she was assistant research professor of women's studies and Institute of Genome Sciences and Policy scholar at Duke University; in 2003–2004 she was a scholar-in-residence at the Pembroke Center for Teaching and Research on Women, Brown University. She is the author of *Race to the Finish: Identity and Governance in an Age of Genomics* (Princeton University Press, 2004).

LAUREL SMITH-DOERR is assistant professor of sociology at Boston University. Her research takes a sociological perspective on organizations and science. Her book *Women's Work: Gender Equality vs. Hierarchy in the Life Sciences* (Lynne Rienner, 2004) examines how organizational context shapes career opportunities for life science PhDs, such that women face fewer constraints and greater flexibility in commercial biotech firms. During a 2004–2005 fellowship at the Robert Schuman Centre for Advanced Studies at the European University Institute, she expanded her investigation of governmental policies and ethics training for life scientists to the European context. The comparison of science ethics training between the United States and Europe will appear in a forthcoming book.

STEVEN P. VALLAS chairs the Department of Sociology and Anthropology at George Mason University in Fairfax, Virginia. His research focuses on questions involving work, culture, and social inequality within both manufacturing and knowledge-based industries. He has written or edited three books on work organizations and published articles in leading sociological journals. He is currently finishing an ethnography exploring the politics of team systems and the conflict between engineering knowledge and indigenous expertise within U.S. manufacturing.

STEVEN WOLF is assistant professor in the Department of Natural Resources at Cornell University. He received his doctorate in environmental studies from University of Wisconsin–Madison and conducted postdoctoral research at University of California, Berkeley and Institut National de la Recherche Agronomique in Toulouse, France. His major research and teaching interests lie in the organizational and institutional resources that shape the pace and nature of innovation applied to the challenges of environmental management and natural resource conservation. He codirects the Cornell University Program on Development, Governance and Nature, and he is engaged in collaborative research on development of forest biodiversity conservation competencies with the Finnish Environment Institute, Helsinki.

EDWARD J. WOODHOUSE teaches political science in the Department of Science and Technology Studies at Rensselaer Polytechnic Institute. He is a democratic decision theorist attempting to understand how to shape a commendable global civilization that governs technologies more wisely and more fairly.

His books include *The Policy-Making Process* (with C. E. Lindblom; Prentice-Hall; 1993), and *The Demise of Nuclear Energy? Lessons for Democratic Control of Technology* (with J. Morone; Yale University Press, 1989). In addition to green chemistry and nanotechnology, his research focuses on overconsumption by the world's affluent, runaway expertise concerning robotics, and the design of democratic practices for governing corporate and consumer decision making.

STEPHEN ZAVESTOSKI is assistant professor of sociology and environmental studies at the University of San Francisco. His current research examines the way health-based social movements use science in settling disputes over the environmental causes of contested illnesses. His work on health social movements appears in *Social Movements in Health* (with Phil Brown; Blackwell, 2005), and in *Sociology of Health & Illness*. His other area of research examines the use of the Internet as a tool for enhancing public participation in environmental decision making.

INDEX

SCIENCE AND TECHNOLOGY IN SOCIETY

Daniel Lee Kleinman
Impure Cultures: University Biology and the World of Commerce

Daniel Lee Kleinman, Abby J. Kinchy, and Jo Handelsman, editors
Controversies in Science and Technology: From Maize to Menopause

Jack Ralph Kloppenburg Jr.
First the Seed: The Political Economy of Plant Biotechnology, second edition

Scott Frickel and Kelly Moore
The New Political Sociology of Science: Institutions, Networks, and Power